Contents

Acknowledgments

Many thanks to the ever-professional team at John Wiley & Sons, especially Jeff Golick and Eric Nelson for inviting me to add *Environmental Science* to the Wiley *Self-Teaching Guide* series, and Kimberly Monroe-Hill for steering the manuscript through the production process. As always, my deep thanks to Professor Brian Skinner, my coauthor on a number of Wiley projects, for his advice, his vast knowledge, and his sharing nature. And thanks to my family for their continued acceptance of the never-ending mess on the dining room table.

A Note to the Reader

The environment is all around us—the solid Earth, the air, and the water, supporting and sustaining people, plants, and wildlife. But the environment is more than this. It encompasses the parts of the Earth system and all the living organisms that inhabit it, as well as the endless variety of interactions between them. The environment isn't really a thing as much as it is an active, complex, ever-changing relationship.

This book focuses primarily on the scientific aspects of the study of the environment. Environmental science is both an interdisciplinary science and an applied science; it draws its fundamental principles from a number of basic scientific disciplines, particularly biology, geology, chemistry, and physics. These and other scientific disciplines such as hydrology, climatology, oceanography, soil science, statistics, and meteorology are applied to the study of the environment, contributing to our understanding of this complex planet we inhabit.

The science of the environment is multifaceted. Because we are talking about a *relationship* rather than a *thing*, the study of the environment extends significantly beyond the realm of science. People are the de facto managers of this planet. We manage the Earth system—its impacts on human society and in turn human impacts on the natural environment—as best we can, through our laws, policies, writings, and other human systems. We draw from disciplines as diverse as philosophy and ethics, art and literature, economics, political science, sociology, management, geography, history, anthropology, and even psychology. Thus, although the scientific aspects of the environment are foremost in this book, you will find that social, political, economic, and philosophic aspects are necessarily interwoven throughout. The book begins with a focus on Earth as a planet and some of the important processes in the natural world that shape and influence our environment (chapters 1 through 6). Later chapters focus more on people and technology and their influences on environmental quality (chapters 7 through 16).

Each chapter in the book begins with a brief list of the main things you will be learning about. Here and there you will find some text set apart like this:

This icon indicates something intriguing to think about, or an activity or experiment you might want to try for yourself.

Throughout the book you will find questions and answers, as well as a Self-Test at the end of each chapter, to test your comprehension and retention of concepts and vocabulary.

You can read the chapters in any order you choose. However, each chapter assumes that you understand the vocabulary and concepts presented in the preceding chapters. The first six chapters, in particular, introduce many of the basic concepts that are applied in subsequent chapters. For example, chapter 12 "Water Resources," chapter 13 "Water Pollution and Soil Pollution," and chapter 14 "Air Pollution" all build upon basic concepts introduced in chapter 3 "The Hydrosphere and the Atmosphere." Boldface vocabulary terms are generally introduced once, although they may be briefly redefined in succeeding chapters.

Since you have purchased this book, you probably fall into one of these categories:

- You studied environmental science a number of years ago in college or university and want to refresh or update your understanding of the subject.
- You are currently studying environmental science in high school, college, or university and want to supplement your understanding of the material covered in lectures and labs.
- You have developed an interest in environmental science, perhaps through watching documentaries or news stories about topics such as global warming, ozone depletion, smog alerts, or the loss of biodiversity.
- You are a professional in a field that has been affected by environmental change, such as the insurance industry, engineering, municipal management, or the health professions.

In any case, you bought this book because you want to enhance your understanding of the science of our environment. Whatever your background, I hope this book will meet your needs and that you will enjoy reading it. I hope it will help you to learn more about our environment and to become a more informed and active participant in the management of the human–environment relationship.

1 Our Unique Planet

We speak for Earth.

—Carl Sagan

Objectives

In this chapter you will learn about:

- the field of environmental science;
- the features and characteristics that make Earth unique;
- the geosphere, atmosphere, hydrosphere, biosphere, ecosphere, and technosphere; and
- the materials and processes of the solid Earth.

The Environment and Environmental Science

Once upon a time, the term *environment* just meant "surroundings." It was used in reference to the physical world separate from ourselves—rocks, soil, air, and water. People gradually came to realize that the organisms that inhabit the physical world—the enormous variety of plants, animals, and microorganisms on this planet, including humans—are an integral part of the environment. It doesn't make sense to define environment without recognizing the fundamental importance of interactivity among organisms, and between organisms and their surroundings. Therefore, we can define **environment** to include all of the components, characteristics,

This satellite image shows North and South America as they would appear from space 35,000 km (22,000 mi) above Earth. The image is a combination of data from two satellites: NASA's Terra satellite collected the land data, and NOAA's GOES satellite collected the cloud-cover data.

and conditions in the natural world that influence organisms, as well as the interactions between and among organisms and the natural world. This definition encompasses the physical–chemical–geologic surroundings of an organism, as well as the other biologic inhabitants of the neighborhood. It implies that there is a range of possible interactions among the various components of the environment, both **biotic** (living) and **abiotic** (nonliving).

Our conceptual understanding of the environment continues to evolve, reflecting the emerging understanding that humans, too, are an integral part. We influence the physical world and other organisms; in turn, we are influenced by them. Therefore, more recent definitions of environment incorporate social, cultural, and economic factors in addition to the components of the natural, biophysical world. For example, human technology is affected by the natural environment through the availability or scarcity of natural resources. Technology, in turn, has profound and sometimes devastating impacts on the natural environment. Understanding the **technosphere**—the built, manufactured, industrialized, and domesticated aspects of the world—is fundamental to our understanding and successful management of the environment.

Environmental science, the focus of this book, is an interdisciplinary combination of basic sciences applied to the study of the environment. By one relatively early definition, environmental science includes "all of nature we perceive or can observe . . . a composite of Earth, Sun, sea, and atmosphere, their interactions, and the hazards they present" (from the U.S. Environmental Science Services Administration, 1968). Environmental science draws its fundamental principles from a number of basic scientific disciplines, particularly biology, geology, chemistry, and physics. These and other scientific disciplines, including hydrology, climatology, oceanography, soil science, statistics, and meteorology, are applied to the study of the environment and contribute to our understanding of this complex planet we inhabit.

Environmental science is a multifaceted discipline, but the study of the environment extends well beyond the realm of science. People are the de facto managers of this planet. We manage the environment—our impacts on the environment, as well as its impacts on human society—through our laws, policies, writings, and other human systems. We draw from disciplines as diverse as philosophy, literature, economics, political science, sociology, management, geography, history, anthropology, art, and even psychology. The application of these disciplines to the environment has given rise to such fields as environmental law, environmental economics, environmental management, and environmental ethics. Although these are not scientific disciplines, they are fundamental to our understanding of the environment.

How would you modify the definition of environment stated above so that it takes into consideration human technology and its interactions with the natural environment? ______________

> *Answer:* One possible definition is that environment encompasses the natural physical, chemical, biologic, and geologic aspects and conditions that influence and are influenced by organisms, including humans; the interactions among them; and social, cultural, and economic factors that influence and are influenced by the natural world. Does this definition include everything that you think it should?

Why do you think it is important to be precise and thorough in defining the concept of environment? ______________

> *Answer:* One reason is that the term *environment* is often used in a legal context, where the wording must be very precise. Another reason for being careful and thorough is that how we define environment reflects, in part, how we view ourselves in

relation to other organisms and to the natural world. The impacts of technologic developments on the natural environment sometimes result in degradation of the social or cultural environment; these need to be included in our definition so that they will be taken into consideration when we undertake activities that may alter the natural environment.

Hot Topics in Environmental Science

If you were to take a poll in which you asked citizens of North America and Europe to name the most pressing environmental issues today, you would likely find significant regional variations. People in the northeastern United States and southern Canada might have concerns about the health of the Great Lakes. In the Northwest, people might have concerns about deforestation in old-growth forests. In the Midwest and dry Southwest, the availability of abundant water and the depletion of groundwater supplies might be of concern. People in Western Europe might be most worried about the effects of acid rain on forests and lakes. Some concerns would be common to all regions, including climatic change, the ozone hole, loss of biodiversity, health impacts of air pollution, toxic contaminants in natural waters, overpopulation, energy shortages, and municipal garbage. These are widespread or even global problems that are part of the legacy of industrialization. Throughout this book, we will be looking at the science that underlies these and other environmental problems.

What are the most pressing environmental issues facing your local neighborhood, city, or region? How do they differ from the environmental issues in other regions, and how are they similar? How do your local concerns differ from global environmental issues?

If you were to conduct the same poll in a less economically developed country, perhaps in South America, Asia, or Africa, the list of pressing environmental issues might be quite different. The environmental concerns of people in the developing world tend to be more local, more immediate, and relate more directly to daily survival. The list might include land degradation and its impact on food production; lack of clean water for drinking, washing, and cooking; and the lack of fuel wood and other energy sources for cooking. Today almost 2 billion of the world's poorest people lack access to sanitation facilities and wastewater treatment. Approximately 1 billion people do not have access to clean water, and almost 1 billion people are chronically hungry. These problems threaten people's survival; they are part of the legacy of poverty. Some of them are not strictly environmental problems—there are underlying political, social, and economic causes—but the impacts of environmental degradation on human health and well-being are immediate, local, and severe in the developing world.

Until fairly recently, most developing countries were not particularly interested in entering the international dialogue about dealing with problems like ozone depletion or global warming. Why? It's partly an issue of responsibility and blame; some of our current global environmental problems were caused—or were at least initiated—by industrialization in wealthier nations. It's also partly because people in developing countries are simply too busy dealing with the immediate problems of daily survival and with getting food, water, and other basic services to people in need. Now, however, it is widely recognized that regardless of the cause everyone in the world is potentially at risk from the impacts of environmental degradation. All nations and all people bear a common responsibility to deal with these problems.

A concept that has become familiar in international dialogues about global environmental issues is the idea of common but differentiated responsibility of nations. What do you think it means? ______________

Answer: Common but differentiated responsibility refers to the concept that all nations must bear responsibility for dealing with global environmental problems, but different nations have different capacities and resources with which to respond to these problems.

Welcome to Our World

Now that we have covered some basic terminology and concepts, let's begin our study of the environment by taking a look at the planet itself. Earth is one of nine planets in our **solar system**—the Sun and the group of objects orbiting around it, which originated as a system approximately 4.6 billion years ago. The solar system also includes more than sixty moons, a vast number of asteroids, millions of comets, and innumerable floating fragments of rock and dust. The objects in our solar system move through space in smooth, regular orbits held in place by gravitational attraction. The planets, asteroids, and comets orbit the Sun, and the moons orbit the planets (Figure 1.1).

The planets can be separated into two groups on the basis of their characteristics and distances from the Sun. The innermost planets—Mercury, Venus, Earth, and

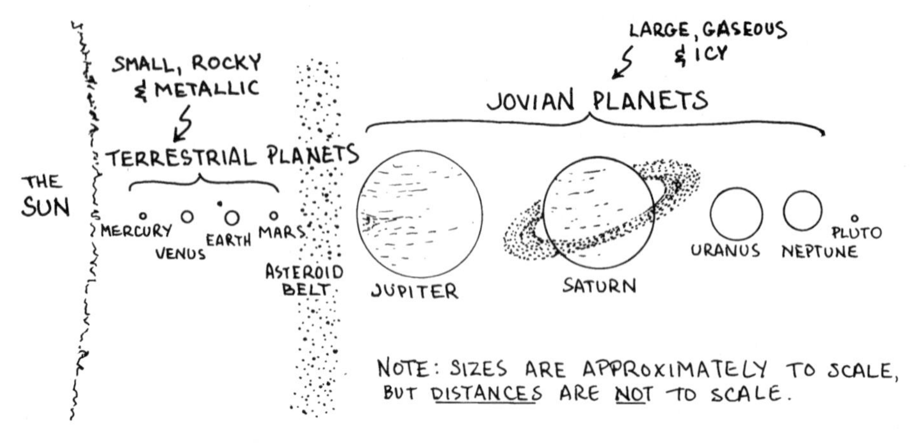

Figure 1.1.

Mars—are small, rocky, and relatively dense. These planets are similar in size and chemical composition. They are called **terrestrial planets** because they resemble *Terra* ("Earth" in Latin). With the exception of Pluto, the outer or **jovian planets**—Jupiter, Saturn, Uranus, and Neptune—are much larger than the terrestrial planets but much less dense, with very thick atmospheres of hydrogen, helium, and other gases. You can learn more about the solar system by reading *Astronomy: A Self-Teaching Guide*, by Dinah L. Moché (John Wiley & Sons, 2004).

The terrestrial planets have many things in common beyond their small sizes, rocky compositions, and positions close to the Sun. They have all been subjected to volcanic activity and intense meteorite impact cratering. They have all been hot and, indeed, partially molten at some time early in their histories. During this partially molten period, the terrestrial planets separated into layers of differing chemical composition: a relatively thin, low-density, rocky **crust** on the outside; a metallic, high-density **core** in the center; and a rocky **mantle** in between. This separation process happened to all of the terrestrial planets, including Earth. In the context of environmental science, the physical Earth—distinct from the organisms that inhabit it—is referred to as the **geosphere**. The term *geosphere* is used in reference to the planet and the whole physical environment—the atmosphere, the hydrosphere, and the solid Earth.

What are the four terrestrial planets, and why are they given this name?

____

Answer: Mercury, Venus, Earth, and Mars. They are all similar to Earth (*Terra*).

What Makes Earth Unique?

In spite of the similarities among the terrestrial planets, the history and specific characteristics of Earth are different enough from those of the other terrestrial planets to make this planet habitable. If you look at a photograph of Earth taken from space, you immediately notice the blue-and-white **atmosphere**, an envelope of gases dominated by nitrogen, oxygen, argon, and water vapor, with traces of other gases. Other planets have atmospheres, but no other planet in the solar system has an atmosphere of this particular chemical composition.

The atmosphere contains clouds of condensed water vapor that form because water evaporates from the hydrosphere, another unique feature. The **hydrosphere** ("watery sphere") consists of the oceans, lakes, and streams; underground water; and snow and ice. Planets farther from the Sun are too cold for liquid water to exist on their surfaces; planets closer to the Sun are so hot that any surface water evaporated long ago. Only Earth has just the right surface temperature to have liquid water, ice, and water vapor in its hydrosphere.

Another unique feature of Earth is the **biosphere**, the "living sphere." The biosphere comprises innumerable living things, large and small, which belong to millions of different species, as well as recently dead plants and animals that have not yet completely decomposed. The **ecosphere** is the physical environment that permits or facilitates the existence of the biosphere. On Earth, the ecosphere extends from the deepest valleys and the bottom of the ocean to the tops of the highest mountains and well into the lower part of the atmosphere. Even the great polar ice sheets host a variety of life forms. Although many new planets have been discovered orbiting distant stars that seem similar to our own Sun, we don't yet know of another planet that offers an ecosphere or hosts a biosphere.

The nature of Earth's solid surface is also special; it is covered by an irregular blanket of loose debris called **regolith** (from the Greek *rhegos*, meaning "blanket"). Earth's regolith forms as a result of **weathering**, the continuous chemical alteration and mechanical breakdown of surface materials through exposure to the atmosphere, hydrosphere, and biosphere. The weathered, broken-down materials are picked up by moving wind, water, and ice, carried downhill under the influence of gravity, and eventually deposited. Weathering and the transport of weathered materials together comprise the process of **erosion**, which is part of the global **rock cycle** (Figure 1.2). Soil, mud in river valleys, sand in the desert, rock fragments, and other unconsolidated debris are all part of the regolith. Some other planets and planetary bodies are blanketed by loose, fragmented material, but in those cases the fragmentation has been caused primarily by the endless pounding of meteorite impacts. Earth's regolith is unique because it forms as a result of complex interactions of

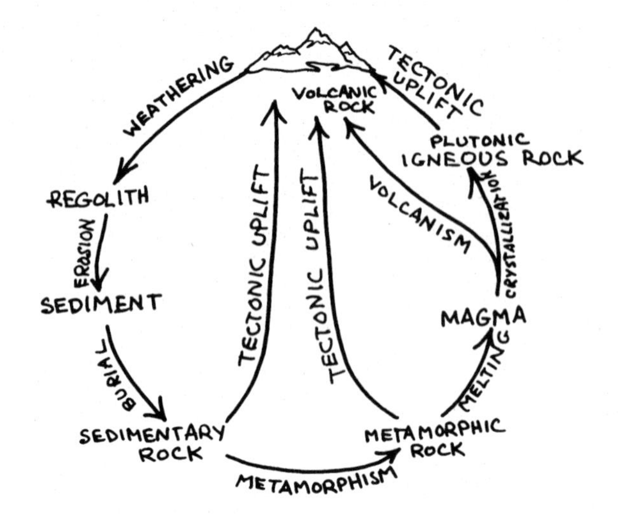

Figure 1.2.

physical, chemical, and biologic processes, usually involving water. It is also unique because it teems with life; most plants and animals live on or in the regolith or in the hydrosphere.

Why does Earth have an ecosphere, whereas all other known planets do not?

Answer: A combination of just the right size and composition (especially the presence of water) and optimal distance from the Sun make the surface conditions on Earth perfectly suited for hosting life.

Earth, Inside and Out

Earth, like the other terrestrial planets, is composed primarily of **rock**, a naturally formed, solid, coherent aggregate. The basic building blocks of rocks are **minerals**—naturally occurring, inorganic elements or chemical compounds that have specific chemical compositions, orderly internal atomic structures, and characteristic physical properties. **Geology** is the scientific study of these and other Earth materials and processes. If you are interested in learning more about the science of the Earth, you can read *Geology: A Self-Teaching Guide*, by Barbara Murck (John Wiley & Sons, 2001).

Three basic families of rocks are recognized. They are:

1. **sedimentary rocks**, which form as a result of the deposition, consolidation, and cementation of unconsolidated rock and mineral fragments (**sediment**) in low-temperature and low-pressure conditions near Earth's surface;
2. **igneous rocks**, which solidify from molten rock (**magma** or **lava**) on the surface (**volcanic rocks**) or deep underground (**plutonic rocks**); and
3. **metamorphic rocks**, which are rocks that have been altered as a result of exposure to very high pressures and/or temperatures.

As everyone knows, rocks are quite durable. However, we live on a planetary surface that is extremely active and ever changing. Winds blow, waves break, streams flow, and glaciers grind away at the surface. These constant, restless energetic forces, driven partly by gravity and partly by solar energy, interact with surface materials, eventually breaking down rocks and minerals to form regolith. Geologic evidence shows that weathering and erosion have been operating throughout most of Earth history—well over 4 billion years. But if these forces are constantly at work, inevitably wearing down and washing away Earth's surface materials, then why are there any mountains left standing? The answer is that other forces are acting on the surface from the inside. Internal forces constantly uplift the surface, creating great mountains and rugged topography that seem to defy the forces of weathering and erosion. Here is how it works.

The outermost, rocky part of Earth is the crust, as mentioned above. **Continental crust** (Figure 1.3) is relatively thick (average thickness 45 km, or 30 mi) and is made mostly of plutonic rocks called granite. **Oceanic crust**, which underlies the great ocean basins, is relatively thin (average thickness 8 km, or 5.4 mi) and is made mostly of volcanic rocks called basalt. Beneath the crust is the mantle, which is also made of plutonic igneous rocks, but they are different from the rocks of the continental crust. At the center of Earth is the core made of iron-nickel metal. Together, the crust and the outermost part of the mantle make up the **lithosphere**, a thin, cold, brittle, rocky layer. The mantle below the lithosphere is very hot, so it is malleable, like putty, even though it is made of solid rock. The part of the mantle immediately beneath the lithosphere is called the **asthenosphere**; it is especially weak and squishy because it is close to the temperature at which rocks begin to melt.

The lithosphere is made of solid rock about 100 km (60 mi) thick, on average. In comparison to the size of the planet as a whole, the lithosphere is an exceedingly

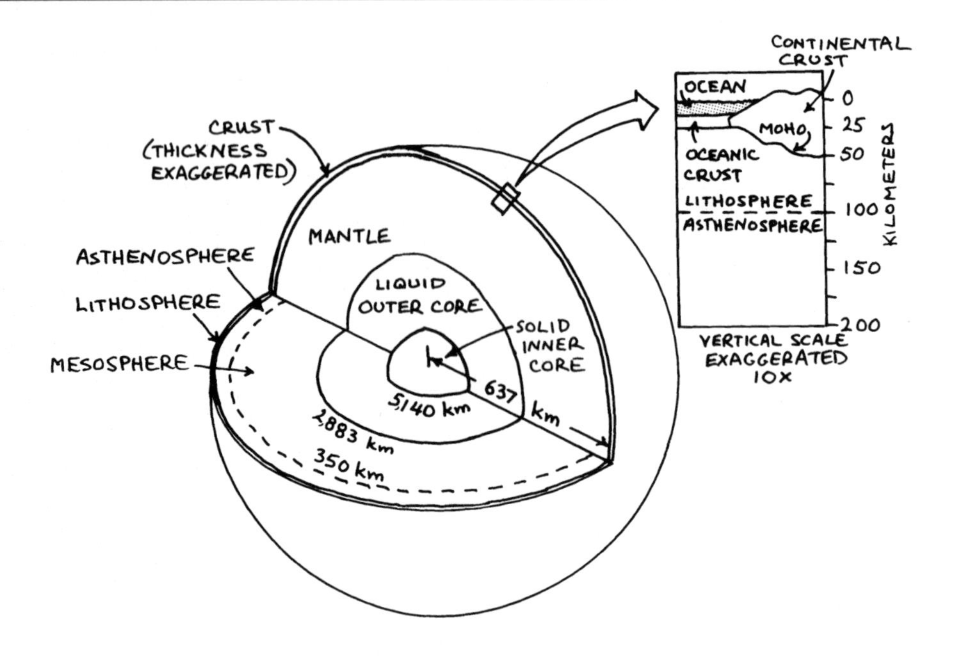

Figure 1.3.

thin shell (Figure 1.3). It has about the same relative thickness as the glass shell of a lightbulb or the skin of an apple.

If you were to do an experiment in which you placed a thin, cool, brittle shell (like the lithosphere) on top of hot, weak material that is rather squishy (like the asthenosphere), what do you think would happen? You might predict that the thin shell would break into pieces. In fact, that is precisely what has happened to the lithosphere—it has broken into a number of large fragments, or **plates**. Today there are six large plates, each extending for several thousands of kilometers, and a large number of smaller plates (Figure 1.4). Note that these are *lithospheric* plates, not crustal plates. The plates are made of the crust *and* the solid rock of the mantle just beneath. Some lithospheric plates, like the Pacific Plate, are capped mainly by oceanic crust; others, like the North American Plate, are capped mainly by continental crust.

You can experiment with plates and plate motion by carefully heating wax in a pan, then letting it cool until it forms a thin skin or crust. If you try this, be careful—molten wax is very hot. Be sure to wear eye protection, and use care in handling hot pans.

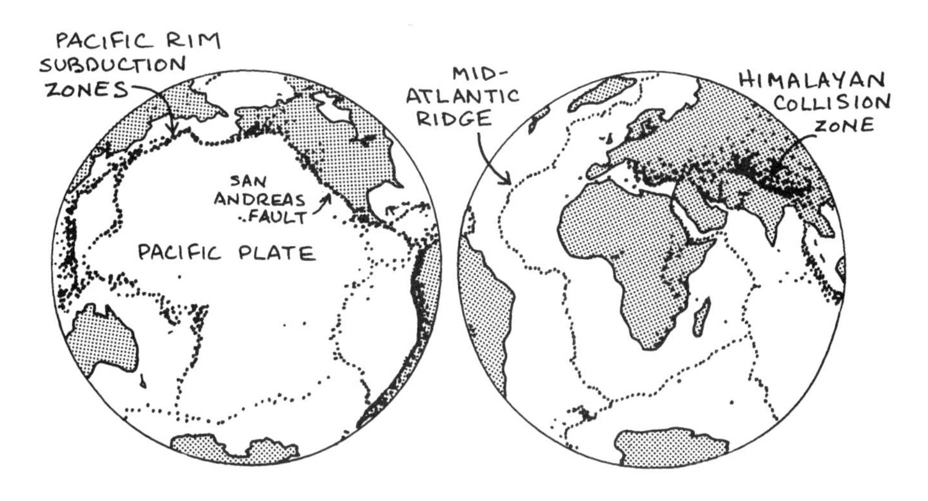

Figure 1.4.

Think again about the expected behavior of thin, brittle fragments floating on top of hot, squishy material. You might expect that movements in the underlying material would cause the brittle fragments to shift about. Again, that is exactly what happens to Earth's lithospheric plates. When movements occur in the hot mantle, the lithospheric plates shift and interact with one another. If a plate happens to be capped by continental crust, the continent moves along with the rest of the plate. You may already be familiar with this process; it is called **continental drift**. The study of the movement and interactions of lithospheric plates is referred to as **plate tectonics** (from the Greek word *tekton*, meaning "carpenter" or "builder").

What is the lithosphere? _______________

Answer: The outer 100 km (60 mi) of Earth; the crust and the upper part of the mantle.

Internal Forces

Lithospheric plates move and shift their positions in response to movements in the mantle beneath. As the plates move, they interact with one another mainly along their edges. Plate margins are where the most violent and intense types of geologic activity originate. Plates can interact in three basic ways: they can move away from each other (diverge); they can move toward each other (converge); or they can slide past each other. Consequently, there are three basic kinds of plate margins (Figure 1.5):

1. **Divergent margins** are huge fractures in the lithosphere where plates move apart from one another, forming great rift valleys on continents and under the oceans. They are characterized by earthquakes caused by the splitting and frac-

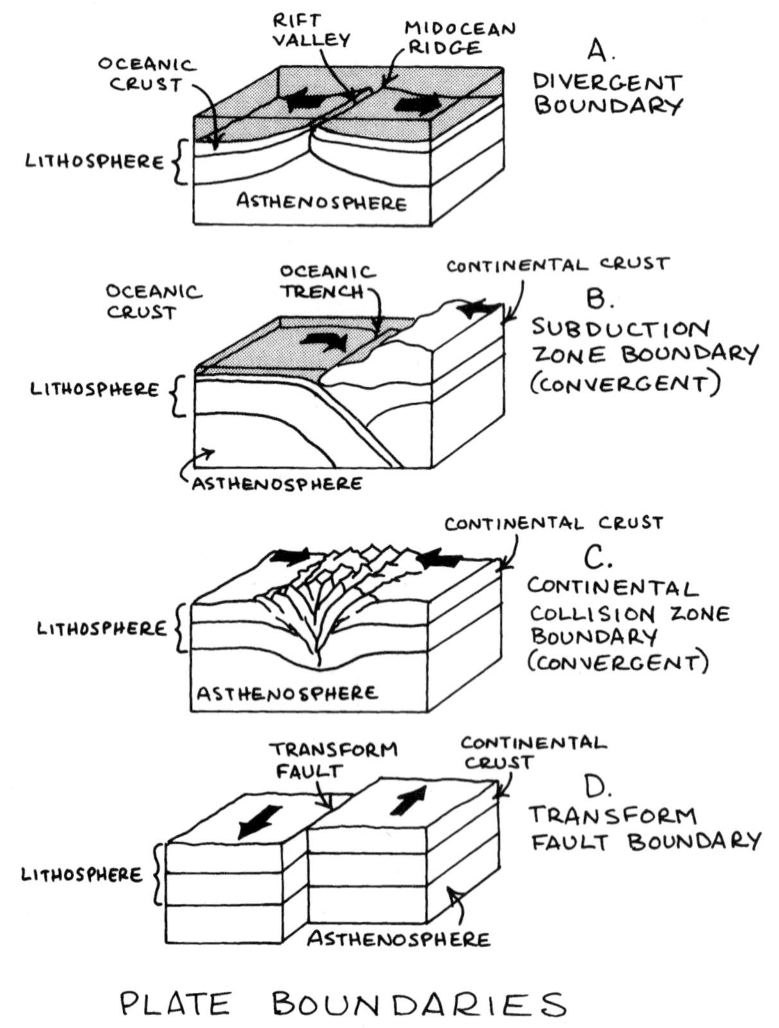

Figure 1.5.

turing of rocks, and by volcanic activity that occurs when melted rock from deep within the mantle wells up through the fractures.

2. **Convergent margins** occur where two plates move toward each other, slowly colliding. At ocean–ocean and ocean–continent convergent margins, a process called **subduction** occurs, in which oceanic crust is forced down into the mantle (Figures 1.5B and 1.6). The downgoing plate melts when it reaches a depth within Earth where the temperature is sufficiently high. Thus, convergent plate margins that involve oceanic crust are characterized by active volcanism. Other convergent plate margins—those that involve only continental crust—are characterized by the uplifting of great mountain chains like the Himalayas (Figure 1.5C). All convergent plate margins are marked by intense earthquakes.

3. **Transcurrent** or **transform fault margins** are huge fractures in the lithosphere where two plates slide past each other, grinding along their edges and causing earthquakes as they go. A famous modern example is the San Andreas Fault in California, where the Pacific Plate is moving north–northwest relative to the North American Plate.

All of these types of plate interactions are occurring today, as they have occurred throughout most of Earth history. We don't often notice plate motion because lithospheric plates move very slowly—usually between 1 and 10 cm (0.4 and 4 in) per year. But we feel the earthquakes and observe the volcanic activity along active

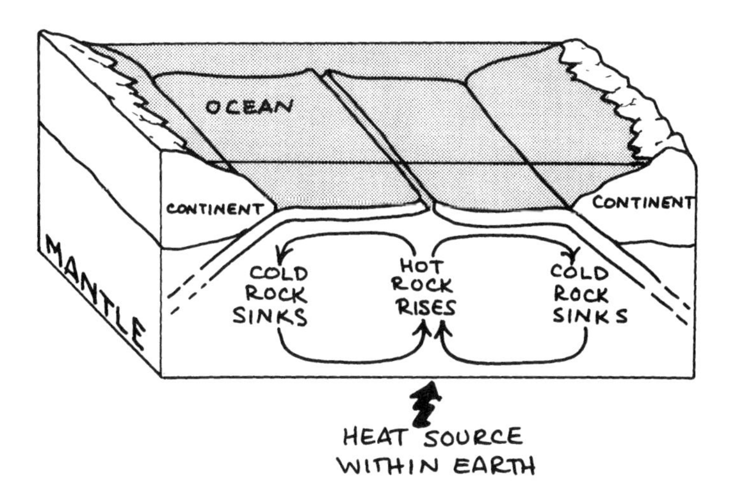

Figure 1.6.

plate margins. The scars and remnants of ancient plate interactions are also preserved in the rock record for us to study.

What causes plate motion? Thermal movement in the mantle is at least partly responsible (see Figure 1.6). Movement in the mantle, in turn, is caused by the release of heat from inside the Earth. The temperature in Earth's interior is high—about 5,000°C (more than 9,000°F) in the core. Some of this heat is left over from the planet's origin, but some of it is generated by the constant decay of naturally occurring radioactive elements. This heat must be released; if it were not, Earth would eventually become so hot that its entire interior would melt.

Some of Earth's internal heat makes its way slowly to the surface through **conduction**, in which heat energy is transferred from one atom to the next. However, conduction is a slow way to transfer heat. It is faster and more efficient for a packet of hot material to be physically transported to the surface. This is similar to what happens when a fluid boils on a stovetop, as in the wax experiment described earlier in this chapter. If you watch a fluid such as wax or spaghetti sauce as it boils, you will see that it turns over and over. Packets of hot material rise from the bottom of the pot to the top. As it reaches the surface, the hot fluid cools, then sinks back down to the bottom of the pot, where it is reheated. The continuous motion of material from bottom to top and down again is called a convection cell, and this type of heat transfer is called **convection**.

Even though Earth's mantle is mostly solid rock, it is so hot that it releases heat by convection. Rock deep in the mantle heats up and expands, making it buoyant. As a result, the rock moves toward the surface very slowly in huge convection cells of solid rock. Near the surface, the hot rock moves along the surface while losing heat to the atmosphere, just like the spaghetti sauce. As the rock cools, it becomes denser (cool rock is denser, or heavier, than hot rock) and sinks back into the deeper parts of the mantle. This convection cycle provides an efficient way for Earth to get rid of some of its internal heat.

The movement of lithospheric plates, formation of new crustal material through volcanism and tectonic uplift, and recycling of plates back into the mantle is called the **tectonic cycle**. Convection, plate motion, and interactions along plate margins create some of the most distinctive geologic and topographic features of the Earth's surface: deep oceanic trenches where lithospheric plates sink back into the mantle; midocean ridges and continental rift valleys where plates split apart; and high folded and crumpled mountain chains that form where continents collide. Plate tectonism is also responsible for generating earthquakes and volcanic eruptions, among other geologic processes that make the surface of Earth such a dynamic, active place.

So, we can now add plate tectonics to our list of unique features. Earth differs from all other known planets because of the unique relationship between its thin, brittle lithosphere and the hotter, weaker rocks that lie below in the asthenosphere. Plate tectonic activity has been an important process throughout much of Earth history. It is responsible for the uplifting of mountains, eruptions of volcanoes, intensities of earthquakes, movement of continents, and formation of deep ocean basins. It has even influenced the formation and chemistry of the atmosphere, the development of climatic zones, and the evolution of life, as you will learn in later chapters.

What is the cause of lithospheric plate motion? _______________

Answer: The release of heat through convection in the mantle.

Why do you think oceanic crust undergoes subduction but continental crust does not? _______________

Answer: Oceanic crust is much denser than continental crust; therefore, continental crust rides up and over a colliding plate, whereas the denser oceanic crust is forced down into the mantle.

So . . . what makes Earth unique? Let's summarize: We know of no other planet where plate tectonics has played and continues to play such an important role in shaping both the land and the physical environment. We know of no other planet where water exists near the surface in solid, liquid, and gaseous forms. No other planet yet discovered offers an ecosphere or hosts a biosphere or would have been hospitable to the origin and evolution of life as we know it. There are billions upon billions of stars in the universe, so it is almost inevitable that there are billions of planets; surely a few of those planets must be Earthlike and therefore capable of supporting life. However, if life does exist on a planet somewhere out in space, we haven't found it so far.

After reading this chapter, you should have a basic understanding of some of the factors that shape our relationship to the physical, chemical, biologic, and geologic world. In environmental science a lot of attention is paid to the interactions and interrelationships among the various spheres introduced in this chapter: the geosphere, atmosphere, and hydrosphere; the ecosphere; the biosphere; and the technosphere. We will revisit many of the topics and ideas introduced in this chapter in greater detail later in the book. In the meantime, you can test your knowledge and retention of this introductory material by trying out the Self-Test.

SELF-TEST

These questions are designed to help you assess how well you have learned the concepts presented in chapter 1. The answers are given at the end. If you get any of the questions wrong, be sure to troubleshoot by going back into that part of the chapter to find the correct answer.

1. Which one of the following is not a terrestrial planet?
   a. Mars
   b. Venus
   c. Mercury
   d. Neptune

2. The irregular blanket of loose debris that covers the surface of Earth is called the _______________.
   a. lithosphere
   b. regolith
   c. asthenosphere
   d. mantle

3. The *plates* in plate tectonics are made of fragments of _______________.
   a. continents
   b. oceanic crust
   c. lithosphere
   d. mantle

4. The weak layer of the mantle immediately underlying the lithosphere is called the _______________.

5. Earth has two fundamentally different types of crust: _______________ crust is made mainly of basaltic rocks, and _______________ crust is made mainly of granitic rocks.

6. Earth formed approximately 4.6 million years ago. (True or False)

7. The abiotic components of the environment are the living organisms that make up the biosphere. (True or False)

8. The term *technosphere* refers to the built, manufactured, industrialized, and domesticated parts of the world. (True or False)

9. Summarize the features that make Earth unique.

10. Briefly describe the processes of weathering and erosion.

11. What is convection, and how does it work inside the Earth?

12. What is the difference between the biosphere and the ecosphere?

13. What are the three main types of plate margins?

ANSWERS

1. d   2. b   3. c   4. asthenosphere

5. oceanic; continental   6. False   7. False   8. True

9. Presence of an ecosphere and biosphere; liquid water at the surface; surface temperature amenable to life; role of plate tectonics; characteristics of the regolith; composition of the atmosphere.

10. The process of weathering occurs when surface rocks and minerals disintegrate, either by chemical alteration or mechanical breakdown. Erosion happens when the weathered fragments are picked up by moving wind, water, or ice, carried downhill, and deposited.

11. Convection is a mechanism of heat transfer in which hot material is physically transported from a hot area to a cooler area. Inside the Earth, hot rock near the core becomes buoyant and rises toward the surface, where it cools. Through cooling, the rock becomes denser and eventually sinks back into the mantle, to be heated once again.

12. These terms are sometimes used interchangeably. *Ecosphere* refers specifically to an environment that is favorable to the existence of life. *Biosphere* refers to the actual living organisms—the life that is hosted by the ecosphere. Sometimes biosphere is used specifically to refer to Earth's ecosphere.

13. The three main types of plate margins are:
    i. divergent
    ii. convergent
    iii. transform or transcurrent fault

KEY WORDS

abiotic
asthenosphere
atmosphere
biosphere
biotic
conduction
continental crust
continental drift
convection
convergent margin
core
crust
divergent margin
ecosphere
environment
environmental science
erosion
geology
geosphere
hydrosphere
igneous rock
jovian planet
lava
lithosphere
magma
mantle
metamorphic rock
mineral
oceanic crust
plate
plate tectonics
plutonic rock
regolith
rock
rock cycle
sediment
sedimentary rock
solar system
subduction
technosphere
tectonic cycle
terrestrial planet
transcurrent or transform fault margin
volcanic rock
weathering

2 The Interactive Earth

When we try to pick out anything by itself, we find it hitched to everything else in the universe.

—John Muir

Objectives

In this chapter you will learn about:

- the storage and cycling of materials and energy within the Earth system;
- interactions among Earth's subsystems and cycles;
- the balance in natural systems and how it can be affected by human activities; and
- the transfer of life-supporting elements from one part of the Earth system to another.

Systems

In environmental science today, there is a strong emphasis on the study of Earth as a unified, interconnected system rather than as a collection of isolated parts. The concept of a system helps us break down large, complex problems into smaller pieces that are easier to study without losing sight of the interconnections between those pieces. The term *system* is often used quite loosely; however, it is instructive to consider the technical definition. The scientific roots of the term date to the 1700s when physical chemists were trying to understand the nature of thermodynamics—heat energy and how its movement causes things to happen.

The river, mountains, forest, and other biota in this photograph of the Alaska Peninsula National Wildlife Refuge are part of a single, large, complex system. The forest alone is a smaller system and so is the river. A tree, a salmon in the river, and an eagle and its nest are all examples of subsystems.

A **system** is any portion of the universe that can be separated from the rest of the universe for the purpose of observing changes within it. By saying that a system is any portion of the universe, we mean that the system can be whatever the observer defines it to be. That's why a system is only conceptual; you define its boundaries for the convenience of your study (Figure 2.1). A system can be large or small, simple or complex. You might choose to study the contents of a laboratory beaker. Or you might observe a lake, a fish, a rock, an ocean, a planet, the solar system, or the human body. A leaf is a system. It is part of a larger system (a tree), which in turn is part of an even larger system (a forest).

The fact that a system can be separated from the rest of the universe means that it has conceptual and physical boundaries that set it apart from its surroundings. The nature of those boundaries is one of the defining characteristics of a system. One important type of system is a **closed system**—its boundaries permit the exchange of energy but not matter with its surroundings. An example would be a perfectly

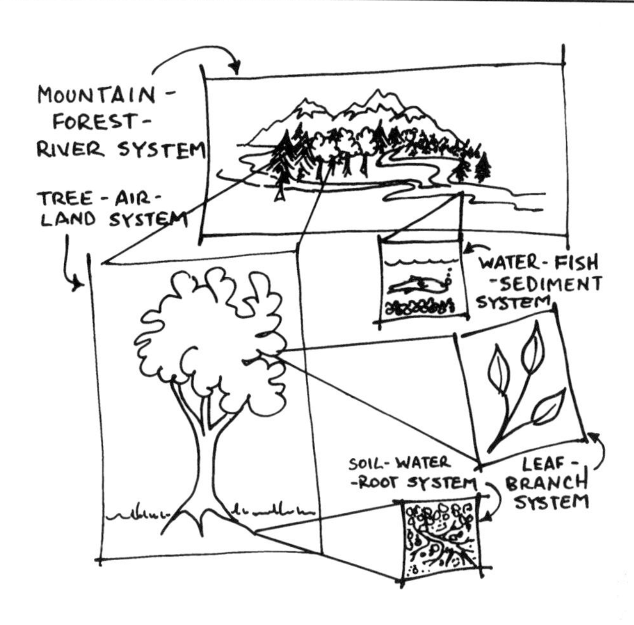

Figure 2.1.

sealed oven, which would allow the material inside to be heated (exchange of energy) but would not allow any of that material to escape (no exchange of matter).

Another important kind of system is an **open system**, which can exchange *both* matter and energy across its boundaries. An island offers a relatively simple example (Figure 2.2). Matter enters the system in the form of precipitation (water) and leaves by flowing into the sea or evaporating back into the atmosphere. Energy enters the system as sunlight and leaves as heat. In the natural environment, open systems are common. Materials and energy enter and leave most natural systems with ease, making them difficult to study because we can't exert the kinds of controls and limitations that we can in a laboratory situation.

We define systems for the purpose of observing changes in them, which brings us back to the fact that Earth is incredibly complex and huge. It has countless parts and interactions among those parts. There are so many variables that it is difficult to study the planet in its entirety. The system concept allows us to study changes within the Earth system, by limiting the size and complexity of the piece we choose to study without sacrificing the concepts of interaction and interconnection.

This approach can be applied to both natural and artificial (that is, human-made) systems. Ecologists have used the systems approach for many years; an *ecosystem* is a system that includes and sustains life—a biologic community, its abiotic

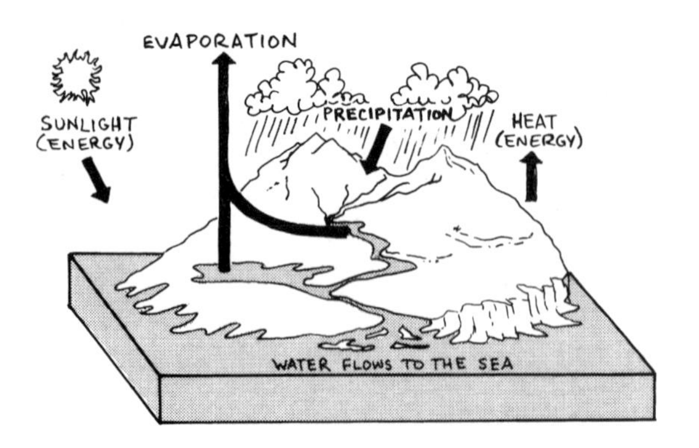

Figure 2.2.

surroundings, and the interactions between and among them. Urban geographers and planners can also apply the systems approach to the study of cities; enormous flows of energy and materials occur in cities, and they are similar to natural systems in many respects.

In the text there are examples of natural and artificial systems—a leaf, a tree, a forest; a city; an oven; and so on. Think of these and some more examples, such as a refrigerator, a school, or a bird's nest. In each case, how does the concept of a system apply? Is the system you are thinking of an open system or a closed system, and why?

The systems approach allows scientists to study Earth as an integrated whole by examining smaller interacting parts, or **subsystems**. The principal subsystems within the Earth system are the atmosphere, the hydrosphere, the solid Earth (both rock and regolith), and the biosphere. They are like huge reservoirs in which materials and energy are stored for a while before moving to one of the other reservoirs. Each of the four main subsystems can be further subdivided. For example, we can divide the hydrosphere into a number of smaller subsystems, including oceans, glaciers, streams, lakes, and groundwater, each of which acts as a reservoir for water.

What term defined in chapter 1 is often used to refer to the physical or abiotic environment—that is, the atmosphere + the hydrosphere + the solid Earth?

Answer: Geosphere.

Earth as a Closed System

Earth itself approximates a closed system. Energy enters the Earth system as short-wave solar radiation. This energy is used in various biologic and geologic processes; then it leaves the system in the form of longer-wavelength energy, or heat. In a perfectly closed system, no matter crosses the system's boundaries, but it is not quite true that no matter crosses the boundaries of the Earth system. We lose a small but steady stream of hydrogen and helium atoms from the outermost part of the atmosphere, and we gain some material every day in the form of incoming meteorites. However, the amount of matter that enters or leaves the system on a daily basis is so minuscule compared with its overall mass that Earth essentially functions as a closed system.

When changes are made in one part of a closed system, the results of those changes will eventually affect other parts of the system. This is sometimes called the **principle of environmental unity**. Earth's subsystems are in a dynamic state of balance; when there is a change in one subsystem, the change spreads throughout the system as a new state of balance, or **equilibrium** is established. Sometimes an entire chain of events ensues. An interesting example is El Niño, a climatic phenomenon in which anomalously warm water accumulates off the western coast of South America. The warm surface water inhibits the normal rise of cold, nutrient-laden waters from great depths. This reduces the available nutrients in the water, causing fish and coastal birds to die off. The unusually warm sea surface temperature initiates a chain of weather-related events around the globe, including abnormally heavy rains, drought, cyclones, unusually cold or mild winters, flooding, and widespread landslides. During an El Niño event, changes in one part of the system are linked to many changes elsewhere in the ocean, in the atmosphere, on the land, and in the biosphere.

The fact that Earth behaves as a closed system has important implications for those of us who live within its boundaries. By definition, the amount of matter in a closed system is finite. The resources on this planet are all we have and, for the foreseeable future, all we will ever have. We must treat Earth resources with respect and use them wisely and cautiously. Another consequence of living in a closed system is that waste materials remain within the boundaries of the Earth system, so we must deal with whatever consequences are associated with the materials we discard. In a closed system, as environmentalists sometimes say, "There is no 'away' to throw things to."

Pesticides have been detected in Arctic polar bears, even though no agricultural chemicals have ever been used nearby. How does this demonstrate the principle of environmental unity? What are some questions an environmental scientist might want to ask about this situation? ______________

Answer: The principle of environmental unity says that when a change is made in one part of the system, the change spreads into other, interconnected parts of the system. If pesticides are introduced in one part of the Earth system, it should not be surprising to find these chemicals appearing somewhere else in the system. Some important questions: How long would it take pesticides to move to the Arctic and into the polar bear from the locations where they were applied? What processes are responsible for the long-distance transport of pesticides? Do pesticides change, break down, or disappear during the course of the journey? How harmful are pesticides to polar bears and other organisms? We will address these and related questions in later chapters.

Cycles and Reservoirs

It is useful to envision interactions within the Earth system as a series of interrelated **cycles**—groups of processes that facilitate the movement of materials and energy among Earth's reservoirs. Much of environmental science is concerned with substances in the environment, how they are stored, how they are cycled among the various subsystems within the Earth system, how they are changed or altered in the process, and how they interact with other substances. You can think of a cycle as a set of two or more interconnected reservoirs, between or among which materials (or energy) move in a cyclic manner. The characteristics of environmental **reservoirs**—the places where materials and energy accumulate and are stored in the Earth system—are of great interest to environmental scientists.

There are two ways to think about environmental reservoirs. One is to think of a reservoir as a holding tank. In this case, a reservoir is like a container with physical boundaries. For example, an ocean is a reservoir for water. So is a lake or a pond. Another more subtle but also more accurate way of thinking about a reservoir is as a mass of material occurring in a particular type of environment. For example, there is a reservoir of salt in ocean water. There is also a reservoir of inorganic silicon, carbon, nitrogen, and many other substances in ocean water. There is a reservoir of ozone in the atmosphere. There is a reservoir of mercury in fish and a reservoir of pesticides in polar bears. And so on.

We can distinguish the reservoir itself from the mass of material contained within it using the concept of burden. The **burden** of a reservoir is its content, the total mass of a particular substance contained within that reservoir. Specifically, the burden or content of a reservoir is the concentration of the substance times the total mass of the reservoir itself:

burden (content) =

mass of substance in reservoir =

concentration of substance in reservoir × total mass of reservoir

You have likely noted from the preceding discussion that some substances are present in high concentrations in certain reservoirs, while other substances are present in very low concentrations. Here are some examples. To figure out the burden or content of sodium chloride (salt) in the ocean, we first need to know the concentration of the substance in the reservoir. The concentration of salt in seawater is roughly 35 grams (g) of salt per kilogram (kg) of ocean water. (Refer to appendix 1 for a review of units and conversions.) Next, we need to know the size of the physical reservoir itself—the mass of the water that makes up the ocean reservoir, which is about $14,000 \times 10^{20}$ g (roughly 1,400,000 km^3 in terms of volume). Plugging into the equation, we find that the burden of salt in the oceanic reservoir—that is, the mass of salt in the ocean—is equal to 35 g/kg × ($14,000 \times 10^{20}$ g), or about 490×10^{20} g.

Now let's look at an example of a substance that is present in much lesser concentrations in seawater. Inorganic silicon in the ocean is measured not in grams per kilogram but in milligrams per kilogram (a milligram is a thousandth of a gram). In fact, the average concentration of inorganic silicon in deep ocean water is about 1 mg/kg. Using the value for the mass of the ocean given above, we get 1 mg/kg × ($14,000 \times 10^{20}$ g) = 0.014×10^{20} g. Both salt and inorganic silicon are stored in dissolved form throughout the ocean, but the salt reservoir is several orders of magnitude larger than the inorganic silicon reservoir because of the difference in concentration.

The concept of reservoirs is extremely important in environmental science. Environmental scientists spend a lot of time measuring the concentrations of different substances in different reservoirs. They study the processes that cause substances to accumulate in reservoirs and the processes that cause substances to move from one reservoir to the next. Sometimes the substances and the processes are natural, such as the sea salt example just given. Sometimes the substances are potentially harmful to humans, such as mercury compounds in the fish we eat. Sometimes the substances are human-made, or **anthropogenic**, such as chlorofluorocarbons (CFCs), a group of anthropogenic chemicals accumulating in the atmospheric reservoir and causing damage to the ozone layer. Sometimes the substances are natural, but anthropogenic activities have altered the processes that cause them to accumulate or move from one reservoir to another. An example is carbon dioxide (CO_2), a naturally occurring substance that is released from long-term rock reservoirs by the burning of fossil fuels. Carbon dioxide is accumulating in the atmospheric reservoir at a rate much greater than the natural rate of accumulation. This causes concern among scientists and policy makers because carbon dioxide has an important role in the greenhouse effect, which keeps the surface of the planet warm. We will return to these topics in subsequent chapters.

What information would you need to calculate the content or burden of mercury in ocean fish? ______________

Answer: You would need to know the size of the physical reservoir—the total mass of fish in the ocean. You would also need to know the average concentration of mercury in ocean fish.

Fluxes

Let's look more closely at the processes that transport substances from one reservoir to another in Earth systems. This includes any process you can imagine that involves the movement of substances in the environment, including physical, chemical, geologic, or biologic processes, or any combination of these. Understanding these processes is key to understanding the dynamic nature of the natural environment, our impacts on it, and our potential exposure to substances within it.

A simple example is the uptake of water by plant roots. Water leaves the geosphere, where it occurs as soil moisture, and is transferred to the biosphere, where it resides in the roots, stems, and leaves of plants. The transfer is accomplished by a process that is both geologic—the movement of water through rock and soil—and biologic—the uptake of water by the plants. In another example, when water falls as rain it leaves the atmospheric reservoir and is transferred to the hydrosphere, where it might reside in a lake or an ocean, or to the geosphere, where it might reside as soil moisture or groundwater. The process is partly physical—rain falls under the influence of gravity—and partly chemical—raindrops condense because of changes in temperature, pressure, or water content of the atmosphere.

Scientists quantify such processes using the concept of **flux**, the amount of material transferred, specified in units of mass or amount (grams, kilograms, cubic meters) per time (seconds, hours, days, years). The flux quantifies the *amount of material*, as well as the *rate of movement*. For example, the flux of water from the oceanic reservoir to the atmospheric reservoir by evaporation from the ocean surface is approximately 425×10^{18} grams per year (or $425 \times 1{,}000$ km^3/yr, by volume). The return flux of water as precipitation from the atmospheric reservoir directly to the oceanic reservoir is approximately 385×10^{18} g/yr (or $385 \times 1{,}000$ km^3/yr, by volume). The flux of carbon dioxide from the solid Earth to the atmospheric reservoir by way of CO_2 emissions from volcanoes is approximately 0.1×10^{12} g/yr. And so on.

Here's something to think about. If 425×10^{18} g/yr of water evaporates from the surface of the ocean, shouldn't 425×10^{18} g/yr of water be returned to the ocean as precipitation? If it isn't, why doesn't the ocean dry up? Find the answer in chapter 3.

It is also useful to consider the forces that *drive* the flux of materials from one reservoir to another in the Earth system: we are not just interested in *how* materials move and *how fast* they move but also *why* they move. Physical and geologic movements of materials are mainly driven by energy from external sources (the Sun and, to a much lesser extent, the gravitational interaction of Earth, Sun, and Moon); by Earth's gravitational effect on objects near its surface; and by Earth's internal heat energy. For example, erosion is driven by wind and water, using a combination of solar energy and gravity. Continental drift, plate motion, and volcanism are driven by Earth's internal energy.

In some cases, the movement of materials from one reservoir to another is driven by chemical energy rather than physical or gravitational energy. For example, if a water body reaches **saturation**—the reservoir is chemically "full" and cannot hold any more of a particular substance—then that substance may precipitate out of the water. This happens in certain zones within the ocean. When seawater is saturated with dissolved inorganic silicon, tiny organisms called radiolaria incorporate the silicon into their shells. If conditions change and the water becomes undersaturated in silicon, the radiolaria shells dissolve, releasing silicon back into the oceanic reservoir.

The rates and directions of fluxes (for example, whether silicon is precipitated in radiolaria shells or dissolved from the shells back into seawater) are controlled by an enormous variety of factors, including the chemical and physical characteristics of the substance itself (Is it chemically reactive? Is it stable? Is it heavy or light? Will it break down or dissolve easily, or is it durable?); the nature of the medium through which or by which it is being transported (Is it soil, or rock, or air? Is it water? Is it fresh or salt water? Is it chemically saturated?); environmental conditions (Is it hot or cold? Is it wet or dry?); and the presence of other substances (Will they react with one another? Will they slow one another down? Will they be chemically or physically altered as a result of their interaction?). The interaction of substances in the environment is particularly important, and we will return to it in subsequent chapters.

What are some processes in the natural world by which materials are moved around? Are they primarily physical, geologic, biologic, or chemical? Think about the driving forces for these processes. ______________

Answer: Here are some examples: precipitation (physical/chemical); wind (physical); flowing water (physical for materials that are carried, chemical for materials that are dissolved); metabolism, the transformation of food energy into body energy (biologic/ chemical); mineral weathering (geologic, chemical, and/or physical); decomposition of organic material (biologic/chemical).

Box Models

It is helpful to portray the burden of a reservoir and the fluxes of material into and out of the reservoir in graph form. Environmental scientists do this with a **box model**, a simplified picture of a cycle as an interconnected set of reservoirs. The boxes in a box model represent the reservoirs, and the arrows between the boxes represent transfer processes and fluxes. For example, we can portray the movement of water between the atmospheric and oceanic reservoirs in a simple box model, shown in Figure 2.3.

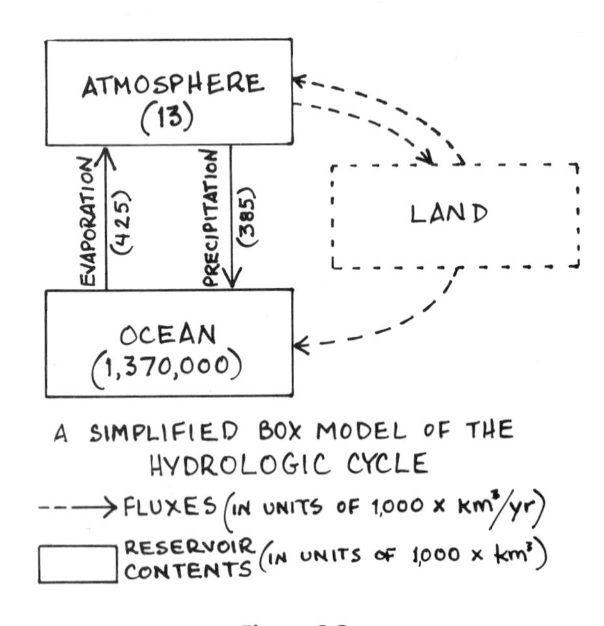

Figure 2.3.

Box models of real environmental cycles are usually significantly more complex than the greatly simplified example shown here. In this example, one box represents the oceanic reservoir, and the other represents the atmospheric reservoir. The numbers in the boxes give the burden or content—the amount of water in each reservoir. The arrows represent evaporation and precipitation—the transfer processes—and the numbers on the arrows show the magnitudes of the fluxes. Note that the flux of water out of the atmospheric reservoir by precipitation is quite large—much larger per year than the entire content of water in the atmosphere. Note, too, that the cycle is not balanced; the amount of precipitation *into* the ocean does not equal the amount of evaporation *from* the ocean, as mentioned above. This suggests that our drawing is missing some crucial parts. (In fact, it is missing all of the land-based reservoirs in the global water cycle, including lakes, rivers, glaciers, groundwater, and the biosphere, which will be discussed further in chapter 3.)

Use the concept of fluxes from one reservoir to another to restate the example given above of radiolaria that build their shells from silicon dissolved in seawater. Then draw a box model to illustrate this cycle. ______________

Answer: When seawater becomes saturated with silicon, there is a flux of silicon from the oceanic reservoir (where it is dissolved in seawater) into a biologic reservoir (where it is precipitated as a solid in the form of radiolaria shells). The opposite flux occurs when seawater is undersaturated with silicon; the radiolaria shells dissolve, releasing silicon from the biologic reservoir back into the oceanic reservoir (Figure 2.4).

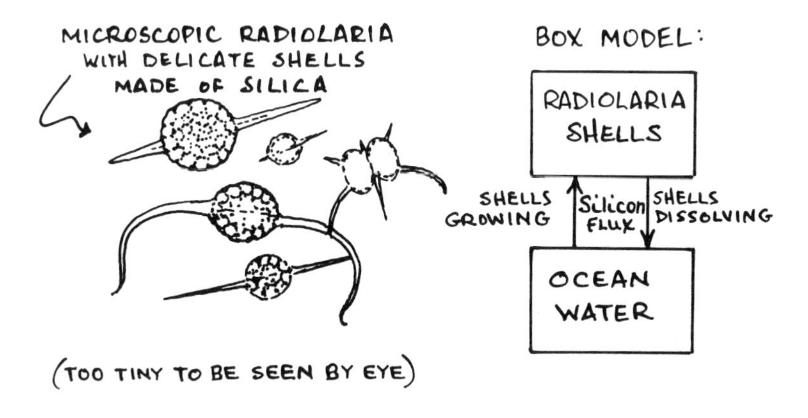

Figure 2.4.

Sources and Sinks

When considering fluxes, it is important to think about where a substance is coming from and where it is going to end up. The reservoir *from* which a substance moves is called a **source**. The reservoir *into* which a substance moves is called a **sink**. The material that moves into a reservoir is called **input** (or inflow, or influx), and the material that moves out of the reservoir is called **output** (or outflow, or outflux).

Consider the case of a laundry tub (Figure 2.5). The tap that brings water into the tub (and the reservoir where the water comes from) is a source. The drain that lets water out of the tub (and the reservoir where the water goes) is a sink. If you turn on the tap but plug the drain, the input of water will exceed the output. The amount of water in the reservoir will increase (the tub will fill up). If you turn off the tap and open the drain, the output will exceed the input. The amount of water in the reservoir will decrease (the tub will empty). If the tap is running and the drain is open to let out just as much water as is coming in, a state of dynamic balance will be achieved, and the water level will remain constant. This state, in which the input and output are equal and balanced, is called **steady state**.

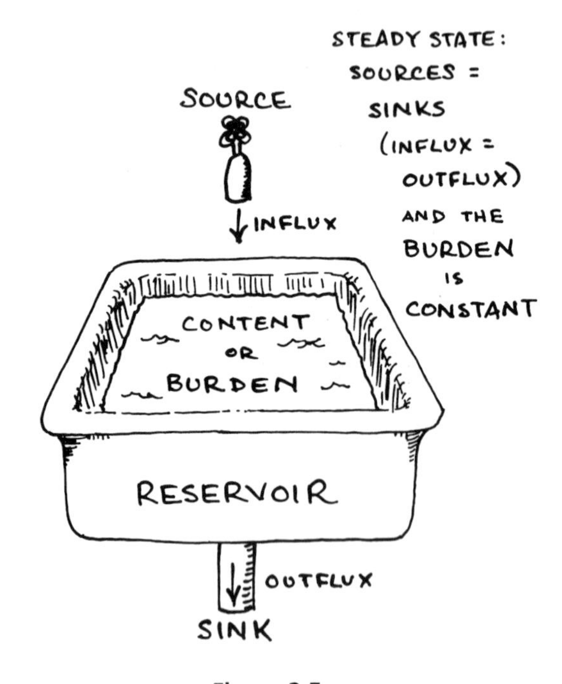

Figure 2.5.

Why are these concepts useful in environmental science? Think of the example of carbon dioxide emissions into the atmosphere from the burning of fossil fuels. An important sink for atmospheric carbon dioxide—one of the "drains" by which carbon dioxide leaves the atmospheric reservoir—is the ocean, which absorbs carbon dioxide into its surface waters. Let's say that atmospheric carbon dioxide is at steady state—the amount of carbon dioxide entering the atmosphere from all sources is equal to the amount that is being removed. Now imagine that we "turn up the tap" by burning more fossil fuels, greatly increasing the flux of carbon dioxide into the atmosphere, but the oceanic sink continues to function at the same rate as before. The predictable result is that carbon dioxide will begin to build up in the atmospheric reservoir. This is exactly what has begun to happen in the atmosphere, to the consternation of climate scientists and many policy makers. Much attention is now being paid in environmental science and policy to the preservation and enhancement of potential "sinks" for atmospheric carbon dioxide.

In the simplified box model of the ocean–atmosphere water cycle, consider the atmosphere to be like a tub that contains about 13×10^{18} g of water. Now imagine that the drain is opened but the tap is not turned on—that is, water is allowed to drain out of the atmospheric reservoir via precipitation (at a rate of 385×10^{18} g/yr) without being replenished by evaporation from the ocean. Under these conditions, how long would it take for the atmospheric tub to be completely emptied of water? ______________

Answer: $(13 \times 10^{18}$ g$)/(385 \times 10^{18}$ g/yr$) = 0.034$ year, or 12.4 days. This answer reveals that there must be sources from which the atmospheric reservoir is replenished with water. Otherwise, the atmosphere would completely dry up in a matter of days.

Residence Time and Turnover Time

If a substance is moving into and out of a given reservoir, how long does it remain in the reservoir? The storage time, or **residence time**—the average length of time an individual grain or molecule of a substance remains in a reservoir—differs greatly from one substance to another and from one reservoir to another, depending on a variety of environmental conditions.

Let's again consider the example of water—an extremely important substance in environmental science, both for its life-supporting characteristics and because it is responsible for transporting many different materials from place to place in the

Earth system. Water is transferred from one reservoir to another via innumerable pathways and processes, both biologic and nonbiologic. The residence times of water in the various reservoirs of the hydrosphere vary widely: an individual molecule of water may spend 100,000 years in a glacier, 1,000 years in an underground reservoir, 7 years in a lake, a few days in the atmosphere, or a few hours in an animal's body.

The concept of residence time can be applied to most substances in environmental reservoirs. For example, the residence times of chlorofluorocarbons (CFCs) in the atmosphere are quite long—forty years or more. (This is one reason why CFCs are problematic.) The residence time of lead (Pb) in the atmosphere, on the other hand, is quite short—about four days. (Lead in the atmosphere comes mainly from automobiles and industrial emissions.) But the residence times of heavy metals like lead, mercury, and arsenic in the bodies of animals can be quite long because the metals accumulate in the tissues of the animal and may remain there for the animal's lifetime. This is a problem because some metals are harmful to organisms.

A final concept related to residence time is **turnover time**, the *average* residence time of a substance in a reservoir. It is an indication of how long it would take to empty the reservoir of that substance if all sources were stopped and all sinks were maintained at a constant rate. The number that you calculated above—the amount of time it would take to empty the atmospheric reservoir of water under certain conditions—is a turnover time. Turnover times and residence times are important because we need to know how quickly or slowly materials cycle through different reservoirs and how long they are likely to remain in those reservoirs.

CFC-12 is a chemical implicated in the destruction of the ozone layer. The CFC-12 burden in the atmospheric reservoir is currently more than 10×10^6 tons. Assuming that CFC-12 breaks down or leaves the atmosphere at a rate of 0.1 million tons per year, and assuming that all anthropogenic inputs of CFC-12 into the atmosphere stop, what would be the turnover time—that is, how long would it take to empty the atmospheric reservoir of CFC-12? ______________

Answer: $(10 \times 10^6$ tons$)/(0.1 \times 10^6$ tons/yr$) = 100$ years.

Feedbacks

In a **feedback loop**, the output from a cycle is fed back into the cycle, becoming an input into the cycle. There are two basic types of feedback loops. In a **positive feedback** loop, the cycle is self-reinforcing. For example, let's say you take a shortcut across the grass to get to your classroom or office from the parking lot. This causes the grass to become beaten down a little bit along the path, which encourages the next person who comes along to take the same shortcut. That person's footsteps wear down the grass a little bit more, making the path even clearer, which encourages the next person to take the same shortcut; and so on, until the path is worn completely clear. This type of feedback loop is sometimes called a *vicious cycle*; it is a growing, self-reinforcing, destabilizing cycle.

In a **negative feedback** loop, in contrast, the cycle is self-limiting. For example, when you are thirsty, you drink water. The result is that you are no longer thirsty, and the cycle comes to an end. In another example, someone who is exercising very hard becomes overheated. The body responds to the overheating by sweating, which cools the person down. Negative feedback loops are self-limiting, stabilizing cycles.

There are lots of examples of negative and positive feedback cycles in the environment. Some particularly instructive examples can be found in Earth's climate system (which we will look at in greater detail in chapters 3 and 16). For example, global warming may lead to increased evaporation of water from the ocean, causing an increase in the amount of water vapor in the atmosphere. Water vapor contributes to the natural greenhouse effect, warming the surface of Earth. More water vapor in the atmosphere would lead to more greenhouse warming, which would cause more evaporation, leading to more water vapor in the atmosphere, causing more warming, and so on. This is a positive, self-reinforcing feedback cycle.

In contrast, increased evaporation of water as a result of global warming may lead to increased cloud cover. Under certain circumstances, clouds block incoming solar radiation, cooling the surface of Earth. This is a negative, self-limiting feedback cycle. Sorting out the relative influences of competing positive and negative feedback loops like these is one of the great challenges in understanding Earth's climate system.

What is the difference between a negative feedback loop and a positive feedback loop? Can you think of some examples that are different from those given in the text? ______________

Answer: Negative feedback loop: self-limiting, self-regulating, stabilizing. Positive feedback loop: growing, self-reinforcing, destabilizing, vicious cycle.

Biogeochemical Cycles

There are many important cycles in the Earth system. You learned about two of them in chapter 1. In the tectonic cycle, rocks are driven across the surface of the planet by internal heat energy, uplifted in great mountain ranges, and recycled back into the mantle. In the rock cycle, solid rock is broken down by weathering, transported by erosion, deposited as sediment, and turned back into sedimentary rock by compaction and other rock-forming processes. In chapter 3, you will learn more about the hydrologic cycle of evaporation, condensation, precipitation, and runoff within the hydrosphere. Another extremely important cycle is the energy cycle (chapter 11), which describes the pathways by which energy enters and leaves the Earth system, the forms and reservoirs in which it is stored, and the processes by which it is transferred from one reservoir to another.

Of particular importance in environmental science are the **biogeochemical cycles**, which trace the movement through the Earth system of chemicals that are essential to life. The most important global biogeochemical cycles are those of carbon, oxygen, nitrogen, sulfur, and phosphorus. Some people consider the water cycle to be a biogeochemical cycle as well. These essential substances move from one reservoir to another by way of both nonbiologic (geologic, chemical, physical) and biologic processes; they occur as gases, as solids, or dissolved in water, and they change from inorganic to organic forms and back again. Let's take a quick look at each of the important biogeochemical cycles.

The **carbon cycle** is complex and poorly understood partly because carbon fluxes vary greatly, both temporally and spatially (Figure 2.6). Carbon is intimately involved in virtually all biologic processes, which can vary dramatically with changes in temperature and other environmental conditions. A further complication is that carbon occurs in a very wide variety of chemical forms, some organic and some inorganic. Some carbon-bearing compounds are **oxidized** (oxygen-bearing, like carbon dioxide [CO_2]), and some are **reduced** compounds (lacking oxygen, like methane [CH_4]). Another complexity is that the processes that dominate the global carbon cycle on a short-term basis (turnover times of hours or days to hundreds of years) are quite different from the processes that dominate on a longer time scale (thousands to millions of years). On the short time scale, the biologic processes of photosynthesis, respiration, and decomposition, which vary daily to seasonally, are the most important controllers of carbon fluxes. On a longer time scale, these biologic processes are just "background noise"; longer-term carbon fluxes are controlled by geologic processes, such as sedimentation, subduction, metamorphism, and volcanism.

Carbon (as carbon dioxide and other carbon-bearing gases) plays a crucial role in the regulation of global climate. Human activities have caused carbon fluxes to increase by releasing carbon from intermediate- and long-term reservoirs through the burning of fossil fuels and biomass (trees, grasses); deforestation and land clear-

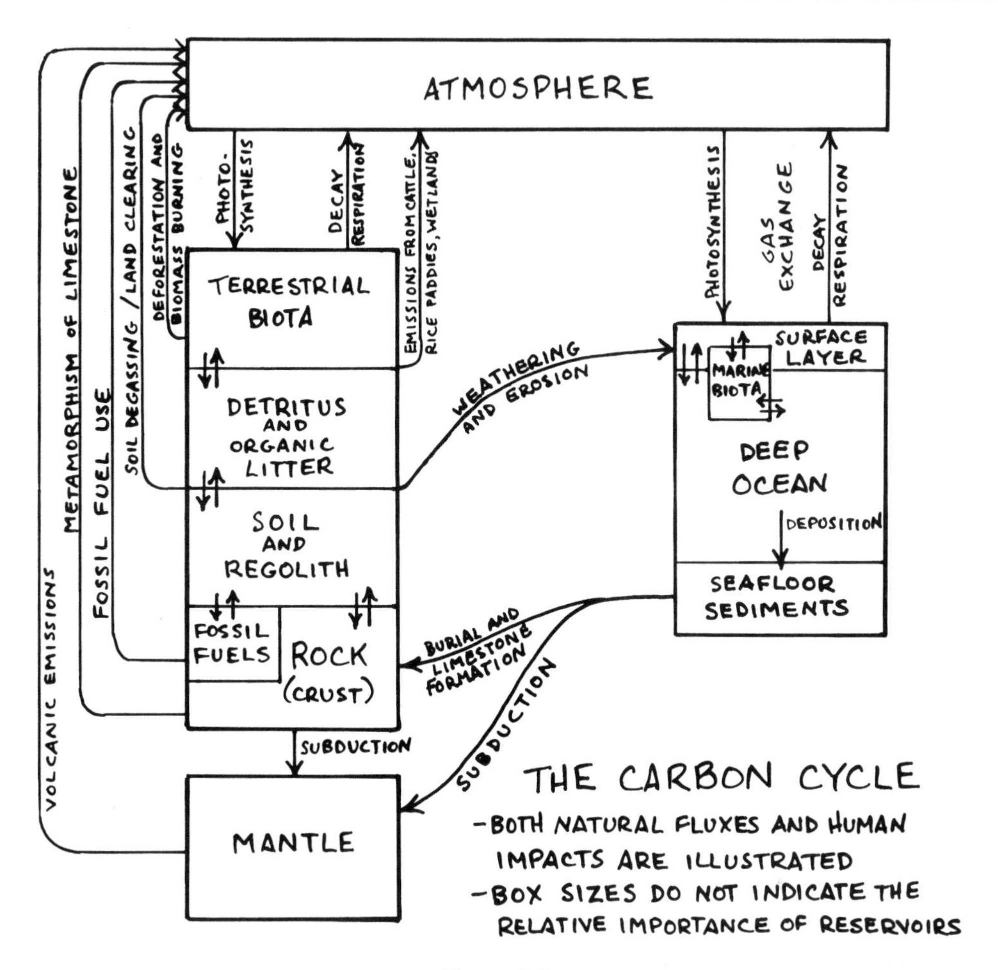

Figure 2.6.

ing; cattle ranching; and rice production. This has led to an increase in atmospheric carbon over the past two hundred years or so. We will reexamine the carbon cycle in the context of global climatic change in chapter 16.

The **oxygen cycle** is strongly coupled with the carbon cycle and, indeed, with the other biogeochemical cycles, since oxygen combines with all of these elements to form a variety of compounds (in other words, oxidizing them). Like the carbon cycle, the oxygen cycle is controlled to a great extent by biologic processes. One of the big drivers of this cycle is the process of decomposition of organic matter, which uses up oxygen and releases carbon dioxide into the atmosphere. For this reason, the oxygen cycle is almost the opposite of the carbon cycle; during periods of Earth's history when atmospheric oxygen has been high, carbon in the atmosphere typically has been low, and vice versa.

The **nitrogen cycle**, like the carbon cycle, is complex (Figure 2.7A). Like carbon, nitrogen moves through the Earth system in many different chemical forms, some oxidized and some reduced. The atmospheric reservoir is of obvious importance in the nitrogen cycle, since 79 percent of the atmosphere consists of nitrogen gas (N_2). In fact, the atmosphere is by far the largest global reservoir for nitrogen. (The largest reservoir in all of the other important biogeochemical cycles, including the oxygen cycle, is the solid Earth.)

Another important aspect of the nitrogen cycle is its role in soil-forming processes and in the provision of nutrients required for plant growth. We will discuss this in greater detail in chapter 9 in the context of soils and agriculture. In the natural cycle, nitrogen is transferred between the atmosphere and the ocean, biosphere, and soils via a number of biochemical reactions (Figure 2.7B). **Nitrogen fixation** reactions convert gaseous N_2 into nitrite (NO_2) and then into nitrate (NO_3), which is usable as a plant nutrient. In turn, **denitrification** reactions release nitrogen gas (N_2) as a by-product. Human activities affect the global nitrogen cycle through

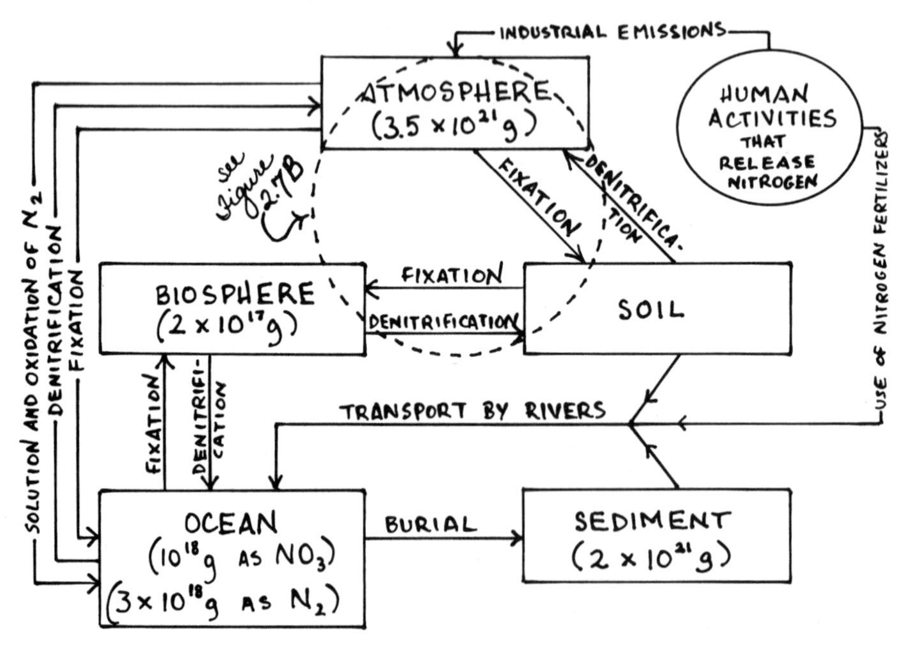

Figure 2.7A.

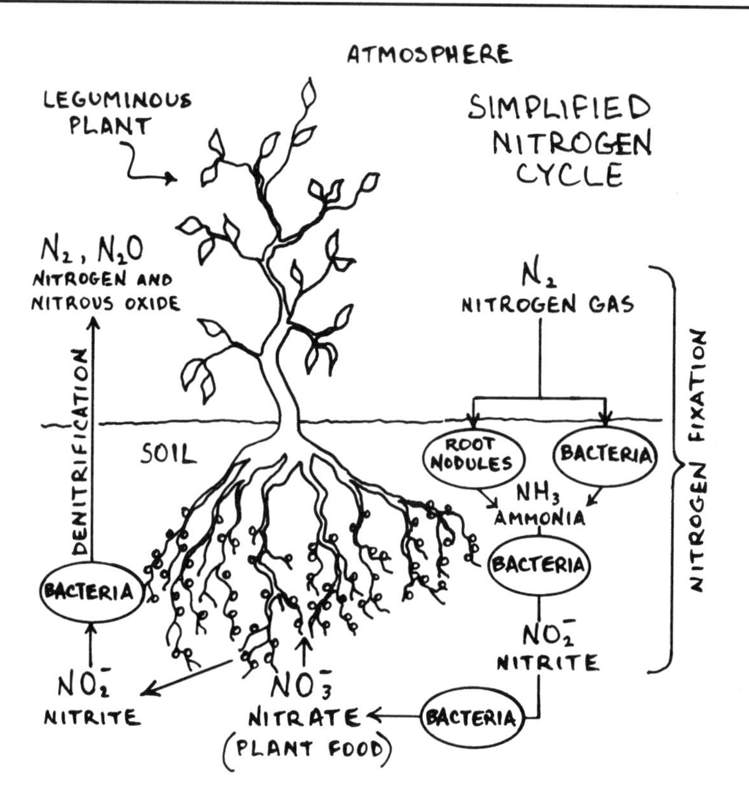

Figure 2.7B.

industrial emissions, automobile exhaust, and the use of nitrogen-rich fertilizers. Impacts include the production of photochemical smog, acid precipitation, and nitrate pollution of surface and groundwater.

Sulfur moves in the global **sulfur cycle** in gaseous, dissolved, and solid forms. In rocks and sediments, sulfur (S) occurs in a number of common minerals, including the mineral pyrite (FeS_2, or "fool's gold"). In the atmosphere, sulfur occurs in the form of reduced sulfur, sulfur dioxide (SO_2), and sulfate aerosols (SO_4). Under certain circumstances, sulfate aerosols in the atmosphere can form a "veil" that partially obscures incoming solar radiation. Thus, atmospheric sulfur plays an important role in the regulation of climate. In water, sulfur occurs in the form of dissolved sulfate. Sulfur compounds in water or in the atmosphere are problematic because they can react with the water, producing sulfuric acid (H_2SO_4). This can cause acid precipitation and acid mine drainage, two challenging environmental problems. Gaseous anthropogenic sulfur emissions come primarily from the burning of fossil fuels, especially coal. Human impacts are significant in the global sulfur cycle; it is now thought that more sulfur is mobilized by human activities than by all natural sources combined.

The **phosphorus cycle** differs from the other biogeochemical cycles because there are no gaseous forms of phosphorus that occur in significant quantity. Furthermore, phosphorus doesn't change its chemical form very readily—it mainly occurs as phosphate (PO_4) dissolved in water or in minerals. Phosphorus is a component of DNA and other basic building blocks of life, as well as being an important constituent of bones and teeth. Phosphorus, like nitrogen, is necessary for plant growth. Human activities affect the global carbon cycle through the use of phosphate-bearing fertilizers and detergents. Because phosphorus is a fertilizing agent, an overload of phosphate in a lake or a river can lead to excessive growth of aquatic plants (algae). This was a difficult problem in the Great Lakes as early as the 1960s. It has been somewhat controlled since then as a result of international agreements between the United States and Canada that limit the use of phosphates in detergents.

In what ways does the phosphorus cycle differ from the other biogeochemical cycles? ______________

Answer: There are no important gaseous forms of phosphorus, and it doesn't change form very readily, being transported mainly as phosphate in dissolved form or in minerals.

Systems, cycles, reservoirs, fluxes, feedbacks—a lot of new terminology and concepts have been introduced in this chapter. They are fundamentally important concepts, and you will find them woven throughout much of the rest of this book. One of the great environmental problems that we must face as a global society is that anthropogenic activities have altered the global fluxes of many substances from one reservoir to another. For some important substances, notably sulfur, the flux from anthropogenic emissions is now thought to greatly outweigh the total flux from natural sources. As you progress through this book, you will build up a more coherent picture of how the different cycles and processes of the Earth system are related to one another, how they work together to create our environment, and how they are altered by our actions. Now try the Self-Test.

SELF-TEST

These questions are designed to help you assess how well you have learned the concepts presented in chapter 2. The answers are given at the end. If you get any of the questions wrong, be sure to troubleshoot by going back into that part of the chapter to find the correct answer.

1. Which one of the following is *not* an example of a system because of the characteristic noted in parentheses?
   a. a rock (not alive)
   b. the universe (too big)
   c. an atom (too small)
   d. a mosquito (living organism)
   e. All of these are examples of systems.

2. Which one of the following is an example of a closed system?
   a. a cup containing water
   b. the human body
   c. a volcano
   d. a bird's nest
   e. None of these is a closed system.

3. Here are two descriptions of feedback loops:
   i. Joe loves reading about science. He reads a lot about science, so he really understands science books. This makes it even more enjoyable for him to read about science, so he reads even more science books.
   ii. Jon hates reading about science. He doesn't do it very often, so he doesn't understand science books very well. This makes it even less enjoyable for him to read about science, so he reads even fewer science books.

   Which one of the following is the correct description of these feedback loops?
   a. Both (i) and (ii) are negative feedback loops.
   b. Both (i) and (ii) are positive feedback loops.
   c. (i) is a positive feedback loop, and (ii) is a negative feedback loop.
   d. (i) is a negative feedback loop, and (ii) is a positive feedback loop.

4. The important global biogeochemical cycles are those of ______________, ______________, ______________, ______________, ______________, and water.

5. The ______________ or ______________ is the total mass of a particular substance that is contained within a reservoir.

6. A mass of material that is transferred during a given period of time is called a(n) ______________.

7. In a system at steady state, there are no inputs and no outputs. (True or False)

8. The atmosphere is the largest reservoir for nitrogen in the Earth system. (True or False)

9. For one of the questions in this chapter, you drew a simple box model of inorganic silicon moving back and forth between the oceanic reservoir and a biologic (radiolaria shells) reservoir. In that box model, which reservoir is the source and which is the sink?

10. What are three important implications of living in a closed system?

11. In the simple model of the water cycle between the ocean and the atmosphere (above), precipitation (water influx into the ocean) and evaporation (water outflux from the ocean) are not balanced. Calculate how long it would take for the ocean to dry up if more water did not come into the oceans from other sources.

12. Define *system*.

13. Would a city be more like an open system or a closed system?

ANSWERS

1. e (None of the reasons given would prevent these objects from being described as systems.)

2. e (They all permit energy and matter to cross their boundaries; hence, they are open systems.)

3. b (They are both positive, self-reinforcing cycles.)

4. carbon; oxygen; nitrogen; sulfur; phosphorus

5. burden; content 6. flux 7. False 8. True

9. Sorry, this was a trick question. Both of the reservoirs can be sources, and both can be sinks. It depends on which way the silicon is moving. From the perspective of the radio-

laria shells, the ocean is a sink for silicon. From the perspective of the ocean, radiolaria shells are a sink for silicon.

10. Resources are limited; wastes remain in the system; changes in one part of the system have an impact elsewhere in the system.
11. The burden of water in the ocean in $14{,}000 \times 10^{20}$ g (or $1{,}400{,}000 \times 10^{18}$ g). The amount of evaporation over and above the amount of precipitation in the model is:

$$(425 \times 10^{18} \text{ g/yr}) - (385 \times 10^{18} \text{ g/yr}) = 40 \times 10^{18} \text{ g/yr}$$

$$(1{,}400{,}000 \times 10^{18}) \div (40 \times 10^{18} \text{ g/yr}) = 35{,}000 \text{ yr}$$

The ocean would completely dry up in 35,000 years if it did not receive extra inputs of water from runoff and from precipitation.

12. A system is any portion of the universe that can be separated from the rest for the purpose of studying changes within it.
13. Cities are open systems. Energy comes into the city in a variety of forms, such as solar energy, and natural gas and oil inputs into the processes of the city, such as lighting, heating, cooling, manufacturing, and so on. Energy leaves the city in the form of heat. Materials, such as food, water, and other resources, flow into the city. Waste materials, including garbage and atmospheric emissions, flow out.

KEY WORDS

anthropogenic
biogeochemical cycle
box model
burden
carbon cycle
closed system
cycle
denitrification
equilibrium
feedback loop
flux
input
negative feedback
nitrogen cycle
nitrogen fixation
open system
output
oxidized
oxygen cycle
phosphorus cycle
positive feedback
principle of environmental unity
reduced
reservoir
residence time
saturation
sink
source
steady state
subsystem
sulfur cycle
system
turnover time

3 The Hydrosphere and the Atmosphere

For Hot, Cold, Moist, and Dry, four champions fierce,
Strive here for mastery.

—John Milton

Objectives

In this chapter you will learn about:

- the hydrologic cycle and the role of water in the environment;
- the ocean, the largest reservoir for water in the hydrologic cycle;
- the characteristics of the atmosphere; and
- the interactions of the ocean and the atmosphere and how they influence and regulate weather and climate.

The Hydrosphere

The hydrosphere is one of the great interacting reservoirs that make up the integrated Earth system. The hydrosphere comprises the ocean, lakes, and streams; underground water; and glaciers. However, water is present in all of the reservoirs of the Earth system, not just the hydrosphere. Water vapor is an important component of the atmosphere, the envelope of gases that surrounds Earth. Water is a constituent of many common minerals in the lithosphere, tightly locked in crystal structures. And, of course, water is a fundamental component of living organisms in the biosphere. The hydrosphere is particularly important in environmental science because of the role of water in maintaining life and regulating other environmental processes, such as weather and climate. Water is also the principal transporting medium by which materials are moved within the Earth system; this includes natural materials, such as soils and sediments, as well as contaminants.

The water cycle, or **hydrologic cycle** (from the Greek *hydro,* meaning "water"), describes the movement of water from one reservoir to another on and just underneath the surface and in the atmosphere. Scientists who study the movement, characteristics, and distribution of water are **hydrologists**. The largest reservoir in the hydrologic cycle is the ocean, which holds 97.5 percent of the water in the hydrosphere (Figure 3.1). The fact that most of the hydrosphere is salty has important implications for humans because we depend on fresh water for drinking, agriculture, and industrial use. The largest reservoir of fresh water is the polar ice sheets, containing almost 74 percent of all fresh water. Subsurface water, or **groundwater**, accounts for almost 98.5 percent of the remaining unfrozen fresh water. A very small fraction of the water in the hydrologic cycle resides in surface freshwater bodies or in the atmosphere, and even less is stored in the biosphere.

The movement of water in the hydrologic cycle is powered mainly by the Sun (Figure 3.2). Heat from the Sun causes **evaporation**, the process by which water changes from a liquid to a vapor. Water evaporates from the ocean, surface water bodies, vegetation, and land surfaces. Water that passes through plants on its way to the atmosphere is said to undergo **transpiration** (or **evapotranspiration**). Depending on local conditions of temperature, pressure, and humidity, some of the

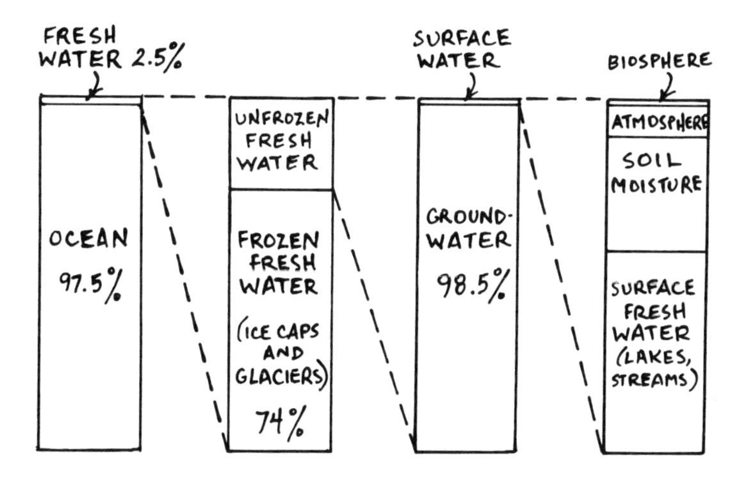

Figure 3.1.

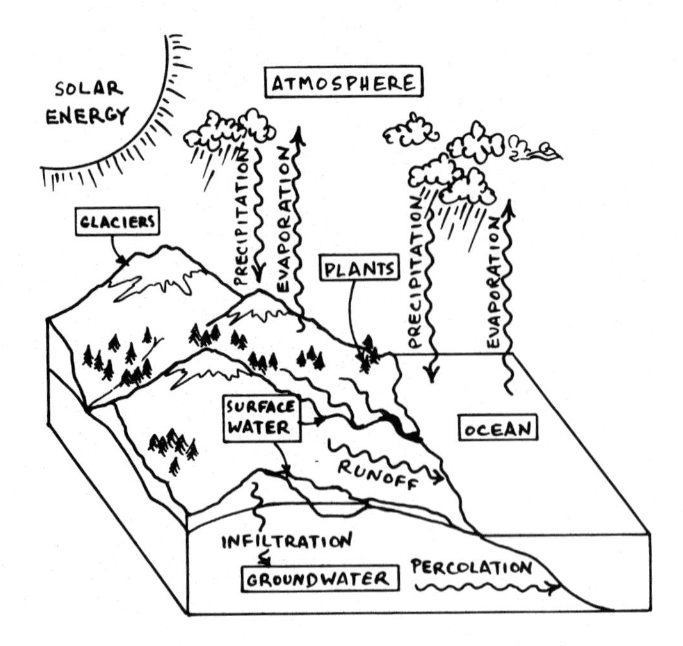

LANDSCAPE DIAGRAM OF THE WATER CYCLE

Figure 3.2A.

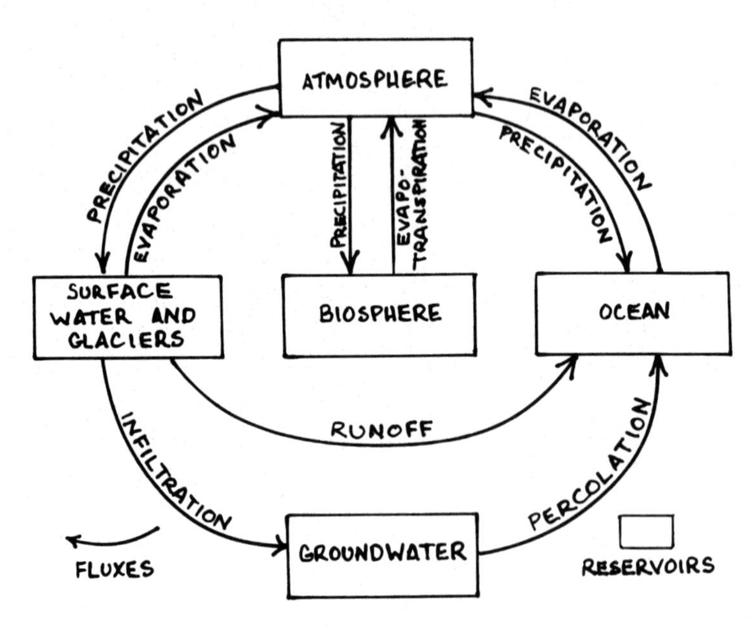

BOX MODEL OF THE WATER CYCLE

Figure 3.2B.

water vapor in the atmosphere may undergo **condensation**—it changes from a vapor into a liquid or a solid. It may then fall as **precipitation** (rain, snow, or hail).

Precipitation that falls on land may evaporate, or be intercepted by plants, or drain over the land, becoming **surface runoff**. It may flow in temporary or long-lasting stream channels, or be stored in surface water bodies such as lakes or wetlands. Some water works its way into the ground by **infiltration**, passing through small openings and channels in soil and rocks. Water that infiltrates deeply enough becomes part of the vast reservoir of groundwater, where it may reside for hundreds of years, migrating slowly toward a discharge point by the process of **percolation**. Water stored in the snow and ice of glaciers may remain locked up even longer, sometimes for hundreds of thousands of years.

Are the water fluxes and reservoirs shown in Figures 3.2A and 3.2B exactly the same? What differences can you spot? Which version of the hydrologic cycle is the "correct" one? _______________

Answer: They are not exactly the same. Figure 3.2A shows glaciers and surface water as separate reservoirs; they are shown together in Figure 3.2B. Figure 3.2A does not give specific information about the transfer of water between the biosphere (plants) and the atmosphere. Both versions are "correct"; they just show two different ways of visualizing and portraying the information, with differing levels of detail. (Note that neither diagram shows soil moisture as a separate reservoir.)

Because Earth operates as a closed system (chapter 2), the hydrologic cycle maintains a **mass balance**: the total amount of water cycling around in the system is basically fixed. In the global hydrologic cycle, the amount of water that evaporates from the surface of the ocean (425×10^{18} g/yr) exceeds the amount that is returned directly to the ocean in the form of precipitation (385×10^{18} g/yr). The ocean doesn't dry up because the amount of precipitation that falls over the land (111×10^{18} g/yr) exceeds the amount of water that evaporates from the land surface (75×10^{18} g/yr). The excess water that falls on the land eventually returns to the ocean as runoff or groundwater flow, balancing the global cycle. Of course, this is an oversimplification; there are many local and regional variations, as well as shorter- and longer-term storage reservoirs that hold water (such as glaciers). But overall a global balance is achieved in the hydrologic cycle.

The processes that control the local and global movements and distribution of water are important in everyday life. During extended periods with below-average precipitation, we suffer from drought. Where infiltration rates are low and precipitation is high, water accumulates and floods may occur. Water also performs important environmental services, such as the dilution and removal of wastes. This sometimes causes problems; for example, rain falling on a landfill may dissolve noxious chemicals, carrying them away to contaminate a stream or groundwater supply. Water transports silt and sediment, which can cause the land to lose its valuable topsoil, a hill slope to become unstable, or a stream channel to become clogged. Large bodies of surface water—especially the ocean—also influence the weather. Even the landscapes around us have been shaped and sculpted by the erosional work of streams, waves, and glaciers. In all these ways and more, the hydrologic cycle influences our everyday lives.

What is the main energy source for processes in the hydrologic cycle?

____

Answer: The Sun.

The Ocean

Seawater covers 70.8 percent of the surface of Earth. Most of the water on our planet is contained in three huge interconnected basins: the Pacific, the Atlantic, and the Indian oceans. (The Arctic Ocean is an extension of the Atlantic.) All three are connected with the Southern Ocean, the body of water that encircles Antarctica. Collectively, these four vast interconnected bodies of water and about twenty smaller seas make up the **world ocean**. Land, which constitutes the rest of Earth's surface, is unevenly distributed around the globe. A view from directly above Russia shows mostly land; a view from directly above New Zealand shows mostly water. The uneven distribution of land and water plays an important role in determining the paths along which water circulates in the open ocean.

What set of processes controls the distribution of continents and ocean basins on Earth? ________________

Answer: Plate tectonics.

The composition of water in the ocean is expressed as **salinity**, or saltiness. The salinity of seawater ranges between 33‰ and 37‰ (‰ = per mil; for comparison, 1% = 1 percent = 1 part in 100; 1‰ = 1 per mil = 1 part in 1,000). The main elements that contribute to this salinity are sodium (Na^+) and chlorine (Cl^-). When seawater evaporates, much of the dissolved matter is precipitated as common salt or sodium chloride (NaCl), the mineral known as halite. Seawater contains most of the other naturally occurring elements as well but in lower concentrations. The elements that make the various salts in seawater come mainly from the weathering of minerals. As rocks of the crust interact with the atmosphere and rainwater during weathering, chemical elements are released to become part of the dissolved material that is carried to the ocean by stream water. Other sources of mineral salts are volcanic eruptions and dust blown out to sea from deserts.

The salinity of seawater is controlled by the interplay between four major processes:

1. Evaporation removes fresh water and leaves the remaining water saltier.
2. Precipitation adds fresh water and makes seawater less salty.
3. Fresh water from rivers makes seawater less salty (river water carries dissolved minerals, but it is still less salty than seawater).
4. Freezing excludes salts (that is, ice crystals are purer H_2O than the salt water from which they crystallize), leaving the unfrozen seawater saltier.

The salinity and temperature of seawater control its density. High salinity causes high density; high temperature causes low density. **Oceanographers** (ocean scientists) define three major density layers or zones in the ocean. The **surface zone**, typically extending to a depth of 100 m (328 ft), consists of water that is relatively warm and therefore low in density. Below the surface zone, to a depth of about 500 m (1,640 ft), the water temperature decreases rapidly with depth, while salinity and density increase with depth. This zone of abrupt change is called the **thermocline**. Below the thermocline is the very cold **deep zone**, which contains the bulk of the ocean's volume.

What are the four major processes that control the salinity of seawater?

Answer: 1. Evaporation removes fresh water, leaving seawater saltier. 2. Precipitation adds fresh water, making seawater less salty. 3. The inflow of fresh water from rivers makes seawater less salty. 4. Freezing excludes salts, leaving unfrozen seawater saltier.

Oceanic Circulation

The salinity and temperature of ocean water are the primary drivers of deep ocean currents. Seawater near Antarctica and in the Arctic Ocean is very cold; it is also quite saline because the formation of sea ice removes fresh water. Low temperatures and high salinity mean high density. In polar regions, the cold, salty, dense water sinks, setting into motion a deep global oceanic circulation system that has a profound influence on climate. The deep, cold water moves away from the poles, resurfaces as it warms while passing through the Pacific and the Indian oceans, and eventually returns to the polar seas. This deep oceanic circulation pattern is called the **thermohaline circulation** (Figure 3.3).

Surface ocean currents are broad, slow drifts of water driven by the wind. Air flowing across the sea surface drags the water forward, creating a current of water as broad as the current of air but only 50 to 100 m deep (that is, confined to the surface zone). The ultimate energy source for surface ocean currents is the Sun, which drives the planetary wind system. In low latitudes (near the equator), surface seawater moves westward along the equator (Figure 3.4) driven by equatorial wind systems. These are called **trade winds** because early exploration and trading vessels took advantage of them in navigating across the ocean.

If Earth did not rotate, oceanic circulation patterns would be controlled by density (that is, temperature and salinity), wind, and the presence of continental landmasses. But Earth does rotate, and the rotation complicates both the deep and shallow circulation patterns in the ocean. The **Coriolis effect**, named after the nineteenth-century French mathematician who first analyzed it, causes anything that moves freely with respect to the rotating Earth to veer to the right in the Northern Hemi-

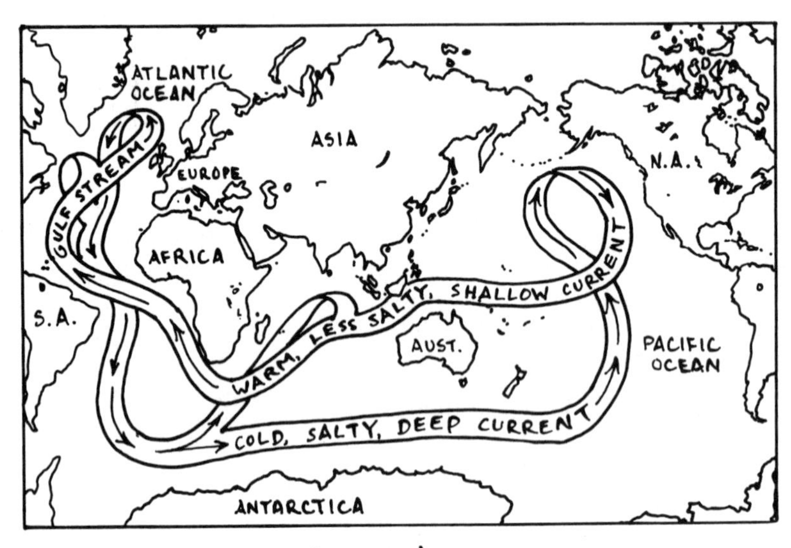

Figure 3.3.

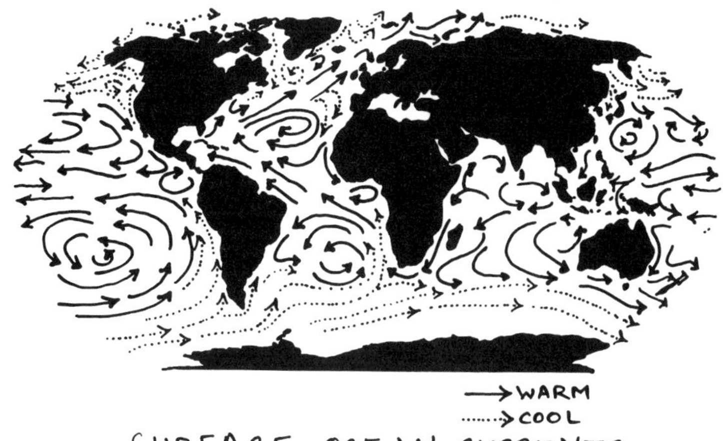

Figure 3.4.

sphere and to the left in the Southern Hemisphere. All fluids, including flowing water and flowing air, respond to the rotation. Thus, the global patterns of both ocean currents and wind systems are influenced and complicated by the Coriolis effect.

To the north and south of the equator, westward-moving currents are deflected wherever they encounter a coast and by the Coriolis effect. Currents of warm surface water originating in the equatorial region thus veer to the north or south along the eastern margins of continents. By the time they reach the middle latitudes, the surface ocean currents turn toward the east as they are driven by the prevailing winds. Eventually, this colder, east-flowing water encounters the western margins of continents and is deflected back toward the equator. The result is a circular motion of water in each major ocean basin north and south of the equator (Figure 3.4).

What is the thermohaline circulation and what drives it? ______________

Answer: The global deep oceanic circulation system and is driven by variations in temperature and salinity.

Tides, Waves, and Shorelines

A majority of the world's population lives within 100 km (60 mi) of the ocean. This reflects our dependence on oceans and the economic benefits afforded by easy access to ocean resources and services, as well as the richness of coastal zone resources. However, the concentration of large numbers of people in coastal areas means that the coastal environment must absorb the impacts of a wide range of human activities. It also means that human vulnerability to hazards is particularly high. Infrequent events such as large storms are hazardous to life and sometimes cause extraordinary damage to property. Slow, continuous processes such as coastal erosion are less dramatic, but they can be equally damaging in the long run.

Two important forces that act upon shorelines in coastal zones are waves and tides. **Tides** are the regular rise and fall cycles of ocean water. The gravitational attraction of the Moon causes ocean water to bulge upward on the side of Earth nearest to the Moon. On the other side of the planet, inertia (the force that tends to maintain a body in motion) created by Earth's rotation also causes ocean water to bulge but in the opposite direction (Figure 3.5). The result is two tidal bulges, one on either side of Earth. (The Sun's gravitational force also affects the tides; however, the Sun is so far away that it is much less effective in producing tides.) While Earth rotates, the two tidal bulges remain stationary beneath the Moon. Thus, any given coastline will move westward through both tidal bulges each day. Every time a landmass encounters a tidal bulge, the water level along the coast rises. A coast passes across both tidal bulges during every complete rotation of Earth, so two high tides and two low tides are observed each day along most coastlines. The shape of a coastline can greatly influence the run-up height (the highest elevation reached by the incoming water), the range (the difference in water level between high and low tides), and the energy of the incoming or outgoing tide.

Waves, like surface currents, receive their energy from winds. The size of a wave depends on how fast, how far, and how long the wind blows across the water surface. A gentle breeze blowing across a bay may ripple the water or form low waves less than a meter high. By contrast, storm waves produced by low-pressure weather centers and intense winds blowing across hundreds or thousands of kilometers of open water may become so high that they tower over ships. Crashing onto the shore, these **storm surges** can cause significant damage.

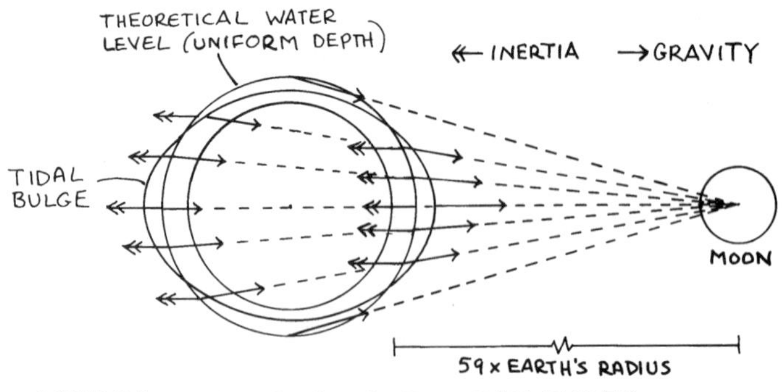

Figure 3.5.

If you visit a shoreline, look carefully at the waves coming onto the beach. You can usually see them arriving from two or more directions. Each set of waves originated from a storm-related wind far beyond your line of sight.

Breaking waves are powerful erosional agents along shorelines. Wave erosion takes place not only at sea level but also below and—especially during storms—above sea level. If you visit almost any coastal zone on two occasions a year apart, you will see changes that result from the constant back-and-forth erosional energy of waves. Sometimes the changes are small, but often they are substantial. Large sand dunes may have shifted. Sand deposits may have built up or eroded away. Steep sections of coastline may have collapsed. Channels may have broken through from the sea to lagoons on the landward side where there were no channels before.

People who build houses, towns, and businesses along beaches and other shorelines sometimes take steps to protect their land and property from erosion. This may take the form of engineered protection, such as concrete seawalls, jetties, or breakwaters; or they may try replenishing the beach by importing sand or stabilizing it by planting dune grasses and other plants. The long-term effectiveness of engineered approaches to the protection of shorelines is a matter of intense controversy among experts. Beaches are in a constant and delicate state of balance with respect to the erosion and deposition of sand by waves and tides. Many experts argue that extensive interference in the natural cycles of shorelines can lead to expensive and possibly irreversible damage in the long run, such as permanent loss of the beach and extensive damage to beachfront buildings.

The two basic approaches to beach protection are called hard stabilization and soft stabilization. What do you think these terms refer to? ______________

Answer: Hard stabilization refers to engineered protection, such as seawalls, jetties, and breakwaters. Soft stabilization refers to sand replenishment and the planting of grasses to hold the sand in place.

The Atmosphere

An atmosphere, defined in chapter 1, is an envelope of gases surrounding a planet. **Air** is the mixture of gases and suspended particles that surrounds one special planet, Earth—Earth's atmosphere is made of air. The composition of air varies from place to place and time to time, mainly because of variations in aerosols and water vapor. **Aerosols** are liquid droplets and solid particles that are so small they remain suspended in the air. Water droplets in fog are liquid aerosols, for example.

Solid aerosols include tiny ice crystals, smoke particles from fires, sea-salt crystals from ocean spray, fine dust stirred by winds, volcanic emissions, and many types of atmospheric pollutants.

The amount of water vapor in air—its **humidity**—is also highly variable. On a hot, humid day in the tropics, as much as 4 percent of the air by volume may be water vapor. On a crisp, cold day, less than 0.3 percent water vapor may be present. Air can become saturated (filled to capacity) with moisture simply by rising, which causes the air to expand and cool. Cool air holds less moisture than warmer air; thus, the moisture in rising, cooling air will condense into droplets and clouds will form. **Clouds** are visible aggregates of tiny droplets of water or ice crystals. They come in a wide variety of shapes that are indicative of the processes by which they form.

Because the water vapor and aerosol contents of air are so variable, the composition of air is usually reported as if the air were dry (free of water vapor) and aerosol-free. The relative proportions of the remaining naturally occurring gases in air are almost constant (Table 3.1). Three gases—nitrogen, oxygen, and argon—make up 99.96 percent of dry air by volume. Other gases are present in very small quantities.

What are the three main components of dry air? What are the two most variable components of air? ______________

Answer: Nitrogen, oxygen, and argon. Aerosols and water vapor.

Another important characteristic that defines the state of the atmosphere in any given place or time is **air pressure**, which is measured with an instrument called a **barometer**. Air pressure is a measure of the weight of overlying air; it therefore decreases with altitude. Anyone who has climbed a mountain knows that the higher you go above sea level, the less oxygen there is and the harder it becomes to breathe.

Table 3.1 Composition of Earth's Atmosphere (in percent, by volume)

Gas	Percent
Nitrogen (N_2)	78.08
Oxygen (O_2)	20.95
Argon (Ar)	0.93
Carbon dioxide (CO_2)	0.035
Neon (Ne)	0.018
Helium (He)	0.00052
Methane (CH_4)	0.00014
Krypton (Kr)	0.00010
Nitrous oxide (N_2O)	0.00005
Hydrogen (H_2)	0.00005
Ozone (O_3)	0.000007

The oxygen supply becomes short not because of a change in the composition of the atmosphere but rather because of the reduction in air pressure. The air that fills our lungs at high altitude is less dense (less matter filling the same volume) than the air at sea level, and therefore it fills the lungs with less oxygen.

The temperature profile of the atmosphere reveals four distinct layers (Figure 3.6), each separated by thermal boundaries called **pauses**. From the bottom, the layers are:

1. the **troposphere**, from the surface up to about 15 km (9 mi);
2. the **stratosphere**, up to 50 km (30 mi);
3. the **mesosphere**, up to 90 km (56 mi); and
4. the **thermosphere**, up to 700 km (more than 400 mi).

We live at the bottom of the troposphere. The troposphere contains 90 percent of the actual mass (that is, the matter) of the atmosphere, including virtually all the

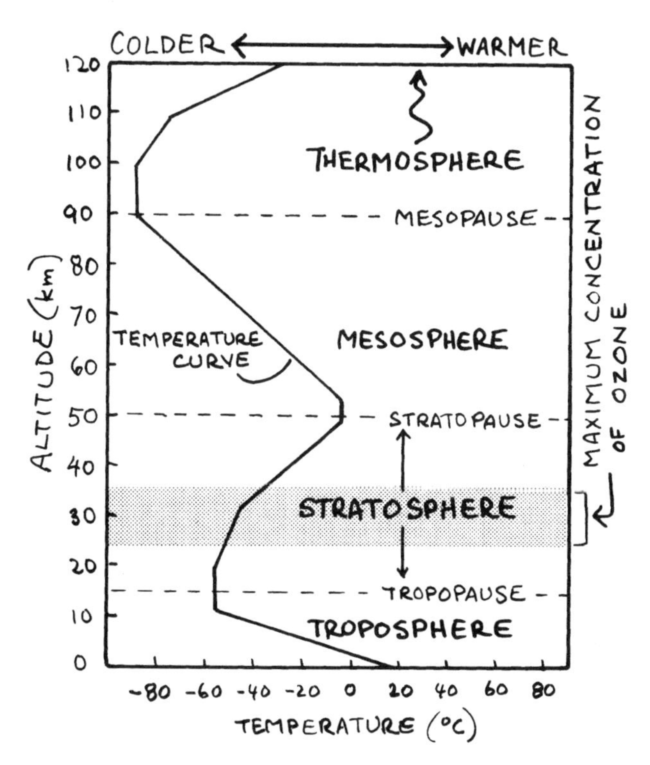

Figure 3.6.

water vapor and clouds. Almost all weather-related phenomena originate in the troposphere. Although little mixing occurs between the troposphere and the stratosphere, the troposphere itself is dynamic, constantly moving, and thoroughly mixed by winds.

The troposphere contains a concentration of naturally occurring **radiatively active** or **greenhouse gases** (**GHGs**) that are responsible for warming Earth's surface. The most important of these are water vapor (H_2O) and carbon dioxide (CO_2). Incoming solar energy is mostly short-wavelength (ultraviolet) radiation. Through interactions with the atmosphere, ocean, land, and biosphere, the short-wavelength radiation is changed into longer-wavelength (infrared) radiation, which is reradiated back into outer space. While passing through the atmosphere, the outgoing energy encounters radiatively active gases that absorb the long-wavelength energy and retain it in the lower atmosphere in the form of heat. This natural process is called the **greenhouse effect**; without it, the surface of Earth would be cold and inhospitable to life. However, the buildup of human-generated (anthropogenic) greenhouse gases in the atmosphere is causing a great deal of concern lately because of the potential for enhancement of the greenhouse effect and resulting global warming. This will be discussed further in chapter 16.

The stratosphere contains another 19 percent of the atmosphere's total mass. This leaves about 1 percent to reside in the upper 250 km (150 mi) of the atmosphere. The stratosphere and its high-speed wind systems play an important role in the global distribution of materials in the atmosphere. If pollutants, dust, or volcanic emissions are ejected high enough to reach the stratosphere, they may be circulated globally within a matter of days or weeks. The stratosphere also contains a natural concentration of the gaseous chemical ozone (O_3). Ozone protects life by absorbing some of the most harmful short-wavelength ultraviolet radiation, preventing it from reaching the surface. This concentration of O_3 in the stratosphere is called the **ozone layer**. Recently there has been concern about the breakdown of stratospheric ozone by anthropogenic chemical pollutants, notably a group of chemicals called chlorofluorocarbons (CFCs), which will also be covered in chapter 16.

What is the scientific term for greenhouse gases? ________________

Answer: Radiatively active gases.

Where is the ozone layer, and what important function does it perform?

____

Answer: In the stratosphere; it prevents harmful short-wavelength radiation from reaching the surface.

Atmospheric Circulation

The amount of solar energy reaching the surface of Earth, or **insolation**, varies from place to place because Earth is a sphere. Where the Sun is exactly overhead, which mainly happens near the equator, the incoming rays are perpendicular to the surface and insolation is high. Everywhere else, because of the planet's curvature, the incoming rays are at an angle to the surface; therefore, less energy reaches each square meter of surface area. Atmospheric circulation is driven by the unequal heating of the surface. Circulation in the atmosphere works to smooth out temperature differences by transporting heat and moisture from one part of the globe to another. The work is done by the movement of large bodies of air called **air masses**, which are relatively homogeneous in temperature and humidity. A zone along which two air masses meet is a **front**.

To understand how atmospheric circulation works, remember that air expands when it is heated, becoming less dense (lower air pressure), and contracts when it cools, becoming more dense (higher air pressure). Thus, a heated air mass near the equator expands, becomes lighter, and rises, creating a low-pressure zone. Near the top of the troposphere it spreads outward in the direction of the poles. As the air mass travels at high altitude away from the equator, it gradually cools, becomes heavier, and sinks, creating a high-pressure zone. Upon reaching Earth's surface, this cool, descending air flows back toward the equator, warms up, and rises. This movement—the rising of hot air, lateral flow, and sinking of cooler, denser air—forms a convection cell.

Do you recall another context in which convection is important in the Earth system?

Answer: Plate tectonics and mantle convection. (In the case of plate tectonics, the material that is convecting is solid rock. In the case of the atmosphere, it is air. The two materials could hardly be more different, but the fluid dynamic principles that govern the process of convection are the same.)

If Earth did not rotate, the convection currents of air would rise at the equator, flow from the equator to the poles at high altitude, then cool and sink, flowing back to the equator along the surface. But the planet does rotate, and simple convection in the atmosphere is complicated by the Coriolis effect, which breaks up the flow of air between the equator and the poles into belts (Figure 3.7).

For example, a warm air mass rising at the equator begins to move toward either the North or the South Pole but is deflected by the Coriolis effect. By the time it reaches a latitude of about 30° (N or S), the high-altitude air mass has cooled and is

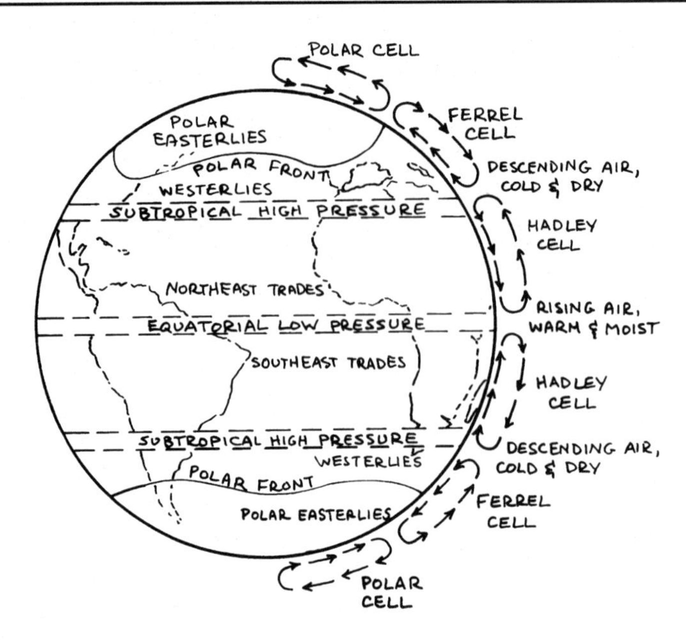

Figure 3.7.

flowing eastward. (Note that a westerly wind blows toward the east *from* the west; an easterly wind blows toward the west *from* the east.) It meets a mass of cool air flowing from the poles toward the equator, and both air masses descend. Upon reaching the surface, some of the descending air flows back toward the equator, where it will become heated and rise again. These circulation cells that dominate the wind systems from 0° to about 30° N and S are called **Hadley cells**. The low-pressure zone at the equator where warm air masses meet (converge) to form the rising limb of the Hadley cells is called the **intertropical convergence zone**. Winds are constantly moving toward this zone because air always tends to flow from regions of high pressure toward regions of low pressure. The low-altitude winds that deliver cool air back to the intertropical convergence zone along the equator flow roughly toward the west (that is, they are easterlies); these are the trade winds mentioned above in connection with surface oceanic circulation.

Hint: If you are trying to understand global atmospheric circulation, it really helps if you follow along on a map or a globe.

In each hemisphere, a second belt of air circulation cells lies poleward of the Hadley cells. These are called **Ferrel cells**. A third set of circulating air masses, called

polar cells, lies over the polar regions. In each polar cell, cold, dry, high-altitude air descends near the pole and moves toward the equator. As this air slowly warms, it rises along the polar front where the polar cells and the Ferrel cells meet, then returns toward the pole. Pressure variations along fronts, especially the polar front, can be quite dramatic. This leads to high-speed winds called **jet streams** traveling in narrow corridors along fronts and exerting a strong influence on the weather.

What are trade winds? ______________

Answer: Easterly winds near the equator, the low-altitude limb of the Hadley circulation.

Weather

The world ocean plays a critical role in regulating the temperature and humidity of the lower part of the atmosphere. Atmospheric circulation, in turn, drives ocean waves and currents and transfers heat. Global patterns of airflow are influenced by uneven heating of the surface, the Coriolis effect, variations in air pressure, and the distribution and topography of landmasses. All of these forces combine to produce weather and climate.

It is important to clarify the difference between weather and climate. **Weather** refers to local conditions in the atmosphere at any given time and is described in terms of temperature, air pressure, humidity, cloud cover and precipitation, and wind direction and velocity. The weather pattern in a given region averaged over a long period is **climate**. The processes that we think of as weather—wind, rain, snow, sunshine, storms, even floods and droughts—are just temporary local variations against the longer-term background of climate.

The interactions among air, land, and water that create weather are not only complex but also highly sensitive to changing conditions. **Meteorologists** (scientists who study weather) sometimes characterize this sensitivity by saying that a change as small as the draft created by a butterfly's wing can become magnified through a network of feedbacks, ultimately developing into a wind or even a storm. They even have a name for this phenomenon: the *butterfly effect*. The variability, complexity, and extreme sensitivity of the atmosphere–ocean system make weather prediction a very tricky business.

To understand more about weather-generating processes, think about the atmospheric circulation patterns described above. When an air mass is heated and rises near the equator, it carries an enormous quantity of moisture that has evaporated from the ocean surface (Figure 3.8). At high altitude, the air mass will cool, and the moisture

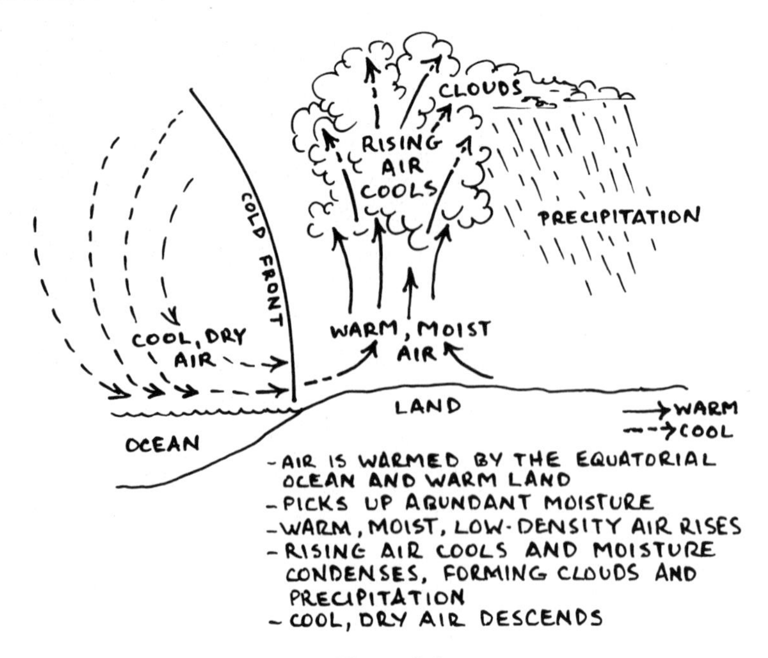

Figure 3.8.

it contains will condense and fall as precipitation. This is why tropical areas are not only hot but usually very cloudy and rainy as well. By the time the air mass has reached high altitude and is moving toward the poles, it has lost most of its moisture. Upon reaching the subtropical front where the Hadley and Ferrel cells meet, the descending air is very dry. Thus, two belts characterized by very low precipitation lie between 20° and 30° latitude, North and South. If you look at a map, you will see that many of the world's great subtropical deserts, including the Sahara, Kalahari, and Great Australian deserts, are associated with these belts, where dry air descends on the downward-flowing limbs of the Hadley and Ferrel cells.

Another important influence on the weather is air pressure. As air near the ground moves toward a low-pressure center, frictional drag across the surface causes the air to develop an inward spiraling motion (Figure 3.9). Similarly, airflow tends to spiral outward from high-pressure centers. Air spiraling inward around a low-pressure center is called a **cyclone**; air spiraling outward around a high-pressure center is called an **anticyclone**. The inward convergence of air in a cyclone results in an upward flow of air at the center of the cyclone. This upward flow carries moisture-laden air, leading to cloud cover and rain. In contrast, high-pressure systems tend to draw dry, high-altitude air downward into the center, leading to clear, cloudless weather. Because of the Coriolis effect, cyclonic systems rotate in a

At the time this image was taken, Typhoon Sudal was off the coast of the Philippines, with sustained winds of 80 km/hr (132 mph) and gusts to almost 90 km/hr (146 mph). Days earlier, the storm lashed Micronesia on its way westward through the South Pacific Ocean, causing extensive damage.

counterclockwise direction in the Northern Hemisphere and clockwise in the Southern Hemisphere; the opposite is true for anticyclones.

Cyclonic storms go by a variety of names, including **hurricanes** (Caribbean and North America), **typhoons** (Western Pacific), and **tropical cyclones** (Indian Ocean). When fully developed, a tropical cyclone is like a huge whirlpool hundreds of kilometers in diameter. Tropical cyclones last up to several weeks and may travel erratically. The intense low-pressure center creates strong winds. Near the center or **eye** of the storm, the inward-rushing winds are drawn upward, so the winds never reach the exact center of the storm. This creates the eerie calm that characterizes the eye. A **tornado** is another type of cyclonic storm, which is characterized by a very intense low-pressure center and extreme winds. Tornadoes are short-lived and local

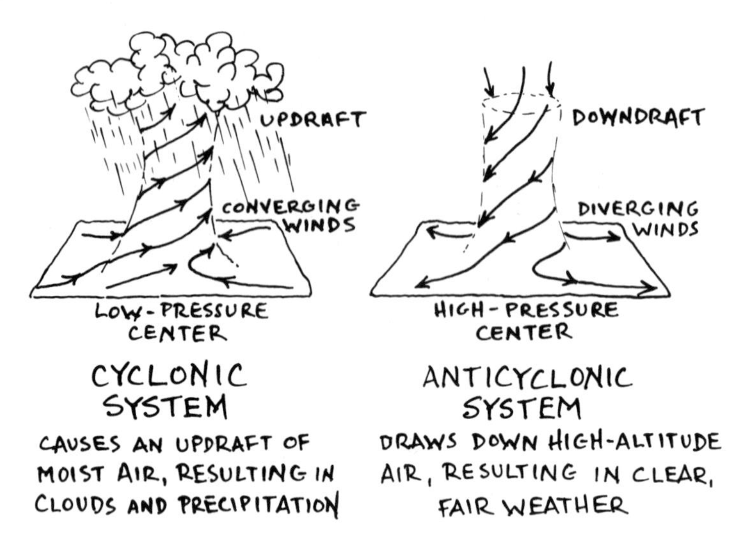

Figure 3.9.

in extent, but they can be very violent. They are particularly common in the southeastern and central United States.

Some weather that we think of as extreme is actually characteristic of the climate zone in which it occurs and can be explained by examining the movement of air masses in the region. An example is the torrential rains typical of the Asian **monsoon**. During the winter months, the wind blows across India from the high, cold plateau of central Asia, bringing clear weather. During the summer, the pattern is reversed. With the Sun overhead on land in the Northern Hemisphere, the landmass of Asia heats up and is covered by cyclones. Warm, moisture-laden winds blow from the Indian Ocean onto the land, and summer is a time of hot, humid weather and heavy rains. The reversing winds that characterize the monsoon are characteristic of this region. They take their name from the Arabic word *mausim*, meaning "change."

Why are low-pressure centers associated with rain and cloud cover, while high-pressure centers are associated with clear, cloudless skies and dry weather?

Answer: The inward convergence of air in a low-pressure system results in the upward flow of warm, moisture-laden air, leading to cloud cover and rain. High-pressure systems draw dry, high-altitude air downward, leading to clear, cloudless weather.

Climate

Climate is the average of weather conditions over a long period. It takes into account the "normal" weather for the area as well as the weather extremes. **Climatologists** (scientists who study climate) use a variety of classification schemes to describe and categorize Earth's climate types. The simplest classification is based on latitude and includes three main divisions:

1. low-latitude, including wet equatorial, wet-dry tropical, dry tropical, monsoon, and trade-wind climates;
2. midlatitude, including dry subtropical, moist subtropical, mediterranean, marine west coast, moist continental, and dry midlatitude climates; and
3. high-latitude, including boreal forest, tundra, and ice-sheet climates.

The most important characteristics of climate are temperature and precipitation. Climate types tend to correspond closely to vegetation zones because vegetation is controlled by temperature and precipitation, and they are also closely related to the movements of large air masses. Climate is further influenced by topography and land–water distribution. For example, deserts often occur on the landward side of large mountain ranges because moisture-laden air tends to cool and lose its moisture through precipitation as it travels over the mountains. By the time it gets to the landward side, the air is very dry, creating a rain shadow with very low precipitation and resulting desert conditions.

The ocean has a moderating influence on climate because water has the ability to absorb and release large amounts of heat with very little change in temperature. For this reason, both the temperature range and seasonal variations in ocean temperatures are much less than on land, and coastal inhabitants benefit from the resulting mild climates. Along the Pacific coast of Washington and British Columbia, for example, winter air temperatures seldom drop to freezing, whereas east of the coastal mountain ranges they plunge to very cold temperatures. In the interior of a continent, summer temperatures may exceed 40°C, whereas along the coast they typically remain below 25°C.

The influence of wind systems, water bodies, and topographic features can be regional, as in the examples cited above, or local. For example, a constantly sunny hillslope may experience a significantly higher average temperature than a slope that faces away from the Sun. Prevailing winds and precipitation can also vary dramatically from one location to another. Differences in climate that are local in scale are called **microclimates**; they can have a profound effect on vegetation.

The complex nature of ocean–atmosphere–land interactions is perhaps best illustrated by the periodic climatic event called **El Niño**. Under normal circumstances, the trade winds blow from east to west, pushing warm surface water across the Pacific and away from the western coast of South America. This allows deep, cold, nutrient-laden waters from the Southern Ocean to well up along the coast. An El Niño begins when the trade winds weaken. Warm surface water collects near the coast of South America, suppressing the upwelling of cold, deep, nutrient-rich water. The fish population declines, accompanied by a great die-off of coastal birds that depend on fish for food. The Peruvian fishery is among the most important in the world, so the occurrence of a major El Niño constitutes a local economic catastrophe. El Niño also causes heavy rains to fall in normally arid parts of Peru and Ecuador. Australia experiences drought conditions; anomalous cyclones appear in Hawaii and French Polynesia; the seasonal rains of northeast Brazil are disrupted; and the Indian monsoon may fail to appear. Unusually cold or mild winters can occur in the northeastern United States, while the Southwest becomes wetter; in California, abnormally high rainfall can produce major flooding and widespread landslides.

When an El Niño event occurs, its effect is felt over at least half of Earth. El Niño not only involves the tropical oceans and atmosphere but also directly affects precipitation and temperature on major land areas, thereby also impacting plants, animals, and people. This is a good example of interactions among the atmosphere, hydrosphere, and biosphere.

What is the difference between a meteorologist and a climatologist?

Answer: A meteorologist studies weather. A climatologist studies climate (weather conditions averaged over a long period).

We have portrayed climate as stable and unchanging in comparison to weather. On a longer time scale, however, a host of natural influences can cause dramatic changes in climate. We will explore some of these, along with potential anthropogenic impacts on climate, in subsequent chapters. Meanwhile, the Self-Test will help you assess your understanding and retention of the material presented in this chapter. Don't forget to test yourself on the vocabulary words as well.

SELF-TEST

These questions are designed to help you assess how well you have learned the concepts presented in chapter 3. The answers are given at the end. If you get any of the questions wrong, be sure to troubleshoot by going back into that part of the chapter to find the correct answer.

1. The three major temperature zones of the ocean are the ________.
   a. thermosphere, mesosphere, and deep zone
   b. surface zone, troposphere, and thermosphere
   c. surface zone, thermocline, and deep zone
   d. convergence zone, thermocline, and surface zone
2. The density of seawater ________ with increasing temperature and ________ with increasing salinity.
   a. increases; increases
   b. decreases; decreases
   c. increases; decreases
   d. decreases; increases
3. The ________ is where most of Earth's weather is generated.
   a. troposphere
   b. stratosphere
   c. mesosphere
   d. thermosphere
4. The Coriolis effect is caused by ________.
   a. unequal solar heating of Earth's surface
   b. tidal interactions among Earth, the Moon, and the Sun
   c. Earth's rotation
   d. All of the above are true.
5. The three major sets of convecting air masses in the global wind system are the ________, ________, and ________ cells.
6. High-speed winds that travel along a front are called ________.
7. The four temperature layers of the atmosphere from the bottom up are the ________, ________, ________, and ________.
8. Approximately 40 percent of the water in the hydrosphere is groundwater. (True or False)

9. Cyclones always spiral clockwise, and anticyclones spiral counterclockwise. (True or False)
10. How does the natural greenhouse effect work?
11. The text states that the hydrologic cycle is powered *mainly* by the Sun. Can you think of another source of energy that drives water movement in the hydrologic cycle?
12. What is condensation?
13. Why do most coastlines experience two high tides and two low tides each day?
14. What is the difference between weather and climate?
15. What are the five factors by which weather is described?

ANSWERS

1. c 2. d 3. a
4. c 5. Hadley; Ferrel; polar 6. jet streams
7. troposphere; stratosphere; mesosphere; thermosphere 8. False 9. False
10. Incoming solar radiation is mostly short-wavelength, ultraviolet radiation. By interaction with various components of the Earth system, it becomes transformed into long-wavelength (infrared) radiation, or heat, which is reradiated from Earth to outer space. As this radiation passes through the atmosphere, it encounters radiatively active (greenhouse) gases, which absorb some of the heat and keep it trapped near the surface.
11. Gravity, which causes both surface water and groundwater to flow downhill.
12. The process by which a vapor changes to a solid or a liquid.
13. Two tidal bulges of water, one on each side of Earth, are caused by inertia and gravitational attraction with the Moon. These bulges remain essentially stationary, while Earth rotates through them. High tide occurs each time a coastline passes through the peak of one of the bulges; low tide occurs in between. A given coastline will pass through each of the two bulges every day. Thus, there are two high tides and two low tides daily.
14. Weather is the local condition of the atmosphere (temperature, air pressure, winds, precipitation, and so on) at any given time. Climate is the weather patterns of a particular area averaged over a long time.
15. Temperature; air pressure; wind direction and speed; humidity; cloudiness and precipitation.

KEY WORDS

aerosols
air
air mass
air pressure
anticyclone
barometer
climate
climatologist
cloud
condensation
Coriolis effect
cyclone
deep zone (ocean)
El Niño
evaporation
evapotranspiration
eye (of a storm)
Ferrel cell
front
greenhouse effect
greenhouse gas (GHG)
groundwater
Hadley cell
humidity
hurricane
hydrologic cycle
hydrologist
infiltration
insolation
intertropical convergence zone
jet stream
mass balance
mesosphere
meteorologist
microclimate
monsoon
oceanographer
ozone layer
pause (atmospheric)
percolation
polar cell
precipitation
radiatively active gas
salinity
storm surge
stratosphere
surface runoff
surface zone (ocean)
thermocline
thermohaline circulation
thermosphere
tide
tornado
trade wind
transpiration
tropical cyclone
troposphere
typhoon
weather
world ocean

4 The Biosphere: Life on Earth

. . . endless forms most beautiful and most wonderful have been, and are being, evolved.

—Charles Darwin

Objectives

In this chapter you will learn about:

- the biosphere and its history;
- the basic functions of living organisms;
- the organization of life into populations, communities, and ecosystems; and
- the flow of energy through ecosystems.

Life on Earth

The history of the biosphere is closely intertwined with that of the atmosphere and the hydrosphere. Without a hospitable atmosphere and hydrosphere, life as we know it could not survive. Without life, the atmosphere and the ocean would not exist in their present forms. More than 4 billion years ago, Earth's atmosphere was very different. There was no "free" oxygen (O_2), a necessity for most forms of life on Earth today. It was also very hot; the early atmosphere consisted primarily of greenhouse gases, which trapped heat near the surface. It was too hot for water to exist as a liquid, so there were no oceans, lakes, or rivers. Atmospheric pressure was much greater than it is today. Altogether, early Earth was inhospitable to life as we know it.

As the planet cooled, water vapor condensed and fell as rain. It collected in low-lying areas, forming bodies of water on the surface, perhaps as early as 4.4 billion years ago. The early hydrosphere reacted with gases in the atmosphere to form acids, and the acids reacted with the rocks of the crust, causing chemical weathering. The compositions of the atmosphere, the hydrosphere, and the lithosphere changed as materials cycled among them. The material that makes up the atmosphere and ocean came mainly from volcanic gases; there is also evidence of contributions of material from comets.

Earth's atmosphere now contains approximately 21 percent oxygen. Volcanic gases (which are typically oxygen-poor) and comets could not have supplied all of this oxygen. Where did it come from? Some oxygen was generated in the early atmosphere by the breakdown of water molecules (H_2O) into hydrogen and oxygen as a result of interactions with solar radiation. This is an important process, but it doesn't come close to accounting for the present level of oxygen in the atmosphere. Another oxygen-producing process was required, and it came from life itself.

How, when, and where life began on this planet is not known, although scientists are getting closer to understanding some of the chemical and biologic processes involved. The oldest fossil remains ever found are 3.55 billion years old, and the chemical signatures of biologic processes have been detected in rocks as old as 3.9 billion years. Some of these very early fossils are deposits left behind by simple, single-celled organisms similar to modern photosynthetic bacteria. In **photosynthesis**, organisms use light energy to make carbon dioxide react with water, producing carbohydrates (food energy) and releasing oxygen. Most organisms in the biosphere depend on photosynthesis either directly or indirectly to obtain food (Figure 4.1). Almost all of the free oxygen currently in the atmosphere originated through photosynthesis. You can learn more about photosynthesis and other biologic processes discussed in this chapter in *Biology: A Self-Teaching Guide*, by Steven D. Garber (John Wiley & Sons, 2002).

Eventually, early photosynthetic organisms generated enough oxygen to permit molecular oxygen (O_2) to build up in the atmosphere. When there was very little atmospheric oxygen, all of the oxygen was used up in combinations with other elements; free oxygen—molecular oxygen that is not in a chemical compound with other elements—could persist in the atmosphere only after oxygen was present in higher concentrations. Along with the buildup of oxygen came an increase in ozone (O_3). When there was sufficient ozone, it began to function as a screen to filter out harmful ultraviolet radiation—the ozone layer. Organisms were then able to survive

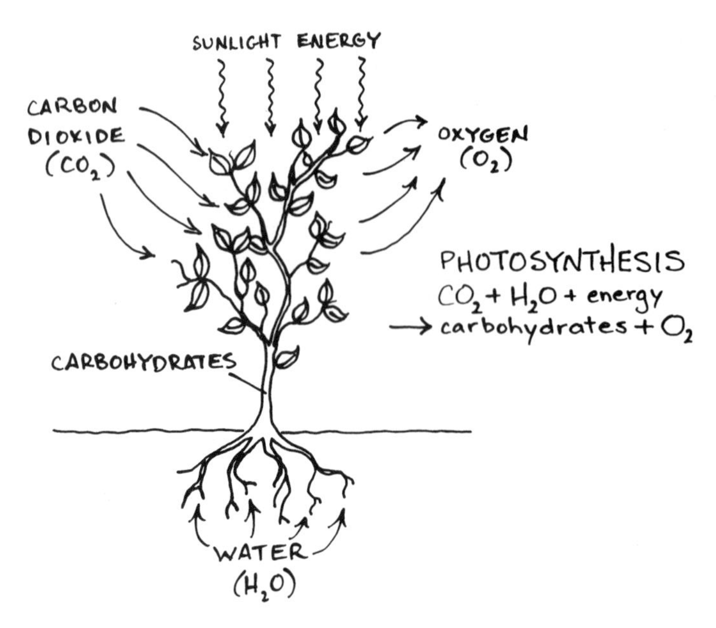

Figure 4.1.

and flourish in shallow waters and eventually on land. This critical stage was reached around 600 million years ago, and the fossil record reveals an explosion of life-forms at that time.

The emerging biosphere in turn had a profound impact on biogeochemical cycling, especially on the carbon cycle. The shells of marine organisms are composed primarily of calcium carbonate ($CaCO_3$), providing a storage reservoir for carbon dioxide ($CaCO_3 = CaO + CO_2$). When these organisms die, their shells are buried by seafloor sediments and are eventually transformed into limestone. In the carbon cycle (chapter 2), limestone is a long-term storage reservoir for carbon dioxide, isolating it from the atmosphere and hydrosphere. If all the carbon dioxide currently stored in limestone and other sedimentary rocks were released, there would be as much CO_2 in Earth's atmosphere as in the atmosphere of Venus, where the greenhouse effect runs rampant and the surface temperature is 480°C (900°F).

Where did the free molecular oxygen in the atmosphere come from?

Answer: Some of it came from the breakdown of water molecules in the atmosphere as a result of interactions with ultraviolet radiation, but most of it came from photosynthesis.

Basic Biologic Processes: An Overview

What distinguishes living from nonliving things is that living organisms can reproduce, metabolize, and grow. Let's first consider reproduction. You may recall from biology class that the genetic plan of a living organism is encoded in its **DNA** (**deoxyribonucleic acid**). DNA consists of two chainlike molecules held together by organic molecules that store genetic information (Figure 4.2). The twisted, ladderlike form of the DNA molecule is called a *double helix*. The information and instructions stored in the DNA are decoded and executed by **RNA** (**ribonucleic acid**). A cell cannot reproduce without RNA because RNA contains the information required to construct an exact duplicate of the proteins that make up the cell.

Another crucial characteristic of life is **metabolism**, the set of chemical reactions through which an organism derives food energy. Organisms that produce their own organic compounds from inorganic chemicals are called **autotrophs**. Most autotrophs produce food in the form of carbohydrates through the process of photosynthesis. A few autotrophs produce food from sulfur-bearing inorganic chemicals in nonoxygenated, or **anaerobic**, environments through a process called **chemosynthesis**.

Organisms that derive food energy by feeding on other organisms or on organic compounds produced by other organisms are called **heterotrophs**. When a heterotroph consumes another organism, the energy stored in the organic compounds is released by one of two metabolic processes. Heterotrophs that cannot tolerate oxygen obtain their energy through the anaerobic process of **fermentation**, in which carbohydrate molecules are partially decomposed to form alcohol, carbon

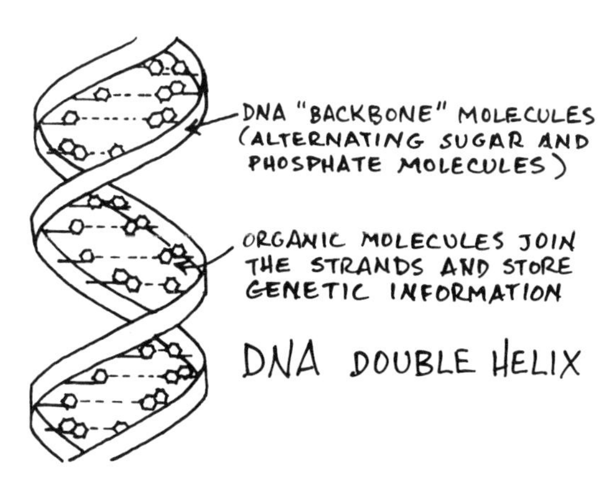

Figure 4.2.

dioxide, and water, releasing energy. Heterotrophs that are oxygen tolerant, on the other hand, obtain their energy through the **aerobic** (oxygenated) process of **respiration**, which means they use oxygen to break down carbohydrates, releasing carbon dioxide, water, and energy. Respiration is a more efficient metabolic process than fermentation.

Growth, the process whereby atoms and molecules organize themselves into larger groupings of molecules, is another characteristic of living organisms. Crystals can grow, but they lack the other characteristics of life: the ability to reproduce and metabolize food energy. Viruses can also grow, but they utilize the reproductive machinery of other organisms to replicate themselves, and they lack the ability to metabolize; hence, viruses are not quite alive—they are somewhere between living and nonliving things.

What is the difference between an autotroph and a heterotroph? ______________

Answer: Autotrophs produce their own organic compounds from inorganic chemical compounds either by photosynthesis or by chemosynthesis. Heterotrophs derive food energy by feeding on organic compounds produced by other organisms.

All organisms are composed of one or more cells. The **cell** is the basic structural unit of life, a complex grouping of chemical compounds and structures enclosed in a porous wall, or membrane. The development of the cell membrane was a crucial step in the evolution of life. The membrane separates the materials and chemical reactions that occur inside the cell from the environment outside it. This allows for complex organizational structures within the cell and facilitates the exchange of materials and energy between the cell and its environment.

The earliest cells were **prokaryotic cells**. Prokaryotes are small, simple cells that lack distinctly defined areas in which the cell's various functions are carried out. Present-day bacteria and related organisms are prokaryotes. **Eukaryotic cells** are larger and more complex than prokaryotic cells. Their DNA is housed in a well-defined area, the nucleus, which is separated from the rest of the cellular material by a membrane. Eukaryotic cells contain a variety of organelles, each of which has a specific role in the functioning of the cell. Humans, animals, plants, and fungi all consist of eukaryotic cells.

What is the difference between prokaryotic and eukaryotic cells? ______________

Answer: Prokaryote: simple cell; lacks distinctly defined organelles; nucleus not separated from the rest of the cell by a membrane. Eukaryote: larger and more complex cell; nucleus separated by a membrane; contains well-defined organelles, each of which has a specific function.

Evolution and Species

We have discussed three characteristics that distinguish living from nonliving things: reproduction, metabolism, and growth. Living organisms also possess a fourth defining characteristic: the potential to evolve and develop new forms, or **species**. It is more difficult to define the term *species* than it might seem. Some types of scientists prefer a definition based on the organism's morphology—that is, its appearance (from the Greek *morph*, meaning "form"). Thus, organisms that look similar to one another and share many of the same structural characteristics are said to belong to one species. Other scientists prefer a definition based on the organism's biologic characteristics; specifically, organisms that are able to breed successfully with one another, producing fertile offspring, are said to belong to the same species.

Paleontologists, who study the fossilized remains of plants and animals, have no choice but to rely on an organism's morphology to define it as a member of one or another species; there is no way to determine whether two organisms that lived millions of years ago would have been able to breed successfully. However, sometimes a morphologic approach doesn't work very well. Consider domestic dogs, for example, which all belong to the same species even though they vary widely in morphology (from enormous dogs like Great Danes to tiny "teacup" poodles). But all dogs are biologically capable of breeding to produce fertile offspring; thus, they belong to a single species. Horses and donkeys, which do look somewhat similar, can breed, but their offspring (mules) are not fertile; hence, horses and donkeys belong to different species.

With the passage of geologic time, the early, simple, single-celled organisms eventually diversified into the vast array of species that inhabit Earth today. But how? The theory of **evolution** basically says that new species evolve from old species and that through a gradual process of adaptation to environmental conditions, all present-day organisms are descendants of different kinds of organisms that existed in the past. Evolution is an essential characteristic of living things, just like growth, metabolism, and reproduction.

Evolution is achieved through the process of **natural selection**, in which poorly adapted individuals tend to be eliminated from a population. Natural selection occurs because the characteristics of individuals are passed from one generation to the next through their **genes**, which are composed of DNA. Sometimes a spontaneous change happens when genes are passed from parents to offspring—a mistake, essentially—that causes variations in the inherited characteristics of offspring. This is called **mutation**; it can happen naturally or as a result of an environmental influence on the organism. All natural populations have individuals with a variety of genetic characteristics. In any given environment, some of these characteristics will be more advantageous than others. For example, certain characteristics might enable an individual to compete more effectively for scarce resources or to escape predators more easily. Individuals that possess advantageous characteristics are more likely to survive and produce offspring with similar characteristics. This is called **adaptation**. When this happens, there are fewer descendants to inherit the genetic characteristics of the poorly adapted individuals and pass them on to the next generation. Over time, the entire population evolves, as natural selection favors individuals that are particularly well adapted to their environment.

A population's characteristics may diverge when part of the group is subjected to new environmental conditions—for example, if part of a population becomes geographically isolated. A mountain chain or an invasion by the sea might provide a physical barrier that separates two groups. Or some individuals might migrate across a large river or to an island and become isolated from the main population. In such cases the separated group must adjust to a new and different environment. Natural selection favors those individuals that are best suited to the new environment. Before the separation occurred, the two groups were part of a single species. After the separation, they may eventually become so different that they can no longer interbreed successfully; a new species has developed.

Given the task of studying the enormous variety of life on this planet, biologists have sought to organize species into a **taxonomic hierarchy** (Figure 4.3). The prokaryotes (bacteria) belong to one of two kingdoms: Archaebacteria or Eubacteria. All living organisms that are not prokaryotes belong to one of four kingdoms of eukaryotes: Protoctista (or protista, single-celled and simple multicellular eukaryotes); Fungi (mushrooms, lichens, and their relatives); Animals (multicellular organ-

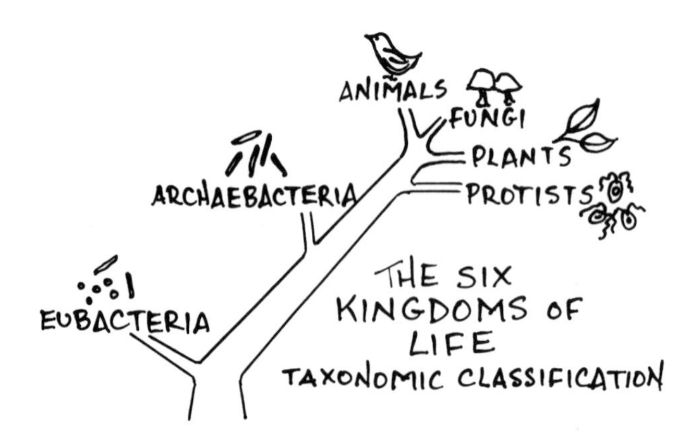

Figure 4.3.

isms that obtain their food by consuming other organisms); and Plants (multicellular, sexually reproducing eukaryotes that produce their own food; more complicated than algae). The kingdoms are further subdivided in a hierarchic manner into successively narrower categories: Phylum, Class, Order, Family, Genus, and Species. For example, a modern human is classified as follows: Kingdom: Animalia; Phylum: Chordata; Class: Mammalia; Order: Primates; Family: Hominidae; Genus: *Homo*; Species: *Homo sapiens*; Subspecies: *Homo sapiens sapiens.*

What is a species? Give both a morphologic and a biologic definition.

Answer: Morphologic definition: organisms that look similar and share many of the same structural features are said to belong to the same species. Biologic definition: organisms that can breed successfully and produce fertile offspring are said to belong to the same species.

Ecology

It should be clear by now that individuals and species do not exist in isolation—there are countless, complex interactions among individuals, among species, and between the biotic and abiotic components of the environment. **Ecology** is the scientific study of these interactions.

Do you recall (from chapter 1) the meaning of the terms *biotic* and *abiotic*?

Answer: Biotic = living; abiotic = nonliving.

As discussed earlier, individual organisms of the same species live together in groups that interbreed and share genetic material; such groups are called **populations**. A population isn't necessarily a group with an interactive social structure; it is simply a number of individuals of the same species living in the same area at the same time with the opportunity to share genetic material. All of the populations of different species that live in an area interact with one another as a **community**. An **ecosystem** consists of a community and its interactions with the abiotic components of the environment, such as water, sunlight, and soil. All of Earth's ecosystems together comprise the biosphere, which in turn interacts with the abiotic environment, the geosphere. In chapter 5 we will look more closely at the major ecosystems, called biomes, that comprise the biosphere.

What is an ecosystem? ____________

Answer: A biologic community and its interactions with the abiotic environment.

Energy in Ecosystems

Through metabolism, organisms derive food energy from the environment by manufacturing it themselves, by eating other organisms, or by eating organic compounds produced by other organisms. Organisms that manufacture their own food energy, either through photosynthesis or chemosynthesis, are autotrophs (from the Greek *auto*, meaning "self," and *troph*, meaning "nourishment"); they are also called **producers**. Organisms that derive their food energy from secondary sources are heterotrophs (Greek *hetero*, meaning "other"); they are also called **consumers**.

Organisms that eat producers are called **primary consumers**; they are commonly **herbivores** (plant eaters). Rabbits and cows are examples of herbivorous heterotrophs; they eat grass, which is autotrophic via photosynthesis. **Secondary consumers** eat the herbivores, and **tertiary consumers** eat the secondary consumers. The secondary and tertiary consumers are commonly **carnivores** (meat eaters) because they eat other organisms or **omnivores** (from the Greek, *omni*, meaning "all") because they eat a variety of foods, including both plants and animals. Lions, owls, and polar bears are carnivores; humans, pigs, and bears are omnivores.

Detritivores are a special kind of consumer that eats **detritus**, the dead bodies, feces, and other organic waste left by other organisms. Examples include earthworms and snails. A particularly important type of detritivore is the **decomposer**, or **saprotroph** (from the Greek *sapro*, meaning "rotten"). These organisms, which include bacteria and fungi, break down dead organic material into simple carbohydrates, mineral salts, and carbon dioxide. They use the carbohydrates to supply themselves with energy. Decomposers and other detritivores are crucial to any ecosystem. Without them, dead organic material and wastes would accumulate endlessly.

All of the digesting and eating of one another that goes on within ecosystems results in the cycling of energy through the system by way of a **food chain**. At the bottom of the food chain are the producers that use sunlight and inorganic materials to make their own food energy. Primary consumers eat the producers, capturing some of their chemical energy and using it to build their own body tissues. Secondary consumers eat the primary consumers, and tertiary consumers eat the secondary consumers. At each step, wastes are emitted, some energy is returned to the environment, detritivores consume the waste, and decomposers are there to break down any remaining organic material.

Each step in the food chain is called a **trophic level**. Producers form the first trophic level, primary consumers are the second trophic level, and so on (Figure 4.4). With each successive level, a significant amount of energy is lost. This is because each organism uses up some of the chemical energy it consumes to do the work of living—moving, growing, breathing, reproducing, and so on. The used-up energy is returned to the environment as heat through respiration, which reverses photosynthesis and returns carbon dioxide to the environment. This loss of energy is why organisms at the top of the food chain—like humans—are much less efficient consumers than organisms closer to the bottom. It's also why there are far more organisms at the bottom of most food chains than at the top.

An example of a simple terrestrial food chain (Figure 4.5) would be: grass (producer, photosynthetic autotroph), eaten by a mouse (herbivorous primary consumer), eaten by a snake (carnivorous secondary consumer), eaten by a hawk (carnivorous tertiary consumer), with wastes and dead bodies broken down by decomposers (bacteria and fungi) and consumed by earthworms and other detritivores. An example of a simple marine food chain would be: algae (producer, photosynthetic autotroph), eaten by krill (tiny, shrimplike animals, primary consumers),

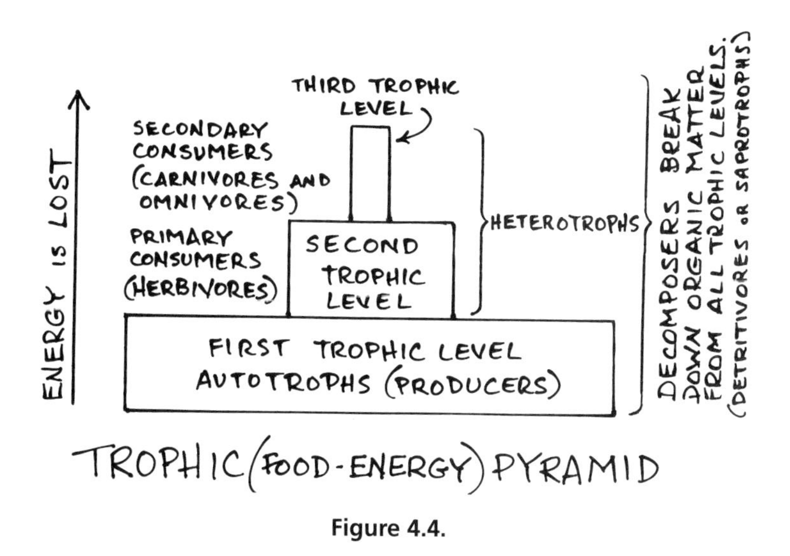

Figure 4.4.

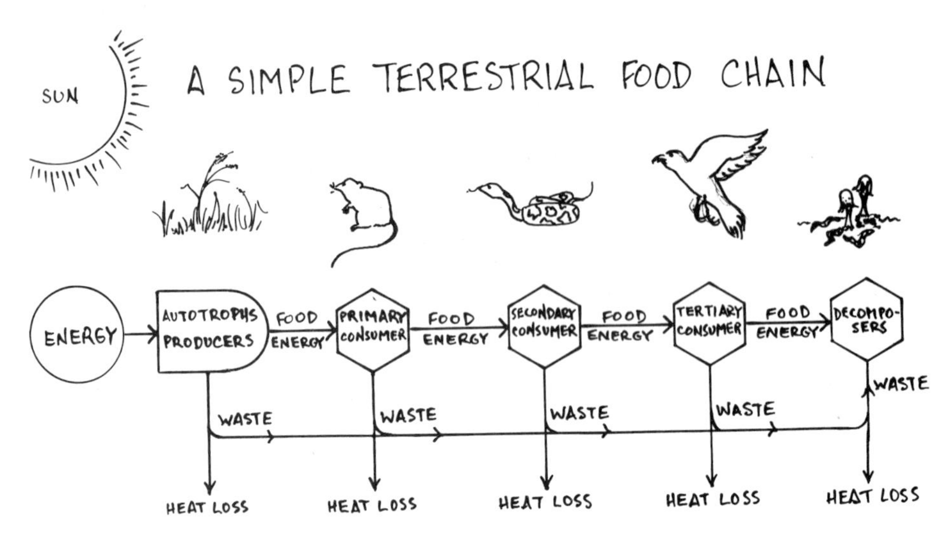

Figure 4.5.

eaten by whales and other marine animals (omnivorous and carnivorous secondary consumers), with wastes and dead bodies broken down by decomposers and consumed by bottom feeders, such as marine snails (detritivores).

Are there food chains in your backyard? Is your yard an ecosystem?

In nature, simple food chains like these are not common; real life is more complex. For example, lots of animals in the ocean, not just whales, eat krill. Hawks eat snakes, but they also eat mice directly, as well as other small animals such as voles or rabbits. And so on. You can see that even in these simple examples there are lots of connections that haven't been discussed. When we begin to consider the full complexity of the various food and energy connections in an ecosystem, we have a **food web**.

Use the appropriate scientific terminology to describe each of the following organisms and their role in a food chain: beetle, human, dandelion, grasshopper, elephant, fungus, oceanic phytoplankton (tiny marine plants). ______________

Answer: *beetle*: detritivore, heterotroph, saprotroph; *human*: secondary/tertiary consumer, omnivorous heterotroph; *dandelion*: producer, photosynthetic autotroph; *grasshopper*: primary consumer, herbivorous heterotroph; *elephant*: primary consumer; herbivorous heterotroph; *fungus*: decomposer, heterotroph, saprotroph; *oceanic phytoplankton*: producer, photosynthetic autotroph.

Biomass and Biologic Productivity

Biomass is a quantitative measure of the amount of organic matter in an ecosystem or part of an ecosystem. Specifically, **phytomass** refers to plant matter and **zoomass** refers to animal matter. Biomass is most often given in terms of the dry weight of organisms at a particular trophic level in a system. For example, the biomass of the first trophic level would consist of the dry weight of all the plant matter in the system. Since biomass is a form of stored energy, and energy is lost at each successive step along the food chain, it makes sense that the amount of biomass at each successive trophic level in the food chain also decreases. The amount of biomass at the top of the food chain is much less than the biomass at the bottom.

We can show this general relationship in the form of an **ecologic pyramid**. A typical pyramid has a broad base representing the first trophic level, with a large number of individuals (producers), a large quantity of biomass, and a large amount of stored energy (Figure 4.4). Moving up the pyramid (and the food chain), the number of individuals, quantity of biomass, and amount of stored energy all decrease with each successive step. This is dramatically illustrated by the fact that 99 percent of Earth's total biomass occurs in the form of phytomass (producers), with only about 1 percent in the form of zoomass (consumers).

Coral reefs are highly productive ecosystems. Here, a school of smallmouth grunt swim among elkhorn coral in the Florida Keys National Marine Sanctuary.

Biologic productivity is a measure of the rate at which biomass is produced. The amount of biomass produced by photosynthesis in a given area during a specific time interval is called **gross primary productivity**, or **GPP**. Some GPP is used up by plant respiratory functions. The remainder, called **net primary productivity**, or **NPP**, is incorporated into organic tissues used for plant growth. Thus, NPP = GPP – plant respiration.

NPP is the plant energy that is available to consumers. It can be measured in units of energy (such as kcal) or units of dry organic matter (such as grams of organic carbon) per unit area (km^2) per unit time (yr). For example, the NPP of tropical swamps is about 3,000 $g/m^2/yr$, while that of a rocky desert might be closer to 3 $g/m^2/yr$. In marine ecosystems, the NPP of a highly productive coral reef might be 2,500 $g/m^2/yr$, while that of the relatively unproductive open ocean would be closer to 125 $g/m^2/yr$. Note, however, that the open ocean is still very important in terms of its overall productivity because the volume of the open ocean is so much greater than the volume of coral reefs and other highly productive coastal near-shore ecosystems.

Some types of ecologic pyramids are inverted, with more individual organisms at the top than at the bottom. Can you think of an example? ______________

Answer: Food chains involving insects, such as thousands of termites feeding on a single tree, provide an example of a pyramid with more individuals (but not necessarily more biomass) at the higher trophic level.

Now that you have mastered some of the basics of ecology, we can go on to consider the geographic distribution of life on this planet, in chapter 5. First, give the Self-Test a try.

SELF-TEST

These questions are designed to help you assess how well you have learned the concepts presented in chapter 4. The answers are given at the end. If you get any of the questions wrong, be sure to troubleshoot by going back into that part of the chapter to find the correct answer.

1. An organism that breaks down and consumes organic wastes left behind by other organisms is called a(n) ______________.
   a. detritivore
   b. decomposer
   c. saprotroph
   d. All of the above are true.

2. The biochemical process that reverses photosynthesis is called ______________.
   a. metabolism
   b. respiration
   c. mutation
   d. chemosynthesis

3. Which one of the following terms refers to the amount of plant matter in a system?
   a. phytomass
   b. biomass
   c. zoomass
   d. sapromass

4. Individuals of the same species that live together and have the opportunity to breed and share genetic material are called a(n) ______________.

5. The three main functional types of organisms in an ecosystem are the ______________, ______________, and ______________.

6. Most of the free molecular oxygen currently in the atmosphere was generated as a result of ______________.

7. The inherited characteristics of individuals are passed from one generation to the next through their genes, which are composed of DNA. (True or False)

8. Evolution is achieved through adaptation and natural selection—that is, the survival of the fittest. (True or False)

9. The oldest signs of life on this planet are almost 3.9 million years old. (True or False)

10. What is ecology?

11. What are the four basic characteristics that distinguish living from nonliving things?

12. What is NPP and GPP, and what is the difference between them?

13. Use the following organisms to construct a hypothetic food web, and identify the functional role of each organism using the appropriate terminology: lion,

antelope, dung beetle, hyena, grass, baobab tree, termite. (Note: antelopes eat grasses and other vegetation; hyenas eat just about everything, especially the carcasses of dead animals.)

ANSWERS

1. d
2. b
3. a
4. population
5. producers; consumers; decomposers
6. photosynthesis
7. True
8. True
9. False
10. The scientific study of the interactions among organisms and between organisms and the abiotic components of their environment.
11. reproduction, growth, metabolism, evolution
12. NPP = net primary productivity; GPP = gross primary productivity. NPP = GPP − plant respiration; that is, NPP is equal to GPP minus the amount of biomass used for plant respiration.
13. lion—secondary consumer; carnivorous heterotroph (eats antelopes)

    antelope—primary consumer; herbivorous heterotroph (eats grass and leaves)

    dung beetle—detritivore; heterotroph; saprotroph (eats lion and antelope dung)

    hyena—detritivore; omnivorous heterotroph (eats dead lions, antelopes, insects)

    grass—producer; photosynthetic autotroph

    baobab tree—producer; photosynthetic autotroph

    termite—primary consumer; heterotroph; detritivore (eats dead wood from trees)

KEY WORDS

adaptation
aerobic
anaerobic
autotroph
biologic productivity
biomass
carnivore
cell
chemosynthesis
community
consumer
decomposer
detritivore
detritus
DNA (deoxyribonucleic acid)
ecologic pyramid
ecology
ecosystem
eukaryotic cell
evolution
fermentation
food chain
food web
gene
gross primary productivity (GPP)
herbivore
heterotroph
metabolism
mutation
natural selection
net primary productivity (NPP)
omnivore
photosynthesis
phytomass
population
primary consumer
producer
prokaryotic cell
respiration
RNA (ribonucleic acid)
saprotroph
secondary consumer
species
taxonomic hierarchy
tertiary consumer
trophic level
zoomass

5 Earth's Major Ecosystems

The web of our life is of a mingled yarn.

—William Shakespeare

Objectives

In this chapter you will learn about:

- the fundamental differences between terrestrial and aquatic ecosystems;
- the geographic distribution of major ecosystems;
- the characteristics of the major terrestrial biomes; and
- the characteristics of the major aquatic biomes.

Biomes

The geographic distribution of living organisms and their communities is the focus of **biogeography**. The most important unit of biogeography is the **biome**, a large geographic area characterized by its environmental attributes and by the plants and animals that inhabit the area. A particular biome can occur in various geographic locations, but it will always display similar characteristics. For example, a desert is basically a desert. There are many variations and many deserts, but all desert biomes wherever they occur have certain characteristics in common—notably, very low precipitation and organisms that are adapted to dry conditions. Different biomes

A mangrove, as shown here, is a coastal swamp, transitional between aquatic and terrestrial biomes.

grade into one another; the boundaries are not sharp or distinct, and there are many variations.

There are two basic types of biomes: **terrestrial** and **aquatic**. Obviously, terrestrial and aquatic environments differ fundamentally, and so do the organisms that characterize these biomes. On land, the most important attributes that define a biome are temperature and precipitation, which in turn influence soil type and other environmental characteristics. In the water, the defining characteristics are temperature, depth, and salinity. On land, plants and animals require structural support in order to survive and hold up against Earth's gravitational pull. These supports mostly take the form of bones and woody parts, and the growth of terrestrial organisms requires an enormous investment in energy for specialized tissue. In aquatic environments, it is more common to find soft forms of plants and animals because water is a more supportive medium than air. Because of gravity, it takes more energy for organisms to move on land than in water, where they can swim or float with ease. Aquatic organisms—especially those that swim vigorously—tend to be streamlined to minimize drag.

There are also differences in the resources available to organisms in terrestrial and aquatic environments. On land, organisms must remain near a source of water. In aquatic environments, light may be in short supply, especially at great depths; therefore, photosynthetic organisms and those that feed directly on them, as well as plants that need to be rooted in bottom sediments, must remain in shallow or nearshore environments where light can penetrate through the water. An adequate supply of light is not a problem in most terrestrial environments, with the exception of a few unusual locations such as caves or the floor of a dense rain forest. The basic types of food energy also differ from terrestrial to aquatic biomes. Land plants mainly produce food energy in the form of carbohydrates; in the ocean, proteins are the main form of organic matter. Organic matter and mineral nutrients also tend to be unevenly distributed and poorly circulated in deep aquatic environments.

How does gravity influence the types of plants and animals that live in terrestrial and aquatic environments? ______________

Answer: On land, plants and animals require structural supports such as bones or woody parts to hold themselves up against the pull of gravity. Water is a more supportive medium than air; therefore, aquatic organisms require less extensive structural supports.

An Introduction to Terrestrial Biomes

Terrestrial biomes are characterized primarily on the basis of variations in temperature and precipitation (Figure 5.1). These in turn drive variations in soil type,

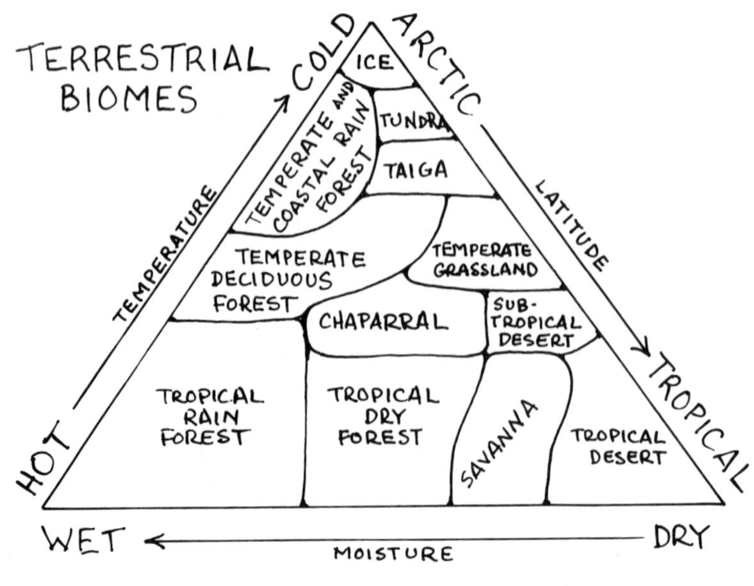

Figure 5.1.

vegetation, and animal life that distinguish one terrestrial biome from another. The major categories of terrestrial biomes include tundra, forest (including taiga or boreal forest, temperate coniferous forest, temperate deciduous forest, tropical rain forest, and tropical deciduous forest), savanna, chaparral, grassland, and desert. Some biomes have **alpine** (mountain) variations (Figure 5.2); elevation exerts a major control on both temperature and precipitation and therefore on vegetation and animal life. Once thought to be essentially barren, ice sheets also present a unique biogeographic environment that hosts a wide range of micro-organisms.

The distribution of Earth's major terrestrial biomes is closely related to climate. Climate (temperature and precipitation) and the resulting plant and animal life in a region are fundamentally influenced by the nature of the air masses that move temperature and moisture around the planet. These in turn are controlled by latitudinal and seasonal variations in solar insolation, the distribution of land and water bodies, topography, and the Coriolis effect. Recall from chapter 3 that convection in the atmosphere combines with the Coriolis effect to create huge cells of rising and falling air masses, forming three belts of high rainfall and four belts of low rainfall. The belts of high rainfall lie along the equator, resulting in tropical climates, and along the two polar fronts, resulting in moist, temperate climates. The belts of low rainfall generate two areas of dry climate in the polar regions and two in the subtropical regions along the 30° N and 30° S latitudes. The effects of the wet and dry belts are most clearly demonstrated by the global distribution of deserts (Figure 5.3). Most of the world's large deserts lie within the belts of dry air near the 30° N and 30° S latitudes.

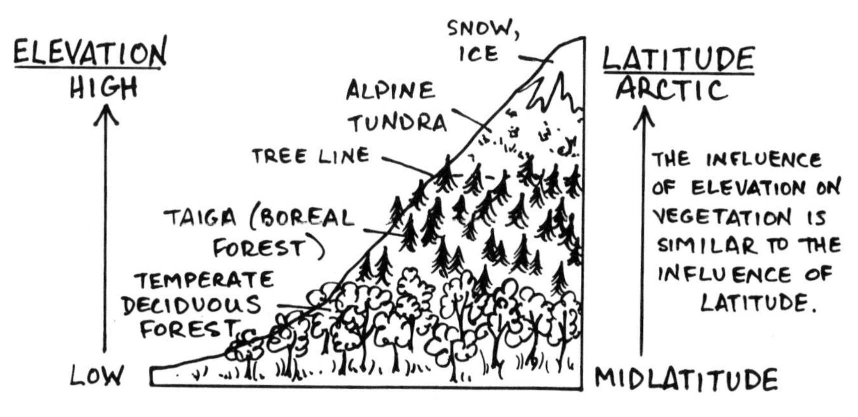

Figure 5.2.

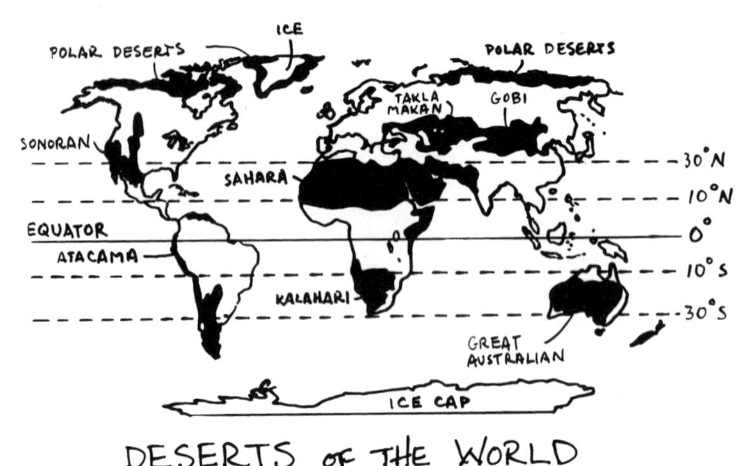

COMPARE TO FIGURE 3.7 - ATMOSPHERIC CIRCULATION

Figure 5.3.

Explain the influence of mountainous terrain on biomes. ________________

Answer: Altitude has an important effect on temperature (cooler temperatures at higher altitudes); steep topography can also have an impact on precipitation. A high-altitude environment in an equatorial region, for example, might be characterized by an alpine biome rather than by a tropical biome.

The Major Terrestrial Biomes

Tundra, or arctic tundra, is a terrestrial biome that occurs in the north at high latitudes. The mountain equivalent, alpine tundra, occurs at high altitudes closer to the equator. Tundra is characterized by treeless plains and by **permafrost**—soil that is perennially frozen except for a thin surface layer that thaws in the summertime. Winters are long, cold, and harsh; summers are short and cool with little precipitation. Plants are typically small because of the short growing season and low light levels during much of the year. Typical plants include mosses, grasses, sedges, lichens, and dwarf shrubs. Small mammals such as voles, marmots, arctic foxes, and snowshoe hares are common, along with large mammals such as muskox, and herds of migratory mammals such as caribou. The tundra is a sensitive environment; because of the short growing season and harsh conditions, tundra ecosystems do not regenerate easily if disturbed or damaged.

What is permafrost? ______________

Answer: Perennially frozen soil characteristic of the tundra biome.

Forests are the most widely varied of the major biomes. They can be roughly categorized into high-latitude, temperate (midlatitude), and tropical (equatorial) forests. We'll look at the major forest types, moving roughly from high to low latitude.

The **taiga** (pronounced *tie-ga*) is the northern **boreal forest** to the south of the tundra in North America and Eurasia. There is no southern equivalent because there isn't much land area in comparable latitudes in the Southern Hemisphere. The taiga is characterized by cold winters (although not as cold as in the tundra), short growing seasons, and low levels of precipitation. The boreal forest of the taiga mainly consists of **coniferous** trees (evergreens with cones), such as spruce, fir, and pine trees, as well as mosses and lichens. Conifers are adapted to conditions of low precipitation. They have needlelike leaves with little surface area, which minimizes moisture evaporation. Animals of the taiga include large, migratory mammals, such as caribou and moose, and many smaller mammals, such as rabbits, foxes, wolves, a wide variety of rodents, birds, and insects, but there are few reptiles or amphibians (because there is not enough water).

The **temperate rain forest** is the coniferous forest typical of the northwest coast of North America (Oregon, Washington, and British Columbia) but also found in other locations such as southeastern Australia. The winters are milder than in the far north and precipitation is high, so the evergreen forests grow thick and tall. These rich wood-producing regions have hosted some bitter confrontations between loggers and environmentalists. Typical trees of the temperate rain forest include pine, fir, redwood, and cedar. **Epiphytes** (ferns, vines, and mosses that grow attached to the branches of trees) also grow in abundance. Bears, mountain lions, wolves, and elk are common, as well as smaller rodents, birds, reptiles, and amphibians.

The **temperate deciduous forest** occurs mainly in the northeastern United States, Europe, and eastern China. The climate is characterized by seasonal changes from summer to winter. The forests consist mainly of broad-leaved **deciduous** trees—trees that shed their leaves each year, including maple, oak, birch, and elm. The soils are rich in organic material and are well suited for agriculture. Deer, bears, and wolves are common, although they are now limited to localities where the forests have not been cleared for agriculture or urban development.

The great **tropical rain forests** of the equatorial regions host an enormous diversity of organisms. Both temperature and precipitation are high, and the growing season lasts all year, so the vegetation is very tall and dense. These are **closed forests**; the **canopy**, or top layer of vegetation, forms an almost continuous cover, so the forest floor can be quite dark (Figure 5.4). There is an astonishing array of plant and animal life in these forests, including a wide range of birds, insects, mammals, reptiles, amphibians, epiphytes, and flowering plants such as orchids and bromeliads. Most of the mammals, such as monkeys and lemurs, live in the canopy and rarely descend to the forest floor. The soils in tropical rain forests tend to be highly weathered and low in organic matter because most of the organic matter resides in the lush vegetation rather than in the soils. With so much warmth and moisture available, there is an abundance of detritivores, including bacteria, fungi, and some types of insects, which consume organic debris almost as quickly as it falls to the ground. Because tropical soils contain so little organic matter, they dry out and are easily eroded when exposed by deforestation. This is one of the reasons why it is problematic when tropical rain forests are cut down.

The equatorial region also hosts large **tropical deciduous forests**, also called tropical seasonal forests or monsoon forests, in which the trees shed their broad leaves during the dry season. The main seasonal variation is in the amount of precipitation, although there is also greater temperature variation than in the tropical rain forest. Tropical rain forests and tropical deciduous forests together are referred to as the **tropical moist forest system**. The largest is Amazonia in South America; other major tropical moist forests are in Southeast Asia (Indonesia) and equatorial Africa.

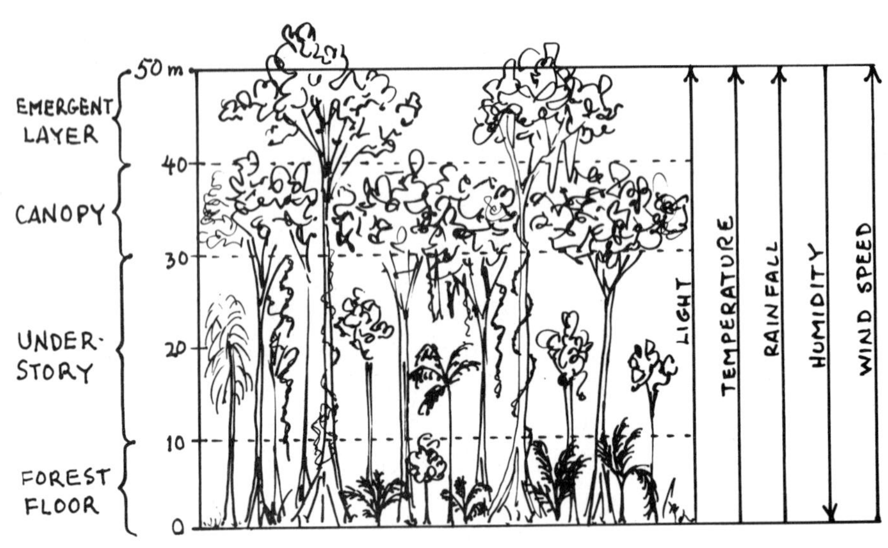

Figure 5.4.

In addition to hosting much of the planet's biologic diversity, the tropical moist forests represent an enormous reservoir for carbon in the global carbon cycle.

Why are soils in tropical rain forests so fragile? ________________

Answer: Tropical soils are deeply weathered; they are also very low in organic matter because most of the biomass in the tropical rain forest resides in the vegetation and detritivores consume the organic debris almost as quickly as it falls to the forest floor. When tropical soils are deforested, they have little organic matter to bind them, so they dry out and are easily eroded.

In contrast to tropical rain forests, the tropical **savanna** is an **open forest** consisting of broad, grassy plains with scattered trees and lacking a continuous canopy. Temperatures are high, but rainfall is low, particularly during the long, dry season. The best-known savanna is the African Sahel (*sahel* is the Arabic word for "border") that lies along the southern margin of the Sahara Desert. The savanna hosts huge herds of migratory mammals such as antelope, zebras, and elephants and large predators such as lions and tigers. This environment is fragile, primarily because of the low precipitation. When overstressed by activities such as intensive agriculture, the savanna can quickly turn into desert.

The biome that is typical of a Mediterranean climate—hot, dry summers and cool, wet winters—is the **chaparral**, which is characterized by low, scrubby evergreen bushes and short, drought-resistant trees. Chaparral occurs in the region surrounding the Mediterranean Sea but is also found in the southwestern United States, as well as parts of Australia, Africa, and South America. Fires are common in chaparral regions, and some plants have fire-resistant adaptations, such as below-ground growth that is able to reemerge following a fire.

Temperate **grasslands** are the vast prairies typical of the midwestern United States and Canada, as well as Ukraine. Grasses have extensive, interconnected root systems. Grasslands are particularly well suited to agriculture because of the rich organic content of the soils. There are two basic types of grasslands. Temperate moist grasslands, also known as **tallgrass prairies**, once hosted great herds of bison and elk and predatory wolves. **Shortgrass prairies** are drier, and the grasses are shorter, more drought-resistant varieties. Typical animals in shortgrass prairies include prairie dogs, snakes, and lizards.

What is the difference between closed and open forests? ________________

Answer: A closed forest has dense foliage and a continuous canopy; an open forest has intermittent trees and a discontinuous canopy.

Deserts round off the low-precipitation extreme of the terrestrial biomes. Although we often think of deserts as being hot, lack of precipitation is actually the main defining characteristic. The term **desert** refers to **arid** lands, where annual precipitation is less than 250 mm (10 in). Four types of hot deserts and one type of cold desert are recognized (Figure 5.3):

1. subtropical deserts in the two subtropical dry belts; examples include the Sahara, Kalahari, and Great Australian;
2. continental interior deserts far from any source of moisture, such as the Gobi and Takla Makan of central Asia;
3. rain-shadow deserts, where a mountain range creates a barrier to the flow of moist air, causing a zone of low precipitation; examples are the deserts to the east of the Cascade Range and the Sierra Nevada in the western United States;
4. coastal deserts, such as the western coastal deserts of Peru, Chile, and southwestern Africa; and
5. the vast, cold, polar ice sheets, where precipitation is extremely low due to the sinking of cold, dry air.

In all deserts, plant cover is sparse. Some desert plants are uniquely adapted to the lack of water, including the many varieties of cacti and succulents, which have few or no leaves and are able to retain large quantities of water. Animals tend to adapt to the extreme temperatures and dryness by storing water, remaining hidden during the day, hunting or foraging at night, and even tunneling underground during extended dry periods.

How can an ice sheet be a desert? ________________

Answer: Although many deserts are hot, a desert is technically defined as a place with extremely low precipitation. Ice sheets fit this definition even though they are cold.

An Introduction to Aquatic Biomes

Aquatic biomes are distinguished primarily on the basis of differences in salinity. There are three major groups—freshwater, marine, and transitional—with further subdivisions based on temperature, depth, and distance from the shore. Saltwater (marine) and freshwater biomes differ from each other in some fundamental ways besides salinity. For example, the ocean is much more strongly affected by tides than even the largest lakes. Waves and deep currents are generally stronger in the ocean than in lakes, and the depth of the water is considerably greater. Low light levels, cold temperatures, and even high pressures can place significant constraints on oceanic organisms.

Aquatic organisms generally fall into one of three categories: plankton, nekton, or benthos. **Plankton** are extremely tiny, free-floating organisms. Of these, **phytoplankton** are tiny, photosynthetic aquatic plants, including algae. Phytoplankton are important because they form the base of the food chain in aquatic ecosystems. **Zooplankton** are tiny, nonphotosynthetic aquatic animals, including the larvae of some larger organisms. In aquatic ecosystems, zooplankton typically eat phytoplankton, which in turn are eaten by larger organisms. **Nekton** are larger animals that are active swimmers, including fish, whales, turtles, and dolphins. **Benthos** are bottom-dwelling organisms, including those that attach themselves to rocks, such as barnacles and mussels; plants that are rooted in bottom sediments; organisms that burrow into bottom sediments, such as octopi; and some that move along the bottom, such as crustaceans.

What are the main differences in the physical environments of freshwater and saltwater biomes? ______________

Answer: Salinity; influence of tides; energy of waves and currents; depth and related factors (low light levels, cold temperatures, high pressures).

The Major Aquatic Biomes: Freshwater and Transitional

Freshwater biomes can be divided into flowing-water and standing-water types. Flowing-water biomes—rivers and streams—are dynamic environments. A river can vary dramatically from its **source** upstream where the river originates to its **mouth** downstream where it empties into another body of water. Upstream or headwater streams tend to be small, cold, and swiftly flowing, whereas downstream rivers tend to be wider, deeper, cloudier, and not as cold. Changes in topography also dramatically affect the character of a stream. Steep topography leads to deeply incised, fast-flowing streams; on flat plains, rivers tend to meander lazily from side to side. Some rivers change seasonally, drying up during one season and overflowing during another. Organisms that live in fast-flowing streams tend to have adaptations that help them survive in the strong current. These include suckers for attaching onto rocks, or a streamlined, muscular build for swimming up-current.

Standing-water environments include lakes, ponds, and wetlands. A **lake** is a standing body of fresh water that occupies a large depression in the ground surface; ponds and pools are smaller standing freshwater bodies. In general, lakes are large enough that terrestrial plants cannot survive in their deepest parts. Lakes contain several zones defined by a combination of depth, temperature, and distance from the shore (Figure 5.5). The warm, shallow, nearshore area is called the **littoral zone**. The littoral zone is the most biologically productive part of a lake, with abundant plant life (cattails, sedge grasses), as well as birds (ducks and other waterfowl), insects, reptiles, and amphibians. The open-water environment is called the **limnetic zone**, which hosts zooplankton and phytoplankton, as well as the larger organisms—mainly fish—that eat them. The limnetic zone extends as far down as light can penetrate. Underneath is the **profundal zone**, which only occurs in the largest and deepest lakes. Organic material in the profundal zone consists mainly of dead organisms that float down and are consumed by bacteria.

The temperature of the water in a large lake can vary greatly from the surface where sunlight heats the water to deeper levels where the water is typically much colder. This is called **thermal stratification**. Recall that water, like air, is denser when it is cool than when it is warm. Cool water sinks to the bottom of a deep lake because it is dense. As sunlight hits the lake, the surface waters become warmer and less dense. This is a stable thermal stratification—cold, dense water at the bottom and warm, less dense water at the top. The depth where the temperature changes very rapidly from warm to cold is called the *thermocline* (defined in chapter 3, in the context of ocean thermal stratification).

You can feel the thermal stratification in a lake if you go swimming in almost any temperate lake in the summertime. Stick your feet down and you will feel a marked temperature difference between the warm surface water and the cold, deeper water. The depth where you feel the sudden temperature change is the thermocline.

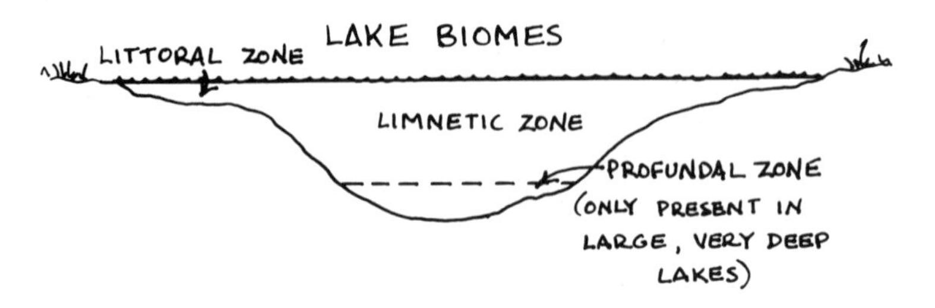

Figure 5.5.

In fall and winter, the surface waters of the lake become cooler and denser. If the surface water becomes dense enough, a turnover can occur, in which the deep and shallow waters mix. In the tropics, the temperature difference between the surface and deep layers is less strong because there is no seasonal cooling of the surface waters; thus, the thermal stratification in tropical lakes is not as pronounced as in temperate lakes.

A freshwater **wetland** is an area that is either permanently or intermittently moist. Wetlands, including swamps, marshes, and bogs, may or may not contain open, standing water, although most wetlands are covered by shallow water for at least part of the year. Wetlands tend to be highly biologically productive, with dense vegetation, migratory birds, reptiles, amphibians, and fish. Some wetlands represent a developmental stage in the natural lifetime of a freshwater body. For example, a lake whose source of incoming water has been cut off will become shallower over time, gradually filling with sediment and organic material to become a peat bog or swamp. Wetlands were once considered to be dirty, mosquito-infested quagmires—prime targets to be drained and developed. More recently, wetlands have been recognized as natural storehouses for a great diversity of plant and animal species. Wetlands also perform many important environmental services, including storing groundwater and removing toxins from the soil.

Wetlands in coastal regions are transitional between freshwater and saltwater environments. These transitional environments include estuaries, salt marshes, and mangroves. An **estuary** is a body of water that is connected to the open ocean but has an incoming supply of fresh water from a river. Salt water and fresh water mix in estuaries, and water levels, salinity, and temperature fluctuate with the rise and fall of the tides. Organisms that inhabit estuaries must be adapted to tolerate these variations. A common feature of temperate estuaries is a **salt marsh**, a coastal wetland dominated by salt-tolerant grasses. Salt marshes typically host abundant shorebirds. The tropical equivalent is a **mangrove forest**, which hosts some of the most productive fisheries in the world. Estuaries, salt marshes, and mangroves, like other wetlands, perform many important ecologic functions, including cleansing the soil and providing homes for many different species. Coastal wetlands also protect the shoreline against the battering energy of oceanic storms. Unfortunately, coastal development is threatening mangrove forests and other coastal wetlands in many localities.

What is the difference between a wetland, a salt marsh, and a mangrove?

Answer: A wetland is an area that is permanently or intermittently moist and may or may not contain open, standing water. A salt marsh is a coastal wetland, in which salt water and fresh water mix. A mangrove is the tropical equivalent of a salt marsh.

The Major Aquatic Biomes: Marine

Marine biomes are divided into several major zones on the basis of depth and distance from the shore (Figure 5.6). The **intertidal zone** is transitional between the shore and the open ocean. This is a dynamic environment where the influence of tidal variations and the energy of breaking waves are strong. Rocky or cliffed shorelines typically host organisms that can attach themselves to rocky surfaces, such as mussels and certain types of algae, and other organisms that can survive periodic drying out during low tide. Some residents of rocky shorelines survive by burrowing and hiding between the rocks. Another common type of shoreline is the sandy beach and barrier island, such as the shorelines that characterize the eastern coast of the United States. Sand-dwelling organisms are often small animals like sand crabs that burrow into the sand to escape the breaking waves and the ebb and flow of the tides.

The **pelagic zone** is the open-water environment of the ocean. It consists of the **neritic province**, which extends from the shore to where the water reaches a depth of 200 m (650 ft), and the **oceanic province**, which encompasses the rest of the pelagic zone—that is, most of the open ocean. Further subdivisions are based on

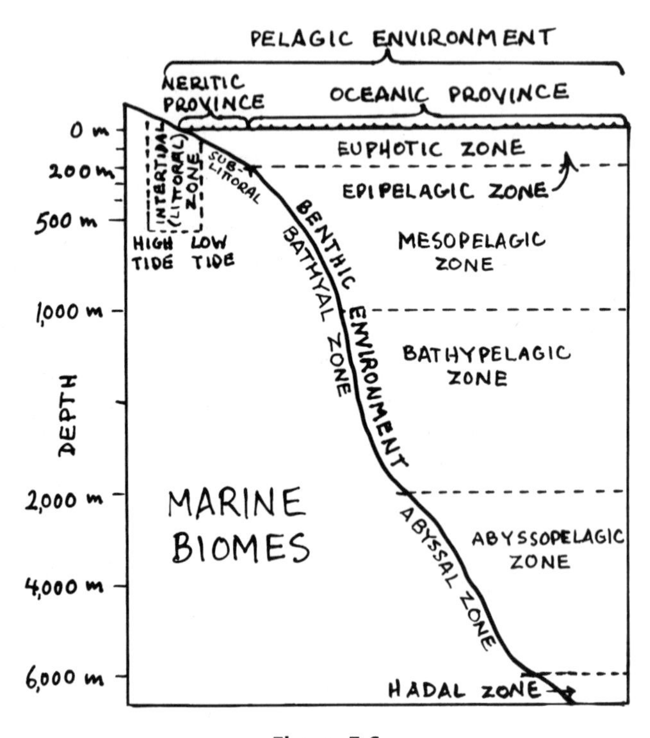

Figure 5.6.

the depth of the water. The **euphotic zone**, or photic zone, is the top part of the pelagic zone, where photosynthesis can occur. The euphotic zone extends from the surface down to the maximum depth of sunlight penetration—about 150 m (490 ft). Below this level are the **bathyal zone**, extending to a depth of 4,000 m (2.5 mi), and the **abyssal zone**, extending to a depth of 6,000 m (3.7 mi). The deepest part of the ocean, called the **hadal zone**, extends into the deepest oceanic trenches 6,500 m (4 mi) below the surface.

Organisms that inhabit the pelagic zone are floating or swimming organisms. The nearshore euphotic zone is the most biologically productive marine environment; the water is relatively warm, and light levels are high. This environment supports abundant phytoplankton and therefore zooplankton and the nekton that feed on them. Among the important zooplankton that inhabit the oceanic province are **foraminifera**, single-celled zooplankton that produce a calcareous (calcium-rich) shell, and **radiolaria**, which produce a siliceous (silica-rich) shell. Nekton encompass the rich diversity of marine life, including large and small fish, sharks, swimming mollusks, and marine mammals such as whales.

Beneath its relatively warm, light surface layer, the oceanic province extends to great depths. No light penetrates the deep waters of the bathyal and abyssal zones. The temperature is constant but cold, and the pressure is high. Organisms must adapt to these extreme environmental conditions. Many of the fish that inhabit these deep waters have light-producing organs that help them find prey or potential mates. Soft bodies prevail, and fish tend to drift slowly rather than swim vigorously. Since photosynthesis is limited by the lack of light, primary biologic productivity is low. Organisms feed on organic debris that rains down from above; thus, many of the larger fish are equipped with adaptations such as scooping jaws.

What are the two provinces of the pelagic zone? ______________

Answer: Neritic and oceanic.

The **benthic zone**, or benthos, is the ocean floor environment. The benthos of the deep parts of the ocean—the bathyal and abyssal zones—consists mainly of sediment, with some burrowing marine animals and bacteria. The sediment is dominated by the tiny skeletal remains of foraminifera and radiolaria, which rain down and accumulate on the bottom. In the deepest parts of the benthos—the deep oceanic trenches—the geologic environment is dominated by **hydrothermal vents**. These are volcanic openings through which hot, mineral-laden waters flow. Unusual life forms inhabit the areas around hydrothermal vents. These organisms are adapted to a completely dark environment; their primary biologic productivity comes from chemosynthesis rather than photosynthesis. Thermophilic (heat-loving) bacteria produce food energy using heat and hydrogen sulfide from the mineral-laden waters of the hydrothermal vents. In addition to albino crabs, clams, and mussels, hydrothermal vents are inhabited by strange creatures such as tube worms, which grow to 3 m (10 ft) in length.

In contrast to the extreme environment of the deep benthos, the shallow-water benthic zone is a warm, light, and highly productive environment. In water up to about 10 m (33 ft) deep, salt-tolerant sea grasses grow in bottom sediments. They provide food and protection for a wide variety of fish, crustaceans, reptiles, and bottom-dwelling detritivores such as mud shrimp. In deeper, cooler water (up to about 25 m, or 82 ft), huge beds of kelp—brown algae—form the bottom of the marine food chain for many organisms, including sponges, sea cucumbers, clams, crabs, fish, and mammals such as sea otters.

A particularly productive benthic environment is the **coral reef**. Coral reefs are worthy of special mention not only because they are a biologically productive biome but also because they are sensitive indicators of environmental stress. Coral reefs are formed by colonies of tiny coral animals that coexist with a photosynthetic bacterium called **zooxanthellae** (pronounced *zoh-zan-thell-ay*). The zooxanthellae require sunlight for photosynthesis, providing the corals with food energy. The corals in turn provide nitrogen—a plant nutrient—to the zooxanthellae. Although corals can exist without zooxanthellae, their presence stimulates the corals to build tiny calcareous shells, which eventually accumulate into reefs; corals without zooxanthellae do not build reefs. Coral reefs come in a variety of forms. *Fringing reefs* closely border the adjacent shoreline. *Barrier reefs* are separated from the land by a lagoon, as in the case of the Great Barrier Reef off Queensland, Australia. An *atoll* is a *circular reef* that forms around the central cone of a submerged volcano.

Reefs are highly productive ecosystems inhabited by a diversity of marine lifeforms. They also perform an important role in the recycling of nutrients in shallow coastal environments. They provide physical barriers that dissipate the force of waves, protecting the ports, lagoons, and beaches that lie behind them, and they are an important esthetic and economic resource. Corals require shallow, clear water in which the temperature remains above 18°C (65°F). Because of their very specific light and temperature requirements, coral reefs are highly susceptible to damage from human activities, as well as from natural causes such as tropical storms. For example, industry and development in coastal zones can lead to soil erosion; the soil clouds coastal waters, blocking the corals and inhibiting photosynthesis by the zooxanthellae. When this happens, the corals die, and all that remains is the white calcareous material of the reef. This process is called *coral bleaching*, and it is an important indicator of many environmental problems.

Why is coral bleaching a sensitive environmental indicator? ______________

Answer: Corals have very specific requirements for water temperature and light levels (that is, water clarity). Coral bleaching occurs as a result of activities that cloud or pollute the water, inhibiting photosynthesis, killing the corals, and leaving only the white calcareous reef structure behind.

Now you know a lot more about the various environments that host life on our planet. Use the Self-Test questions to see how well you have retained the concepts and terminology covered in this chapter.

SELF-TEST

These questions are designed to help you assess how well you have learned the concepts presented in chapter 5. The answers are given at the end. If you get any of the questions wrong, be sure to troubleshoot by going back into that part of the chapter to find the correct answer.

1. Which one of the following is not an example of a biome?
   a. desert
   b. land
   c. wetland
   d. tundra
2. Which one of the following is not a type of zooplankton?
   a. foraminifera
   b. radiolaria
   c. algae
   d. All of the above are zooplankton.
3. Which one of the following is not a characteristic of the taiga?
   a. low precipitation
   b. coniferous trees
   c. cold winters
   d. All of the above are characteristic of the taiga.
4. Tundra is also known as ______________; its mountain equivalent is called ______________.
5. ______________, ______________, and ______________ are examples of standing freshwater biomes.
6. The four zones of the oceanic province, as defined by depth, are the ______________, ______________, ______________, and ______________ zones.
7. In the deepest, darkest parts of the ocean, no living organisms can survive. (True or False)

8. When a lake has cold water on the bottom and warm water on the top, its thermal stratification is stable. (True or False)
9. What is the difference between coniferous and deciduous trees?
10. What type of biome do you live in?
11. What are two terrestrial and two aquatic biomes?
12. Where are the world's most extensive tropical moist forests?
13. What is biogeography?

ANSWERS

1. b 2. c 3. d 4. arctic tundra; alpine tundra

5. lake; ponds; wetlands 6. euphotic; bathyal; abyssal; hadal 7. False 8. True

9. Coniferous trees are cone-bearing evergreens, which often have needlelike leaves that do not fall off in the winter. Deciduous trees have broad leaves, which are shed during the winter season.
10. (For you to find out.)
11. Terrestrial biomes discussed in the text include: tundra, taiga, temperate (coniferous) rain forest, temperate deciduous forest, tropical rain forest, tropical deciduous forest, tropical savanna, mediterranean chaparral, temperate grassland, and desert. Aquatic biomes discussed in the text include: freshwater—flowing-water (rivers and streams) and standing-water (lakes, ponds, and wetlands); transitional—salt marshes, estuaries, and mangrove forests; and marine—intertidal zone, pelagic zone (neritic and oceanic provinces), euphotic zone, bathyal zone, abyssal zone, hadal zone, and benthic zone.
12. Amazonia (mainly Brazil), Southeast Asia (mainly Indonesia), and equatorial Africa
13. The study of the geographic distribution of living organisms and their communities.

KEY WORDS

abyssal zone
alpine
aquatic biome
arid
bathyal zone
benthic zone
benthos
biogeography
biome
boreal forest
canopy
chaparral
closed forest
coniferous
coral reef
deciduous
desert
epiphyte
estuary
euphotic zone
foraminifera
grassland
hadal zone
hydrothermal vent
intertidal zone
lake
limnetic zone
littoral zone
mangrove forest
mouth (of a river)
nekton
neritic province
oceanic province
open forest
pelagic zone
permafrost
phytoplankton
plankton
profundal zone
radiolaria
salt marsh
savanna
shortgrass prairie
source (of a river)
taiga
tallgrass prairie
temperate deciduous forest
temperate rain forest
terrestrial biome
thermal stratification
tropical deciduous forest
tropical moist forest system
tropical rain forest
tundra
wetland
zooplankton
zooxanthellae

6 Habitat and Biodiversity

To keep every cog and wheel is the first precaution of intelligent tinkering.

—Aldo Leopold

Objectives

In this chapter you will learn about:

- the interactions that occur among species;
- the factors that influence diversity in organisms;
- the importance of preserving biodiversity; and
- current threats to biodiversity.

Ecologic Niche and Species Interactions

One of the central environmental concerns today is the preservation of species and the protection of the natural environment in which they live. To begin to address this issue, we need to look at some fundamental concepts of ecology, including the roles and interactions of organisms within ecosystems. The physical environment where a particular species lives or could live is called its **habitat**. This is different from the *role* of the species within that environment, its **ecologic niche**. In his classic book *Animal Ecology*, published in 1927, the ecologist Charles Elton said that where a species lives is its habitat; what it does for a living—its profession—is its niche. The niche concept opens up all sorts of questions about the possible interactions among species in a particular habitat. What roles can species play in their interactions with one another? Can two species occupy the same niche? Can species cooperate with one another to fulfill a particular role within a habitat? This chapter will attempt to answer these questions.

There are three basic categories of interaction among species: competition, symbiosis, and predation. In **competition**, species vie for scarce environmental resources, including food, water, space, and light. Competition for resources is one of the main factors that controls population in a community. In its simplest form, competition is related to the principle of natural selection, which says that the organism best adapted to a particular environment will survive and prevail. This can be restated as the principle of **competitive exclusion**, which says that species in direct competition cannot coexist—one of them will inevitably win out over the other (Figure 6.1). Note that the principle of competitive exclusion applies to an organism's niche rather than its habitat. Another way of stating this principle is that two species cannot occupy exactly the same niche within a particular habitat.

Competition can occur within a species (*intraspecific* competition) or between different species (*interspecific* competition). Within a population of animals, intraspecific competition might take the form of territorial aggression, in which an individual stakes and defends its territory against other individuals of the same species. This is common in many species of birds and among predatory animals such as wolves. Plants also compete with one another for resources such as nutrients, water, and light. You have seen this if you have ever planted seeds too close together in a garden; at some point, the density of the foliage becomes so great that access to light will favor an individual plant. The plant that grows the tallest or broadest blocks the light, preventing it from reaching adjacent seedlings.

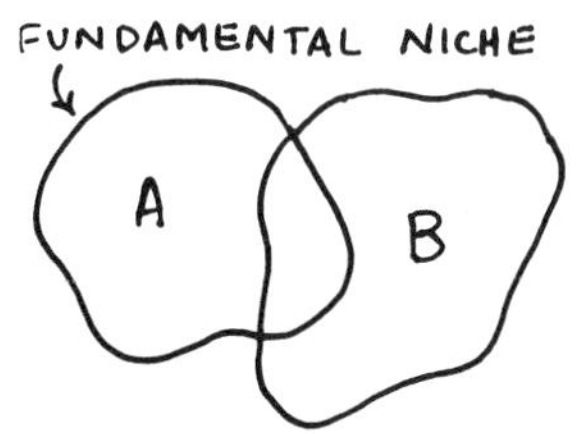

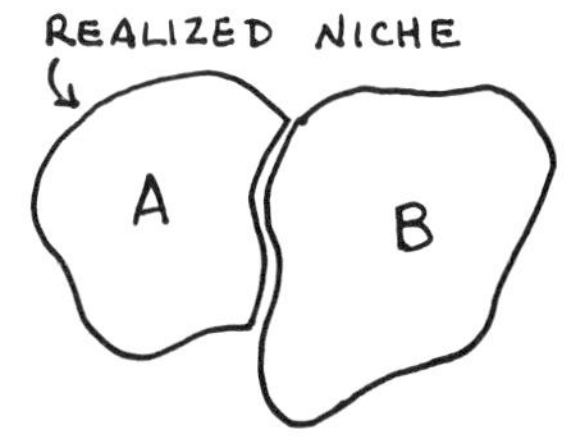

Figure 6.1.

What is the difference between intraspecific competition and interspecific competition? ________________

Answer: Intraspecific: competition among individuals of the same species. Interspecific: competition among individuals of different species.

Competition, in which the outcome is invariably negative for one or both species, is not the only form of species interaction. Sometimes species interact in ways that are beneficial to one or both; this is called **symbiosis** (from the Greek words *sum* and *biosis*, meaning "life together"). Symbiosis is very common; most organisms, even humans, participate in some form of symbiotic relationship. Humans, for example, host a wide range of microorganisms (about a hundred common ones). Some of them are of benefit to us, such as the organisms that reside in our intestines and help us digest our food. Some of them are just along for the ride, such as the mites that inhabit our eyelashes. And some of them can be annoying or harmful if they grow out of control, such as the fungus that causes athlete's foot.

There are three basic forms of symbiosis. In **mutualism**, the symbiotic relationship is beneficial to both partners. An example is the relationship between photosynthetic zooxanthellae and the corals they inhabit (chapter 5). The zooxanthellae provide food energy for corals; the corals in turn provide nitrogen—a plant fertilizer—for the zooxanthellae. Sometimes mutualism is expressed by one species offering shelter or protection to another. For example, ladybugs and some kinds of ants "farm" aphids; they corral the aphids, offering them protection from predators in exchange for being allowed to harvest the sugary liquid they produce.

In some cases, such as that of the coral-zooxanthellae relationship, the symbiotic partnership is mutually beneficial, but the partners can still exist without each other. In other cases, one or both of the organisms is completely dependent on the symbiotic relationship and could not exist without it. An example occurs in ruminants, animals such as cows, deer, moose, and reindeer, which have specialized four-chamber stomachs. These animals are herbivores; they eat grasses that contain cellulose, a relatively indigestible plant material. Specialized bacteria that live in the guts of ruminants help them digest the cellulose, making the plant nutrition available; the animal in turn supplies the bacteria with food. This type of mutualism is an example of **coevolution**, in which two species evolve together into their dependency on each other.

Another form of symbiosis is **commensalism**, a symbiotic relationship in which one partner benefits without affecting the other. An example is seen in epiphytes, a category of plant that is common in temperate and tropical rain forests. Epiphytes include mosses, vines, ferns, lichens, and other plants that grow by attaching themselves to the branches of trees. The epiphytes use the tree as an anchor, but otherwise they do not obtain any nutrients directly from the tree. The trees in turn apparently do not suffer any ill effects from the relationship, nor do they benefit from it.

In the third type of symbiosis, one partner benefits, while the other is harmed by the relationship. This is called **parasitism**. The partner that benefits is the parasite; the other is the host. Some parasites live outside the body of the host; for example, ticks attach themselves to a mammalian host and feed on its blood. Other parasites live inside the host's body; examples are human intestinal parasites such as tapeworm. Many human diseases are caused by parasites: Bubonic plague is caused by a bacterium that lives as a parasite in rats and the fleas that feed on them and is then transferred to people. Malaria spends part of its life cycle in the intestines of mosquitoes before being passed to a human host. A parasite that either causes or transmits disease or kills its host is called a **pathogen**. Killing the host is not a particularly good strategy, especially if the parasite depends exclusively on the availability of one particular type of host. In most cases, parasites just weaken their hosts without actually killing them; alternatively, they may develop strategies whereby they are easily passed to a new host if the original host dies.

What are the three main types of symbiosis? ______________

Answer: Mutualism, commensalism, and parasitism.

So far we have discussed competition and symbiosis. The remaining basic type of species interaction is **predation**, in which one organism, the predator, eats another organism, the prey. Any organism that directly feeds on another—whether the prey is killed—is a predator. Thus, even parasites and pathogens that derive nutrients from their hosts can be considered predators. However, we usually think of predators as organisms that specifically hunt and feed upon a certain type of prey. Wolves, lions, owls, and alligators are examples of large carnivorous predators. Some predators, such as caterpillars and pandas, are herbivores.

Predators develop characteristics that help them to be efficient hunters, such as speed, a venomous attack, a disguise, or a clever means of attracting prey. Prey species also develop adaptations that help them evade capture or defend themselves. In plants, these include spines, thorns, or poisons that discourage animals from eating them. Animals display these and a host of other defenses, including warning postures, colors, behaviors, and noises. Warning colors may advertise the poisonous or otherwise unsavory nature of the prey. Camouflaging colors or patterns allow the animal to hide by blending in with the background. Some animals puff up so that they look fierce; others shoot barbed spines or smelly chemicals at potential predators. All of these tactics—both predatory and defensive—are genetic adaptations. Thus, in its own way, predation has led to the evolution of a greater diversity of species.

What is the difference between predation and parasitism? ______________

Answer: In predation, an organism feeds directly on another organism. Parasitism, a type of symbiosis in which the parasite derives nutrients from the host, can also be considered a type of predation.

Conservation and Keystone Species

The examples used here to illustrate interactions among species—competition, symbiosis, and predation—seem simple and straightforward. In real ecosystems, of course, the interrelationships among species are anything but simple. Ecosystems are open systems, in which species interact in immeasurably complex ways with one another and with the physical environment. It may be tempting to try to disentangle one species from this complex web and examine its role in isolation from the rest of the system. What is the ecosystem role of a particular species of toad, for example, or a certain type of insect, or a flowering plant?

In some cases, it is possible to remove a species from an ecosystem with no discernible short-term effect on the rest of the system. In the long run, however, ecologists do not fully understand how ecosystems respond to the removal of one or more species. This speaks to the central theme of **conservation**—conserving and protecting as many species and as much of the natural environment as possible. If one or another species is lost from an ecosystem, how can we really be sure of the long-term impact on the rest of the ecosystem?

Sometimes a particular species is known to play a central role in maintaining ecologic balance in a given ecosystem. For example, alligators burrow into the mud of subtropical wetlands, creating depressions ("gator holes") that retain water during dry periods. This creates a habitat for aquatic organisms where none would otherwise exist; without alligators, the entire food chain of the ecosystem would be altered. Many large predators, such as wolves, and some smaller, less conspicuous species play similarly crucial roles in their particular ecosystems. A species that plays a fundamental role in an ecosystem, or whose influence is much greater than might be expected given its abundance, is called a **keystone species**. We know that removing a keystone species from an ecosystem will be damaging or even disastrous. What is less well known is the ultimate or cumulative effect of removing other, less central species. Given this uncertainty, it does seem wise, as suggested by the famed ecologist Aldo Leopold, to make our best effort to "keep every cog and wheel."

It has been suggested that the best way to conserve species is to focus efforts on the conservation of keystone species. Why does this make sense? ______________

Answer: If you preserve the keystone species, the other species that depend on it will be more likely to survive. If a keystone species is lost, all species that depend on it will be negatively impacted.

Biologic Diversity

The principle of competitive exclusion tells us that species compete against one another for scarce resources, and no two species can occupy exactly the same niche within an ecosystem. This might lead to the conclusion that a single species would eventually and inevitably come to dominate a given ecosystem, to the exclusion of others. Instead, as discussed above, competition, symbiosis, and predation, along with constant adaptations to changing environmental conditions, have led to increasing richness and variety of species. This variety of life-forms is called **biodiversity** (or biologic diversity).

There are different types of biodiversity. **Habitat diversity** (or **ecosystem diversity**) refers to the variety of habitat types in an ecosystem and the biologic richness of those habitats. Some experts believe that preserving habitat diversity is the key to preserving biodiversity in general. **Genetic diversity** refers to the amount of variability or heterogeneity that is available among the DNA of individuals within a population or species. This is important, as you will recall from the discussion of natural selection in chapter 4, because genetic variability gives a species a better chance to compete, adapt, and survive. If there are only a few individuals left in a population, even if there are still breeding pairs, the population may be doomed if their genetic diversity is too limited to allow them to adapt to changing environmental conditions.

Species diversity encompasses the concepts of species richness and evenness. **Species richness** refers to the number of species in a community. This number alone is not always a good indicator of diversity. Think about two ecologic communities that have the same number of species (Figure 6.2). In one of the communities, 90 percent of the individuals belong to the same species. In the other community, each species is equally represented with individuals. Which community is more diverse? In the first, a single species is overwhelmingly **dominant**; this

SPECIES DIVERSITY

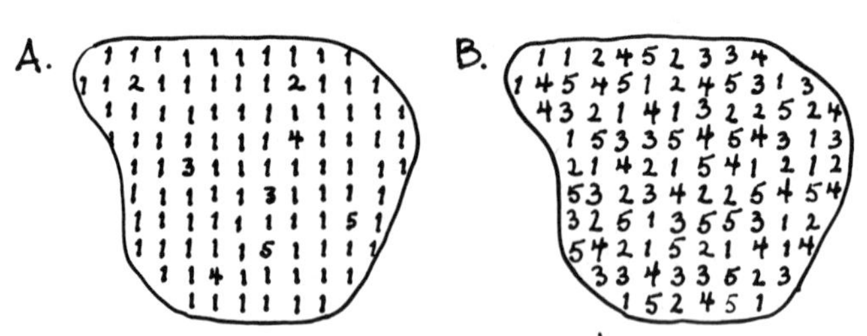

~ TWO IDENTICAL HABITATS ~

SPECIES RICHNESS: BOTH HABITATS HOST THE SAME NUMBER OF SPECIES (5).

SPECIES EVENNESS: SPECIES 1 IS DOMINANT IN HABITAT A; SPECIES ARE MORE EVENLY REPRESENTED IN B, AND OVERALL DIVERSITY IN B IS HIGHER.

Figure 6.2.

limits the species diversity of that community. In the second, the species richness is the same, but the **species evenness**—the relative abundance of individuals within each species—is much higher, the chance of encountering a variety of species is higher, and the overall diversity is greater.

No one knows for certain how many species there are altogether. Scientific estimates range anywhere from 10 million to 100 million. So far, scientists have identified about 1.8 million species, and about 10,000 new ones are identified each year. The majority of species identified so far are insects (around 500,000) and plants (around 400,000), but other types of organisms are being discovered all the time. For example, in the early 1990s, a previously unknown, relatively large (goat-size) mammal was discovered living in a forest in Vietnam. How did there come to be so many different species on this planet?

Evolution results from genetic variability, mutation, and natural selection. Under certain circumstances, a new species may emerge. This can happen, for example, when individuals are forced to adapt to a new or changing environment. Those individuals whose characteristics allow them to adapt successfully to the changes will survive and will be more likely to have offspring with the same characteristics. Eventually, the genetic character of the whole population changes, perhaps to the point where the individuals can no longer breed successfully with individuals of the original population. The emergence of a new species, whether by this or other mechanisms, is called **speciation**.

You can imagine a number of circumstances that could lead to the isolation of a group of individuals and the subsequent emergence of a new species. For example, a group of individuals could migrate across a river during a dry spell; later, when the water is higher, those individuals might be cut off from the main population. Near the end of the last ice age, Siberia and Alaska were connected by a land bridge that permitted the migration of animals (and possibly early humans). After the land bridge was covered by the rising ocean, animals that had migrated into North America were isolated from their original populations. A similar process happened in Australia, which was once joined to Antarctica and India as part of a supercontinent that geologists call *Gondwanaland*. As a result of plate tectonics (discussed in chapter 1), Australia split from Gondwanaland and drifted off on its own; the communities it hosts have been on their own evolutionary path for the past hundred million years. This process is called **divergent evolution**. Australia is now home to some very unusual species, notably the monotremes, which are egg-laying mammals such as platypuses (Figure 6.3).

Organisms are opportunists. If resources are available, there is an opportunity for a species to make use of those resources and fill a specialized niche within the ecosystem. For example, there are many different species of finch on the Hawaiian

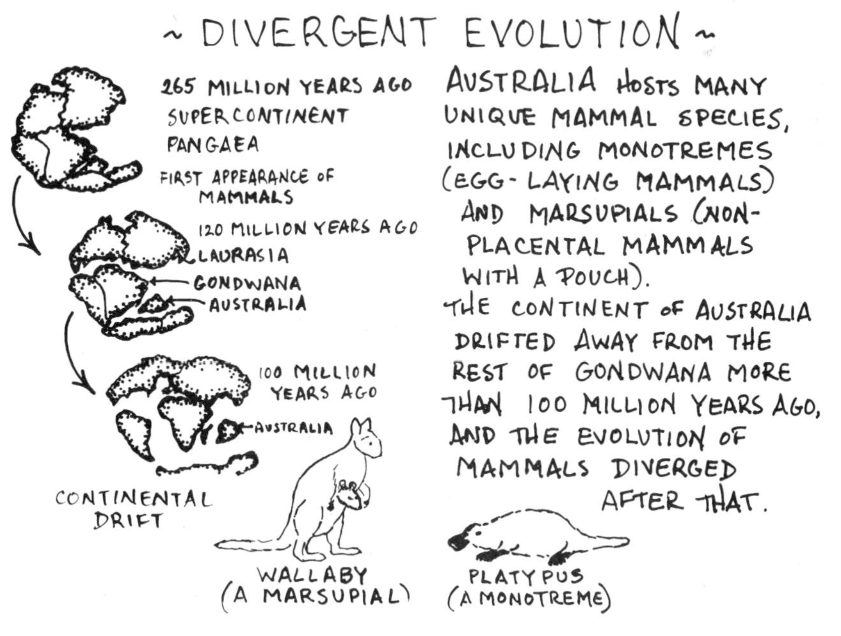

Figure 6.3.

Islands. They evolved from a common finch ancestor into a wide variety of species, each with its own specialized, uniquely shaped beak suited to a particular type of food—fruit, seeds, nectar from flowers, insects from tree bark, and so on. The process of a species evolving into a number of new species, each adapted to fill a particular, specialized niche, is called **adaptive radiation**.

Sometimes the evolutionary response to environmental opportunities and constraints leads to adaptations that are similar, even in communities that are widely separated geographically (Figure 6.4). For example, finches that sip nectar from flowers tend to have similarly shaped beaks—long, thin, curved, and pointy—no matter where in the world they live or evolved, or how long ago they were separated from their ancestral species. This phenomenon is called **convergent evolution**. It is also seen in desert plants, many of which have evolved similar adaptations in response to a particular environmental stress (lack of water). Thus, even if they are genetically unrelated, desert plants often display features such as spines, tough outer layers, and the ability to retain water in their stems.

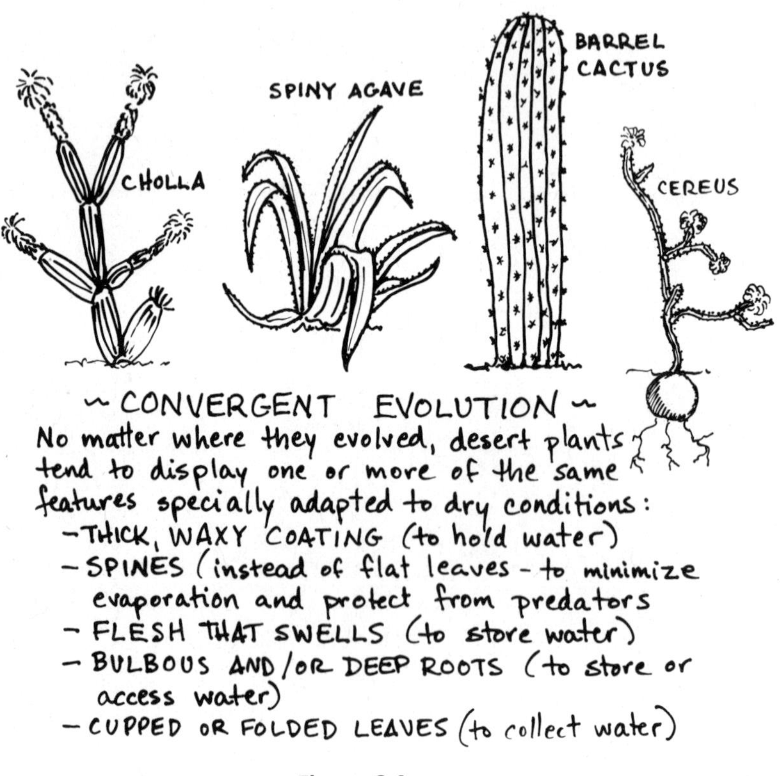

Figure 6.4.

Why do some experts say that preserving habitat diversity is the key to preserving biodiversity? ______________

Answer: If no appropriate habitat is available, nothing can help a species survive (short of maintaining it in a zoo or another type of conservatory).

Why Is Biodiversity Important?

Why is biodiversity important? Write down all the reasons you can think of, then check your list again after you finish reading this section. ______________

Most people would begin their list of reasons why biodiversity is important with something like, "Scientists may find new drugs to cure cancer from undiscovered species in the rain forest." It is fundamentally a **utilitarian** argument for the importance of biodiversity—an argument based on the usefulness or potential usefulness of species. Indeed, new substances are being tested and discovered every day not just from the tropical rain forest species but from other ecosystems as well. For example, penicillin—which completely revolutionized health care—comes from a natural fungus. What would the world be like today if that fungus had gone extinct before its medicinal use had been discovered? It was recently discovered that a substance called paclitaxel, which is derived from the Pacific yew tree, may be useful in the treatment of ovarian cancer. Several hundred-year-old trees must be harvested to yield enough paclitaxel for one treatment. Unfortunately, yew trees are an integral part of the old-growth forests of the Pacific Northwest that provide habitat for the northern spotted owl and other endangered species.

Would you choose to save the old-growth forest that provides habitat for the northern spotted owl, or would you harvest the Pacific yew for its medicinal potential? In any case, if the yew trees are harvested, will they grow back quickly enough for the supply of paclitaxel to be sustainable? Can you think of possible compromises?

Aside from medical and pharmaceutical purposes, another utilitarian argument for the value of biodiversity is that many species have direct economic value. Virtually all of the food we eat and many other products we depend on come from the natural environment. Recently, people concerned with preserving the tropical rain forests have begun to capitalize on the economic argument. By marketing rainforest products such as nuts without damaging the forest, they are emphasizing the economic value of the intact rain forest rather than its value in terms of cut lumber.

Other economic arguments are based on the recreational value of biodiversity. Tourism in the natural environment—including hunting, fishing, hiking, boating, enjoyment of scenery, and many other outdoor and wilderness activities—is a major contributor to the world economy. Many developing countries are beginning to capitalize on **ecotourism**—tourism that focuses on the natural world with minimal environmental impact—as a way to both preserve and profit from their natural assets.

Ecotourism brings up another reason why biodiversity is important, which has to do with the aesthetic value of the natural environment. The concept of aesthetics—appreciation of the innate value of beauty—may seem somewhat vague in relation to the natural environment. However, imagine a home overlooking a wild ravine with songbirds and small animals in the yard, a pond with fish, and a mountain in the background. Now imagine a similar home overlooking a shopping mall. Which home would have greater real estate value? We can appreciate the beauty of a mountain view, but it may be more important for the long-term preservation of biodiversity to be able to attach a specific dollar value to it.

Something that is more difficult to evaluate quantitatively is the spiritual and cultural importance of the natural environment, especially for aboriginals. Many indigenous populations depend on the natural world for life support, as well as for their spiritual fulfillment and sense of identity. Their intimate knowledge of the natural environment, called **traditional ecologic knowledge** (**TEK**), may have accumulated over many generations living in a particular environment. As aboriginal communities around the world lose access to their lands and natural resources, their traditional knowledge is lost. Ultimately, it may be up to the court system to place a dollar value on TEK. For example, some pharmaceutical companies have been sued for having unfairly profited from the traditional ecologic knowledge of indigenous peoples with regard to the healing properties of various plants.

Aside from the utilitarian and economic value, biodiversity provides ecologic benefits. For example, wetlands perform a variety of environmental services such as storing groundwater and cleansing toxins from soils and water. Coral reefs, in addition to hosting extremely productive biologic communities, protect shorelines from storm damage. Rain forests provide an enormous storage reservoir for carbon, without which our global climate would suffer massive changes. Bees are crucially important in the pollination of many crops. Bacteria and other detritivores protect us from drowning in our own waste products. And so on—we cannot even begin to comprehend the value of the environmental services performed by species and ecosystems.

Note that *every one* of these reasons for valuing biodiversity is related directly or indirectly to meeting the needs of humans. This is an **anthropocentric**, or human-centered, view of the value of biodiversity. The economic and utilitarian justifications are obviously anthropocentric, but even the spiritual, aesthetic, and ecologic arguments ultimately represent a human perspective. Yet many people believe that biodiversity should be protected and preserved for moral reasons because species have a value of their own—an **intrinsic value** that is wholly separate from any economic value they might have or service they might provide to humans. The philosophy that all species, including nonhuman species, have intrinsic value and thus the right to continue to exist, and that humans have a moral obligation to protect other species, is called **deep ecology**.

Finally, it is worth noting that the concept of biodiversity speaks to a **holistic** or integrated vision of the environment in which the cumulative value of species and ecosystems together is worth incalculably more than individual species considered separately. This vision is akin to the idea of marginal value in economics; the marginal value of a single, obscure species of salamander or insect, taken on its own, may be next to nothing. The true value of a species may lie in the role it plays within the whole. One way to think of this concept is that species are like individual cards in a house of cards, or like the rivets in an airplane; they are being removed, one by one, but the airplane is still flying. How many rivets can we remove without causing the structure to fail?

After having read this section, would you change your list of reasons why biodiversity is important? Why or why not? ______________

Threats to Biodiversity

Biodiversity is increasingly under threat. Just as new species arise as a result of evolution, through **extinction** they disappear forever. Extinction is not a new phenomenon; throughout the history of life on this planet, species have gone extinct at a rate of roughly one per year. Superimposed on this are several catastrophic episodes—**mass extinctions**—in which many species became extinct within a geologically short time (Figure 6.5). For example, approximately one-quarter of all known animal families, including almost all of the dinosaurs, became extinct 65 million years ago. There have been at least five great mass extinctions in Earth history. Some evidence suggests that the mass extinction that killed the dinosaurs may have been caused by a giant meteorite impact; other extinctions were likely caused by climatic change, volcanism, or other catastrophic environmental changes.

Overall, in the geologic history of life on this planet, the evolution of new species has outpaced extinction. Today, however, we are in the midst of an extinction that rivals or exceeds the great mass extinctions of our geologic past. This

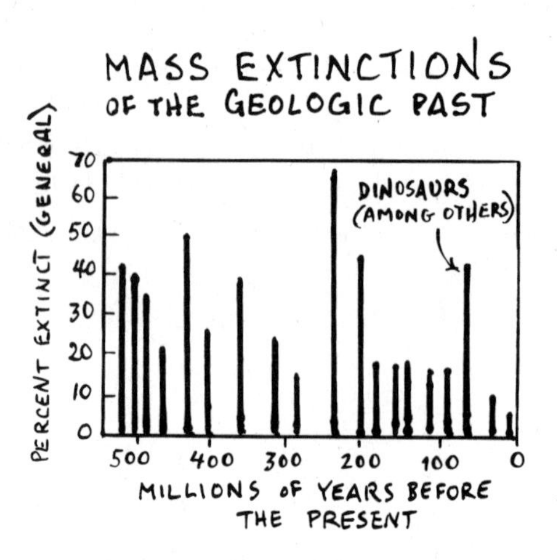

Figure 6.5.

extinction is being caused by human actions rather than by a natural environmental catastrophe. Scientists estimate that in the past 150 years the rate of species extinctions has soared to thousands per decade—more than a hundred times greater than the natural rate of extinction.

An **endangered species** is a species in imminent danger of extinction. Examples include the black rhino, Siberian tiger, giant panda, mountain gorilla, and many others. Some, like these, are well known; others are obscure. There are currently about 1,500 endangered species in North America. A species that has shown a significant decrease in population or **range** (the natural area in which a species is found), or shows signs of imminent local extinction, or **extirpation**, is a **threatened species**. Examples include the gray wolf and sea otter, which are abundant in some parts of their natural range but are locally endangered.

One of the interesting things about biodiversity is that it tends to cluster in certain, highly productive areas that are sometimes referred to as megadiversity "hot spots." These areas, mostly tropical, represent just 2 percent of the world's land surface but collectively hold at least 44 percent of all vascular plant species and 38 percent of all vertebrate animal species (Figure 6.6). Endangered species can be protected by national laws, such as the Endangered Species Act in the United States, and by the United Nations Convention on International Trade in Endangered Species (CITES). However, only about 2 percent of the world's wilderness land area is protected by international agreements; unfortunately, it's not the same 2 percent that hosts the biodiversity hot spots.

Certain biologic or lifestyle factors can make a species particularly vulnerable to extinction. For example, different types of organisms employ different reproductive strategies. Some organisms invest their reproductive energy in producing a very large

Figure 6.6.

number of offspring, only some of which are expected to survive. These are called ***r*-strategists**; they tend to have a small body size and short life spans, and they produce large numbers of eggs, offspring, or seeds on a frequent basis. Examples include mosquitoes and milkweed. ***K*-strategists**, in contrast, tend to be larger, with longer life spans, but they produce a small number of offspring on an infrequent basis. *K*-strategists devote considerable energy to the parental care of their offspring. Examples include elephants, owls, whales, and humans. *K*-strategists (with the exception of humans) tend to be more vulnerable to extinction because of their low reproductive rate.

Other biologic and lifestyle factors that can lead to species vulnerability include highly selective feeding habits and a limited natural range. For example, the giant panda lives only in a few locations in China and eats only bamboo. The panda is a **specialist**—an organism that has evolved to fill a narrow niche and is completely dependent upon the availability of a specific habitat or resource. A **generalist**, however, adjusts readily to changes in its habitat or available resources. A familiar example is the raccoon, which eats almost anything. A small population also renders a species more vulnerable, partly because of lack of genetic variability and partly because it is difficult to find a mate. An example is the California condor, which was brought back from the brink of extinction by a captive breeding program and subsequently released into the wild. The condor is still at risk because so few are alive in the wild, and none of them is yet known to have bred successfully. Extremely selective breeding habits can also cause vulnerability. The green sea turtle breeds only in a few isolated locations on a specific type of sandy beach. Salmon do not breed in the wild unless they can make their way upstream to their original spawning ground.

A significant cause of extinction is **poaching**, the illegal harvesting of wild

This endangered gray wolf in Yellowstone National Park will be tranquilized and given a radio collar for tracking by U.S. Fish and Wildlife Service scientists.

species for commercial exploitation. Animals are killed for everything from their fur to their horns, feathers, or any of a variety of body parts used as trophies or for reputed medicinal treatments. A rhinoceros or an elephant is likely to be killed by poachers who are only interested in the horn or tusk of the animal. Many exotic fish and birds are captured live and sold to collectors and pet owners. Often these animals are caught and transported illegally under poor conditions; many animals die in the process. Plants, too, are harvested illegally, usually either for medicinal or aesthetic purposes. The popularity of cacti for arid landscaping (called *xeriscaping*) has led to the poaching of cacti from public lands; in some cases, the plants are hundreds of years old and may be rare or endangered. Poaching is difficult to control because it requires intensive policing of wilderness areas. Poachers have strong incentives to continue because their harvest may bring a staggering price on the black market.

Legal hunting and fishing (for sport but more often for commercial purposes) can also lead to the extinction of species if overexploitation is allowed to occur. The passenger pigeon and the bison are among the species that have been hunted to extinction (or nearly so, in the case of the bison) within the past 150 years. Overexploitation of fisheries (chapter 8) is a problem in many parts of the world and resulted in the collapse of the Canadian Atlantic cod fishery in the early 1990s.

One of the greatest threats to biodiversity is the introduction of **exotic** (or **alien**) **species** into an ecosystem. Although it can happen naturally, accidental or purposeful importation is common. Lacking its natural predators, an exotic species may take hold and flourish in the new environment, becoming an **invasive species** and crowding out native or **endemic species**. There are many unfortunate cases. The African "killer" bee was imported to South America intentionally in an effort to increase the honey production of domestic bees; this aggressive insect has now spread far into the United States. Purple loosestrife was originally grown as an ornamental flower, but it "escaped" from gardens and has spread into wetlands across the northern United States and southern Canada, choking native plants. Zebra mussels are thought to have hitched a ride on a cargo vessel across the Atlantic Ocean and into the Great Lakes, where they have proliferated, covering all available solid surfaces and clogging intake pipes. The northern snakehead, a fish native to northern China, was imported for medicinal purposes. A pair of snakehead released into a pond in 2002 by a Maryland man caused widespread concern. The fish is very aggressive and eats just about everything in its path; worse, the snakehead can survive and travel, pulling itself along on its fins, for up to three days on dry land.

Some populations have been intentionally reduced or brought to near-extinction by pest-eradication programs. For example, government animal control officers in the United States kill approximately 100,000 coyotes each year to prevent them from preying on domestic animals. This brings up some interesting questions.

What constitutes a pest? Is it justifiable for humans to intentionally cause the extinction of another species?

Before answering these questions, consider the case of smallpox (a virus, not quite alive, but a good example nonetheless). Through vigorous inoculation programs in the 1960s and 1970s, the smallpox virus was essentially eradicated; only a few populations remained protected in laboratories. Some argued for its complete annihilation; others argued that if the virus were to reemerge, for whatever reason, scientists would need laboratory populations from which to manufacture a vaccine. Which do you think was the correct course to take?

The introduction of exotic species is a form of **biotic pollution**; so is the genetic contamination or genetic assimilation of wild populations. This happens when domesticated species escape and breed with wild species, contaminating or diluting the gene pool of the wild population. Salmon and trout raised in captivity may escape and breed with wild fish. Another example is the accidental contamination of organically grown canola, a grain raised for its oil, by genetic material from biologically engineered canola.

Chemical pollution also contributes to the loss of biodiversity. Acid precipitation and ozone depletion, in particular, are known to have detrimental effects on organisms. Many animals have choked on discarded plastic and other debris, especially in the marine environment. The shifting of natural climatic zones and biomes in response to global warming ultimately may prove to be one of the most harmful impacts of climatic change. Above all, however, the single most important cause of extinction and loss of biodiversity is the destruction of habitat. It has been estimated that as many as 85 percent of endangered species have been affected by the destruction of their natural habitat. A significant part of the problem is habitat fragmentation, in which large tracts of natural area are broken up into smaller patches by roads and other developments. We will discuss this further in chapter 8 in the context of wildlife conservation.

What is the most important cause of loss of biodiversity? ________________

Answer: Habitat destruction.

SELF-TEST

These questions are designed to help you assess how well you have learned the concepts presented in chapter 6. The answers are given at the end. If you get any of the questions wrong, be sure to troubleshoot by going back into that part of the chapter to find the correct answer.

1. The process whereby two species evolve together into mutual dependence is called ________________.
   a. coevolution
   b. convergent evolution
   c. natural selection
   d. genetic assimilation

2. When a species evolves into a number of new species, each adapted to fill a particular, specialized niche, it is called ________________.
   a. divergent evolution
   b. species diversity
   c. adaptive radiation
   d. species richness

3. Which one of the following is not a utilitarian argument for the value of biodiversity?
   a. economic value of products
   b. intrinsic value of species
   c. medical or pharmaceutical potential
   d. ecotourism

4. The three basic categories of species interaction are ______________, ______________, and ______________.

5. An organism that plays a fundamental role in an ecosystem or whose influence is greater than would be expected on the basis of its abundance is called a(n) ______________.

6. A parasite that causes or transmits disease or kills its host is called a(n) ______________.

7. Although new species are being discovered all the time, all of the large mammals were discovered by the turn of the twentieth century. (True or False)

8. The principle of competitive exclusion says that only one species can be dominant in any given ecosystem. (True or False)

9. What concepts are encompassed by the term *biodiversity*?

10. What are the main threats to biodiversity?

11. What is the difference between an organism's habitat and its niche?

12. What is the difference between *K*-strategists and *r*-strategists? Give examples of each.

13. Some people strongly object to the idea of placing a dollar value on the spiritual, aesthetic, cultural, or ecologic functions of the natural environment. Others feel that this is the best way—perhaps the only way—to ensure that biodiversity is preserved. What do you think?

ANSWERS

1. a 2. c 3. b

4. competition; symbiosis; predation/parasitism 5. keystone species 6. pathogen

7. False 8. False

9. Habitat or ecosystem diversity; genetic diversity; and species diversity, which includes species richness and species evenness.

10. *Natural causes,* including meteorite impacts, climatic change, volcanism, and other natural catastrophes; *biologic causes* that lead to vulnerability of species, including certain reproductive strategies, selective feeding habits (specialization), limited range, small population, and selective breeding habits; and *anthropogenic causes,* including poaching,

live capture, hunting or fishing that leads to overexploitation, introduction of exotic species (biotic pollution), genetic contamination or assimilation, pest eradication, chemical pollution (acid precipitation, ozone depletion, litter, climatic change), and habitat destruction and fragmentation.

11. Habitat is where the organism lives; niche is its "profession"—what it does for a living, or its role in the habitat.

12. *K*-strategist: typically large body size; long life span; produce small number of offspring on infrequent basis; considerable parental care of offspring; includes elephants and humans. *r*-strategist: small body size; short life span; produce large number of eggs or seeds on a frequent basis; little or no parental care of offspring; includes insects and weeds.

KEY WORDS

adaptive radiation
anthropocentric
biodiversity
biotic pollution
coevolution
commensalism
competition
competitive exclusion
conservation
convergent evolution
deep ecology
divergent evolution
dominant species
ecologic niche
ecosystem diversity
ecotourism
endangered species
endemic species
exotic (or alien) species
extinction
extirpation
generalist
genetic diversity
habitat
habitat diversity
holistic
intrinsic value
invasive species
K-strategist
keystone species
mass extinction
mutualism
parasitism
pathogen
poaching
predation
r-strategist
range
specialist
speciation
species diversity
species evenness
species richness
symbiosis
threatened species
traditional ecologic knowledge (TEK)
utilitarian (argument for biodiversity)

7 People, Population, and Resources

The Lion looked at Alice wearily. "Are you animal—or vegetable—or mineral?" he said, yawning at every other word.

—Lewis Carroll

Objectives

In this chapter you will learn about:

- the factors that influence population growth in ecosystems;
- the socioeconomic, cultural, and technologic factors that influence human populations;
- the impacts of human activities and lifestyle on the natural environment; and
- the basic concepts of resource utilization and management.

Population Dynamics

You have learned about populations, communities, and ecosystems, and how individual organisms interact with one another and with the abiotic environment. In this chapter we look more closely at the factors that control the size and growth of populations. We begin with some of the basic biologic controls on population and move on to consider the factors that influence human population growth and resource utilization.

Resources are very important in the population dynamics of biologic systems. The growth and reproductive success of organisms depend on the availability of life-supporting resources. In an environment where unlimited resources are available and where nothing interferes with growth or reproduction, **exponential growth** can occur (Figure 7.1). In exponential growth, population increases geometrically at a given rate of increase per unit of time—the **growth rate**. Let's say that you have one penny and each day you double it. On the second day you will have two pennies, the third day you will have four pennies, and the fourth day you will have eight pennies. At this rate of growth, by the twenty-first day you will have over a million pennies, the twenty-second day you will have over two million pennies, and so on. This is also called **geometric growth**. It differs from **arithmetic growth**, in which the *amount* of growth per unit of time, rather than the *rate*, is a constant (Figure 7.1). If you were to simply *add* one penny each day instead of doubling, the growth would be arithmetic; on the second day you would have two pennies, three pennies on the third day, twenty-one on the twenty-first day, and so on.

The mathematic equation that describes exponential growth is $dN/dt = rN$, where N is the number of individuals, t is the unit of time, and r is the growth rate (like the interest rate in a bank account). The term d is a mathematic term that means "change." Therefore, the equation can be read as "The change in the number of individuals per change in unit of time is equal to the number of individuals times the growth rate."

Try this formula with a handheld calculator using the example of an organism that reproduces quickly, such as a bacterium that reproduces itself by dividing in half every 30 minutes. Start with $N = 1$, $t = 0.5$ hr, and $r = 1.0$ (that is, 100 percent). After each calculation, the result gets recycled back into the equation as the new N. When does N reach 1,000? When does N surpass 1 million? One billion? Try it again

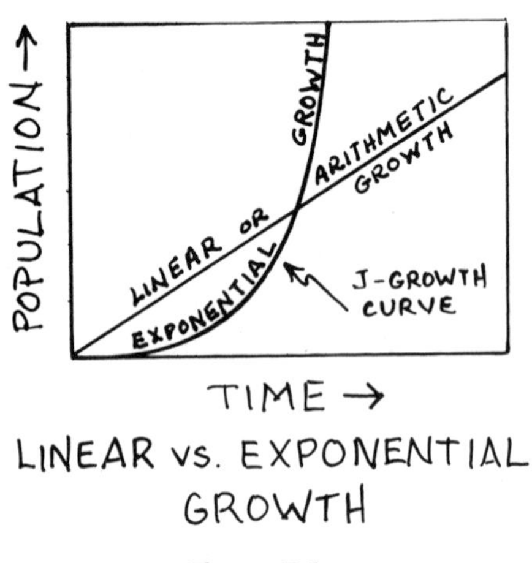

Figure 7.1.

using a typical reproductive rate for another organism, such as fruit flies, which might reproduce at a rate of 150 percent ($r = 1.5$) per day, or for elephants, which might reproduce at a rate of 100 percent ($r = 1.0$) per decade. ____________

Answer: For the bacteria, N is greater than 1,000 by the time 5 hours have passed; $N >$ 1,000,000 by 10 hours, and $N >$ 1,000,000,000 by 14.5 hours.

If you plot the results of the penny or bacteria calculations on a graph using N as the vertical axis and t as the horizontal axis, you will see that a graph of exponential growth is shaped like the letter J. This **J-curve** is characteristic of exponential growth. (If you plot an arithmetic increase, however, you should get a straight-line graph.)

Clearly, unlimited exponential growth could be problematic. If a bacterial population can grow from one to over a billion individuals in less than fifteen hours, why isn't the whole world completely awash in bacteria (not to mention other organisms that reproduce rapidly, such as cockroaches, fruit flies, or mosquitoes)? The answer, of course, is that growth in biologic populations in the real world is *not* unlimited; in fact, the reproduction of organisms in the real world is limited by a variety of factors.

Do you recall from chapter 6 what scientists call organisms whose reproductive strategy is to produce many offspring on a frequent basis, as opposed to organisms that produce few offspring infrequently but provide them with abundant parental care and protection? ____________

Answer: Numerous, frequent offspring: *r*-strategists. Few, infrequent offspring, with parental care: *K*-strategists.

An obvious oversight in the calculation is that we have only considered growth as a function of **birth rate**, the number of births (new individuals added to the population) per unit of time. However, organisms also die. The **death rate** (or **mortality**) in a population represents the number of individuals that have died during the same period and must be subtracted from the population. In the natural environment, organisms can also enter a population through immigration or leave through emigration. These adjustments to population must also be made.

Furthermore, as mentioned above, in the real world the growth and reproduction of organisms is limited by the availability of resources. The growth that a population theoretically could experience in the absence of any limitations is called the **biotic potential** of the population. As a population grows, however, many factors begin to place limitations on its growth. Some of these are related to **population**

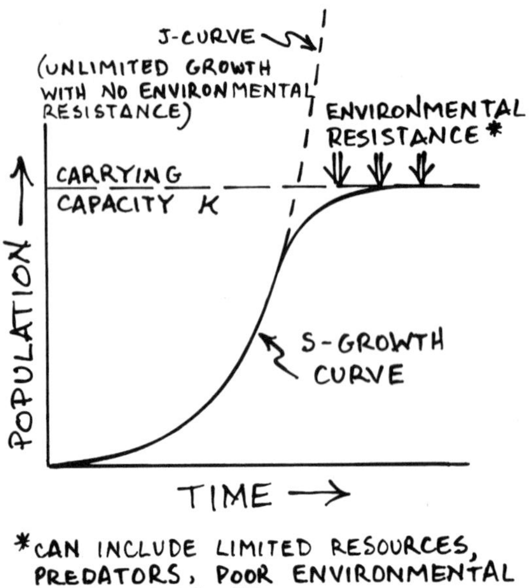

Figure 7.2.

density, the number of individuals per unit of space. If a system is very densely populated, individuals may have a difficult time gaining access to the resources they need, such as light, water, food, shelter, or living space. Resources may become depleted or run out altogether. In a community with more than one species, additional factors come into play. For example, with increasing population density, an individual might be more likely to encounter a predator. Environmental risk factors, such as drought, a cold winter, or a population explosion among predators, also influence the growth of a population.

Anything that acts to limit or control population growth is called a **limiting factor**. In a test tube crowded with bacteria, poor access to food might be the limiting factor. On the ground floor of a densely vegetated rain forest, lack of light could be a limiting factor. In a desert, the lack of water might be the most important limiting factor for organisms. For pandas, which eat only bamboo, the poor availability of bamboo forest habitat is a limiting factor. For snowshoe hares, an abundance of lynx—an important predator—might be a limiting factor. And so on. Some limiting factors are related to biologic characteristics of the species itself, such as slow reproductive rates. Other limiting factors, such as scarcity of resources or environmental risk factors, are external.

These factors—limited food, water, space, or light, slow reproduction, compe-

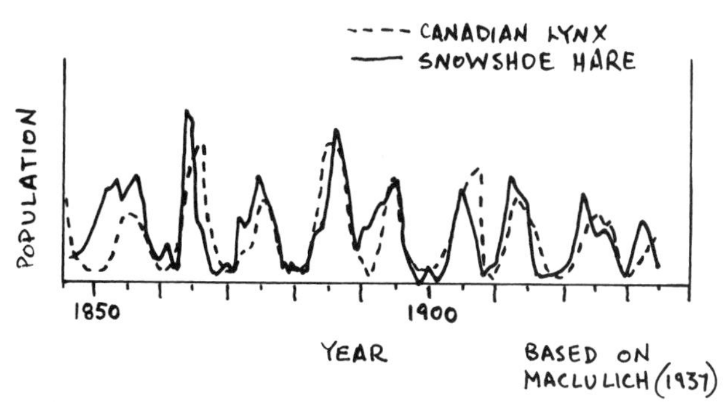

Figure 7.3.

tition, predation, disease, and other risk factors—provide **environmental resistance** to unlimited growth in biologic populations. Typically, a population will start out growing rapidly on an exponential J-curve. As population density increases, environmental resistance comes into play, and the population will begin to level off, resulting in a curve that is S-shaped (Figure 7.2). The leveling-off point represents the maximum number of individuals in a population that can be supported by the ecosystem on a long-term basis; it is called the **carrying capacity** (*K*).

Note that the concept of carrying capacity explicitly states that the ecosystem can support the population *on a long-term basis*—that is, without incurring serious or permanent damage. This is referred to as **sustainability**. Sometimes organisms—especially *r*-strategists that reproduce and grow very rapidly (*r*, as in growth *rate*)—overshoot the carrying capacity of the ecosystem; their number increases so rapidly that resources are quickly depleted. In this scenario, what often happens is a sudden dieback resulting either from a lack of resources or an abundance of predators. The dieback returns the population to a sustainable level for the carrying capacity. In particular, *r*-strategists tend to go through boom-and-bust population cycles, with huge growth spurts followed by massive diebacks repeated the following season or year (Figure 7.3). Populations of *K*-strategists (*K*, as in carrying capacity) tend to approach the carrying capacity of the ecosystem more slowly, leveling off without the dramatic boom-and-bust oscillation.

Explain the graph in Figure 7.3 based on a classic study by D. A. MacLulich (*Fluctuations in the Numbers of the Varying Hare* (Lepus americus). Toronto: University of Toronto Press, 1937, reprinted 1974), which shows variations in the populations of the snowshoe hare and lynx in northern Canada. Use the concepts of carrying capacity, environmental resistance, and boom-and-bust population cycles.

Answer: The hare, an *r*-strategist, reproduces quickly in the absence of predators (see the year 1850, for example). The hare population increases, overshooting the carrying capacity of the ecosystem. After each boom, the hare population encounters environmental resistance and experiences a dieback. The diebacks come from a lack of resources and from environmental conditions such as cold, as well as from predation, which is shown on the graph by the concomitant increases in the number of lynx. With each decrease in the hare population, the lynx population, lacking its prey, also decreases, and the cycle repeats itself.

Human Population Growth

The study of human population is called **demography**. Humans are fundamentally biologic organisms, so the human population of this planet is subject to many of the constraints discussed in the previous section. However, humans are complex creatures, and two major factors greatly complicate the population scenario: our socioeconomic and cultural structures, and our technologies. Socioeconomic and cultural factors are important because humans are intelligent and are generally capable of understanding the consequences of their actions. Humans do not mate strictly by instinct; instead, they make conscious decisions that influence mating and reproductive behaviors. Social policies and laws are very important, too. For example, China's strict one-child policy has led to a drastic reduction in the birth rate. As a result, India will soon surpass China as the most populous nation on Earth.

Some socioeconomic and cultural forces tend to promote, rather than limit, population growth; these are **pronatalist** influences. An example is religious beliefs that prohibit birth control. Poverty can also promote population growth. In the poorest nations of the world where **infant mortality** (the death rate for children under the age of five) is very high due to hunger, poor health, and difficult living conditions, having a large family is seen as an economic necessity. A family requires a certain number of hands to work in the fields; if half of the children can reasonably be expected to die before the age of five, couples are likely to produce more children to ensure that enough survive to help with food production.

Technology is the other major factor that influences human population growth. The story of population growth from the emergence of the first modern humans is really a story of technologic advancement and innovation. An early milestone was the Agricultural Revolution about 12,000 years ago, when people first developed the ability and technology with which to cultivate crops and domesticate animals. Until that time, the entire population of the world was probably no more than 250,000 people, in widely scattered, small, seminomadic groups of hunter-gatherers. From studying modern nomads, we know that supporting a small population with a hunting-and-gathering lifestyle requires access to a very large resource base. In contrast, agriculture increases the effective carrying capacity of a system by making food resources more readily available. Thus, the Agricultural Revolution opened up the possibility of a sedentary, or settled, lifestyle and ultimately led to a significant increase in both population density and total population.

Subsequent technologic advances have introduced human populations to more modern agricultural techniques, new materials, sanitation, and other medical and technologic innovations. Even literacy and printing were innovations that have served to enhance and preserve human life. Each of these advances in technology

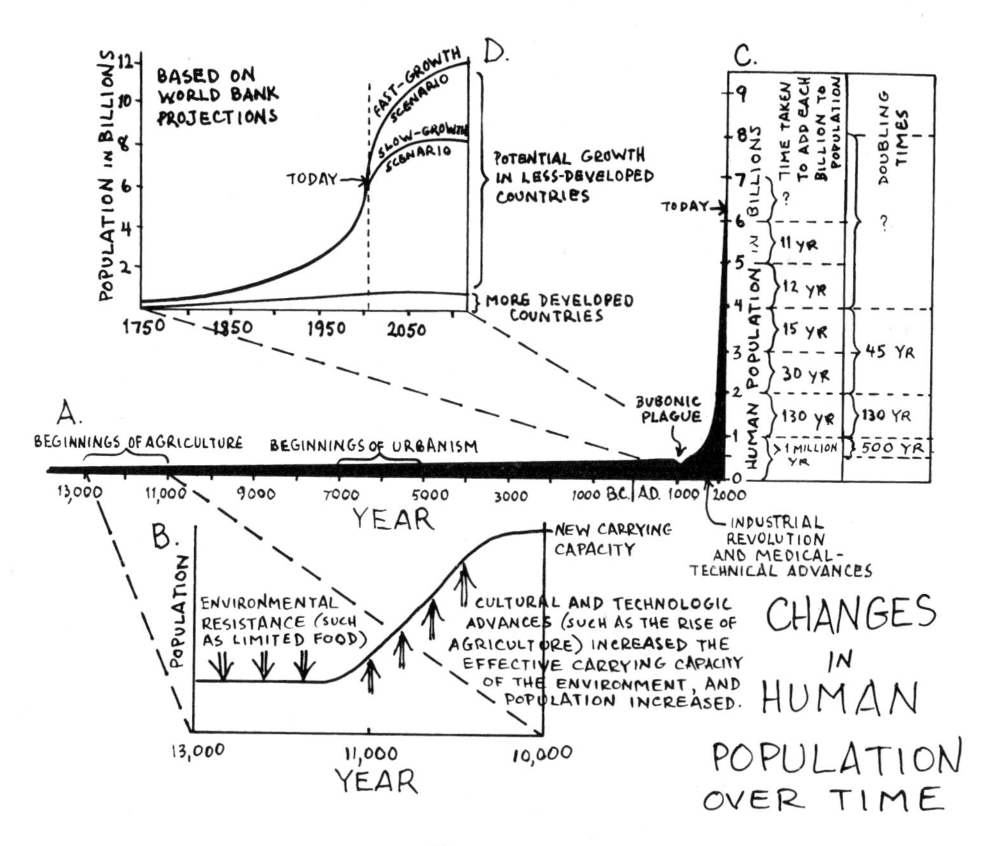

Figure 7.4.

removed some constraints, such as disease, hunger, lack of information, or lack of shelter, thereby lessening environmental resistance, increasing the effective carrying capacity of the environment, and causing or permitting growth in the human population. The overall growth curve for the human population is J-shaped, representing exponential growth (Figure 7.4A). However, if we look closely at the growth patterns and correlate them with technologic advances, we can see that human population growth is not like the exponential growth of bacteria in a test tube. There have been many constraints on human population growth, but some of them have been lessened or removed by our own technologic ingenuity (see Figure 7.4B).

The low-density, scattered human population increased to roughly 10 million shortly after the Agricultural Revolution, then reached 50 million by about 5000 B.C. and 500 million by about A.D. 1300. Between A.D. 1300 and A.D. 1800, the population doubled, reaching about 1 billion. It took about 130 years to double again, to 2 billion (by 1930); 30 more years to reach 3 billion (1960); 15 more years to reach 4 billion (1975); and just 12 more years to reach 5 billion (1987). The current world population is about 6.2 billion. (You can check the up-to-the-minute human population at several Web sites, including the population clock at www.probeinternational.org.)

Note that the **doubling time** for the human population (from 500 million to

These are slum apartment buildings in Calcutta, India. India will soon be the most populous country in the world (surpassing China).

1 billion, from 1 to 2 billion, from 2 to 4 billion, and so on) has decreased over the past several decades (see Figure 7.4C). This is because of an increase in the annual growth rate of the human population. You can calculate approximate doubling time (in a population, or anything with compounded interest such as a bank account) by dividing 70 by the annual percentage growth rate. For example, a city that is growing at 3 percent per year—not an unusually rapid growth rate for cities in some parts of the world—will double its population in 23.3 years. A population with a growth rate of 0.5 percent per year will double in about 140 years. Following many years of increase, the annual growth rate of the world population has recently dropped somewhat, from a high of 2.2 percent in 1960 to the present level of about 1.4 percent per year.

What is the doubling time for a population with an annual growth rate of 1.4 percent? ______________

Answer: 50 years.

Can we expect to have a human population of 12 billion—twice what we currently have—by the year 2050? Maybe, but not necessarily. It is notoriously difficult to project global population trends; much depends on the technologic, social, economic, and cultural aspects of our health and reproductive behavior. Trends in death (mortality) rates (the number of deaths per thousand people per year) and **life expectancy** (average age at death) obviously influence population trends. Access to reproductive health care and education—especially of girls—plays an important role in limiting **fertility rates** (the average number of children born to each woman during her reproductive lifetime). A wide variety of environmental risk factors can contribute to higher death rates; significant factors include natural disasters, wars, pollution, and infectious diseases such as AIDS.

The first important predictions about human population were made more than two hundred years ago by Thomas Malthus. He concluded that the human population inevitably will overwhelm available food resources, resulting in famine, disease, and war over resources. Modern population theorists who predict catastrophes and conflicts as a result of uncontrolled population growth are sometimes referred to as **neo-Malthusians**. Malthus, of course, could not have foreseen the technologic, agricultural, and medical advances that have made resources more available and have increased longevity since his time. A more optimistic view, therefore, suggests that human population has not grown out of control but has grown in response to the expansion in carrying capacity created by human technologic advancements, and this will continue to be the case. A final important conceptual perspective suggests that poverty, poor conditions, inequitable opportunity, and lack of social justice—rather than just population growth—should be at the root of our concern, and that if we solve those problems, population will stabilize in response.

What is a neo-Malthusian? ________________

Answer: A population theorist who, like Malthus, argues that human population growth is out of control and will eventually exceed available resources, resulting in catastrophes such as famine, war (competition for resources), and disease.

The Demographic Transition

Organizations that make predictions about population trends—notably, the World Bank and the United Nations—predict that the global population will begin to level off or even decline slightly sometime during the present century (see Figure 7.4D). Estimates of the leveling-off time vary from about 2030 to 2060. Estimates of the final population also vary, ranging from low, optimistic estimates of about 8 billion to high, pessimistic estimates of 12 billion or more. The populations of some countries have already leveled off. Japan and some European countries currently have negative growth rates. Other European countries, Australia, New Zealand, the United States, and Canada have nearly stable fertility rates (although their populations continue to increase partly as a result of immigration and partly because of baby-boomer fertility). Demographers say that these countries—the established, industrialized, wealthy nations of the world—have passed through a **demographic transition** to attain population stability (Figure 7.5); most of the population growth in the coming decades will take place in the poor countries of the developing world, which have not yet made this transition (see Figure 7.4D).

The demographic transition has several stages. In the *preindustrial* stage, birth rates are high, but death rates are also high; the resulting population is stable. In the *industrial* stage, death rates fall as a result of higher incomes and better health care, but birth rates remain high. The result is a dramatic increase in population; many countries of the developing world are currently in this stage. Next is the *transitional* stage, in which birth rates decline in response to low death rates. However, during this stage, which also characterizes many developing countries today, much of the population is under age fifteen. As a result, even if fertility rates drop, the population will continue to grow under its own momentum while these children grow up and have children of their own. Finally, with low birth and death rates, the population can begin to stabilize and may even decline slightly; this is the *postindustrial* stage that

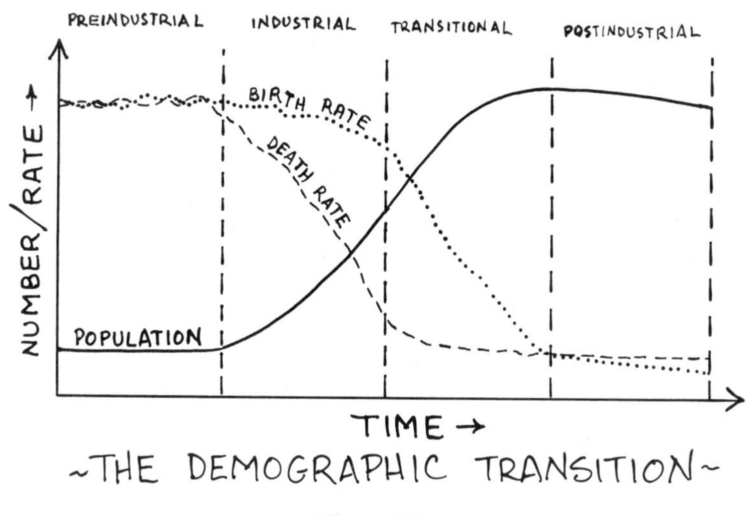

Figure 7.5.

characterizes Japan, Australia, North America, and some of the wealthy European countries today. Some countries have moved very slowly through the demographic transition; it has taken the United States more than 150 years. Others, such as Peru, Mexico, Thailand, and China, are moving very quickly, with fertility rates dropping dramatically in just a couple of decades. Many other developing countries are still in the rapid-growth stage.

A population that is stable with a growth rate of zero—neither increasing nor decreasing—is said to have achieved **zero population growth**, or **ZPG**. It takes a couple of generations of reproduction at or below replacement levels for a population to stabilize to ZPG. The effect of age distribution can be seen in **population pyramids** for fast-growing and slow-growing populations (Figure 7.6). In a

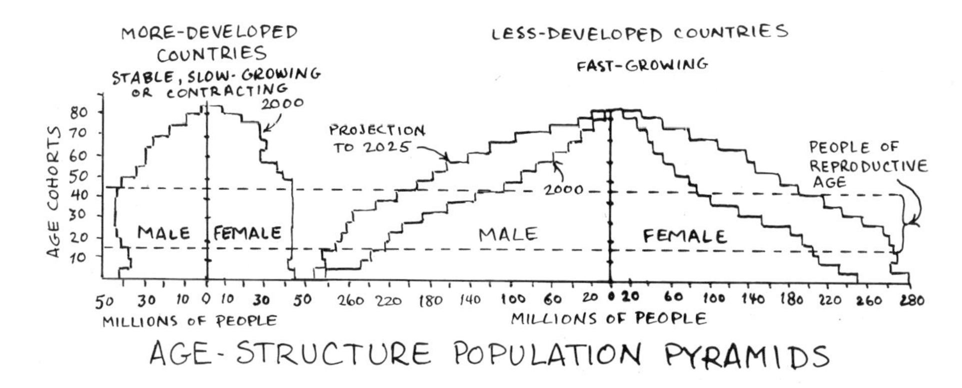

Figure 7.6.

fast-growing population, a large proportion of the population is under the age of fifteen; in a slow-growing population, the age distribution of the population is more even. Some economists worry that ZPG will mean that the population will age too quickly and there will be too many old people and not enough young, working people to support the nation's productivity. Others argue that ZPG and the demographic transition are necessary to stabilize the world's population and keep it within the carrying capacity of the planet.

What is the demographic transition? Try to draw the three curves of the demographic transition from memory—birth rate, death rate, and population versus time. Be sure to label all of the axes. Also name and label the four stages of the demographic transition. ______________

Answer: Demographic transition: change in population trends in a society undergoing development. Stages: (1) preindustrial (high birth rate, high death rate, stable population); (2) industrial (high birth rate, falling death rate, very rapid population growth); (3) transitional (low death rate, falling birth rate, population still growing because of momentum); (4) postindustrial (low death rate, low birth rate, population stable or declining slightly).

Human Impacts on the Environment

Is the carrying capacity of this planet sufficient to support a population of 12 billion, or even 8 billion? Are we already overshooting the planet's carrying capacity even at just over 6 billion? Unfortunately, there is no simple answer. Certainly, human use of environmental resources already outstrips carrying capacity in some local areas, where significant long-term or permanent environmental damage has occurred.

One of the interesting issues related to human population is the question of how our impact on the environment through sheer numbers is related to the impacts of lifestyle and levels of resource utilization. The poor countries of the world—those with the least access to financial resources—still have high birth rates and are where most of the remaining population growth will occur. These countries seem most likely to push our global population over the sustainable limit. Furthermore, poverty itself can cause environmental degradation. As quoted in the Brundtland Report, "Those who are poor and hungry will often destroy their immediate environment in order to survive: They will cut down forests; their livestock will overgraze grasslands; they will overuse marginal land; and in growing numbers they will crowd into congested cities. The cumulative effect of these changes is so far-reaching as to make poverty itself a major global scourge." (*Our Common Future* [1987], commonly known as the Brundtland Report, was the report of the United Nations World Commission on Environment and Development, in which the term *sustainable development* was first popularized.) Poverty, in combination with a breakdown of traditional social and cultural patterns, contributes to environmental degradation, and the poor are often the most seriously affected victims.

It is tempting to think that those of us who can "afford" to take care of our children and provide for them should have the right to have more children. But what if we measure environmental impact on the basis of intensity and rate of resource use? On a per capita basis, Americans use about four times as much steel and twenty-three times as much aluminum as people in Mexico. The average Japanese consumes nine times as much steel as the average Chinese. Canadians routinely top the list in per capita energy consumption. In overall use of nonreplenishable resources, the average Swiss consumes as much as forty Somalis. The point is not that the lifestyle of the Swiss or Somalis or Americans is right or wrong. Nor is it that everyone's living standard should be lowered in order to fit within Earth's carrying capacity. Rather, the point is that there are many choices to be made, many resource issues to consider, many ways to use resources wisely, and many ways to do more with less. When you consider the fact that a Somali family could have thirty-nine children and still not be consuming as much as a Swiss or North American family, it gives you a different perspective on population and resource use.

The United Nations Conference on the Human Environment was held in Stockholm in 1972. In her address to the conference, Indira Gandhi, at the time the prime minister of India, said, "How can we speak to those who live in the villages and in the slums about keeping oceans, the rivers, and the air clean, when their own lives are contaminated? Are not poverty and need the greatest contaminators?" Do you agree that poverty and need are the greatest global environmental problem?

The idea that people may have a greater or lesser impact on the environment as a result of their lifestyle and resource consumption is encapsulated in the concept of the **ecologic footprint**, which is a measure of the land and water area required to support a person or a community at a specified material standard of living. Most people in the industrialized world have a footprint much larger than the actual plot of land they occupy. For example, the ecologic footprint of the average American is 10 ha (25 acres). The average Canadian lives a bit more lightly, at 7 ha (17 acres), and the average Italian's ecologic footprint is 3.6 ha (9 acres). The calculation takes into account the land and water requirements for six categories of activities that support our material lifestyle: growing crops for food, feed, and fiber; grazing animals; fishing; harvesting timber; building infrastructure (roads, houses, and so on); and burning fossil fuels. According to the ecologic footprint calculation, the current human population of 6.2 billion has already exceeded Earth's carrying capacity. It would take about 1.25 Earths to meet our current resource needs sustainably and 3 or more Earths for everyone on the planet to live at the level of resource consumption of the average North American.

What is your ecologic footprint? You can find out more about ecologic footprints and calculate your own footprint by visiting www.redefiningprogress.org.

Basic Concepts in Resource Use and Management

Our hunter-gatherer ancestors once lived entirely on renewable resources—the animals they caught and the plants they gathered. Eventually, our ancestors crossed one of the great thresholds of human development. They picked up stones to use as hunting tools and became the first and only species to use nonrenewable resources routinely. It was not long before they discovered that some stones are tougher and make better spear points than others. Because the best stones could only be found in a few places, trading started. As time passed, people started gathering and trading salt. Hunting communities could satisfy their dietary needs for salt by eating meat, but when farming started, diets became grain based, and extra salt was needed. We don't know when or where salt mining started, but long before recorded history, trade routes for the exchange of nonrenewable resources—beginning with spearheads and salt—crisscrossed the globe.

Nonrenewable resources, such as minerals (including salt), fossil fuels, and soils, cannot be renewed or replenished. The geologic processes responsible for the formation of most types of mineral and energy resources are still operating today, but they may take hundreds of millions of years to complete. As a society, we can't wait that long to replenish a resource that has been used up. Therefore, it is more accurate to say that nonrenewable resources are not renewable or replenishable *on a humanly accessible time scale.*

The key feature of nonrenewable resources is that the more we use, the less remains in the **stock**. When we use up nonrenewable resources, we do so not just for ourselves but for future generations as well. Wherever possible, we must seek to extend the availability of these resources through conservation, substitution, reuse, and recycling. A wise management strategy will also look toward replacing reliance on nonrenewable resources with reliance on renewable or inexhaustible ones. For example, we might replace nonrenewable fossil fuels with solar, geothermal, tidal, or biomass sources of energy. **Inexhaustible resources**, such as solar and tidal energy, can never be used up. **Renewable** or **replenishable resources** are those that can be regenerated, such as fish and trees.

Renewable resources require a different management style from nonrenewable resources. Most living resources are renewable—*if* they are managed properly. Some nonliving resources, such as groundwater and soil, are also renewable, but people may not be willing or able to wait long enough for the depleted resource to be replenished. Increasingly, the guiding principle in the management of renewable resources is **sustainable development**—cautious, planned utilization of Earth resources to meet current needs without degrading ecosystems or jeopardizing the future availability of those resources.

Specifically, the sustainable management of renewable resources must ensure that the **flow** of material from the resource—the rate at which the resource is being used or withdrawn from the stock—does not exceed that at which it is replenished or regenerated (Figure 7.7). Fish should not be harvested faster than new fish are

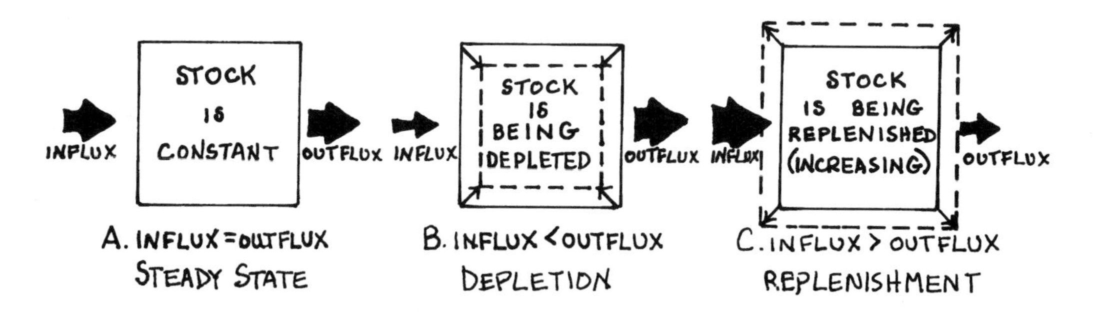

Figure 7.7.

hatching, and trees should not be cut down faster than new trees can grow. If the stock of a renewable resource becomes too severely depleted, it may never regenerate, no matter how long we wait. For example, fishing industries in many parts of the world, including the eastern coast of Canada, have collapsed (both economically and biologically) because the fish were overharvested.

It is also possible for nonliving renewable Earth resources to become depleted to the point at which they cannot regenerate naturally. For example, groundwater reservoirs can be replenished by rainfall, although the process may take many years or even centuries. However, if too much water is withdrawn from an aquifer too quickly, the sediments of the aquifer may become compacted. When this happens, the underground storage reservoir can never be replenished, no matter how much it rains and no matter how long we wait.

Summarize the basic management strategies required for renewable and nonrenewable resources. _______________

Answer: Renewable: balance stocks and flows so that the rate of withdrawal does not exceed the rate of renewal or regeneration. Nonrenewable: conservation, reuse, recycling, replacement with renewable substitutes.

From an environmental perspective, an important type of resource is the **common property resource**, or "commons," a resource that is jointly or publicly (rather than privately) owned and accessed. Parks, town squares, and national, state, and provincial forests are examples of common property resources. The Great Lakes are a common property resource; they are jointly managed by the United States and Canada, but they are publicly owned. The atmosphere and the ocean are common property resources; their scale and importance to all people is so great that they are referred to by a special name, the **global commons**. Managing common property resources is somewhat trickier than managing privately owned resources. The reason is explained by the concept of the **tragedy of the commons** from a classic 1968 paper entitled "The Tragedy of the Commons," published in the journal *Science*, by the ecologist Garrett Hardin.

Imagine a goat pasture to which all the people of a certain town have equal access. The pasture has a carrying capacity—a specific number of goats that is the maximum that can be sustained indefinitely without causing damage to the pasture ecosystem. Let's say that the carrying capacity of the pasture allows for the grazing of five goats from each family. No one polices the resource; it is taken on good faith that everyone will put only five goats out to graze. Then imagine that one person gets a bit greedy and decides to put fifteen goats out to graze; after all, who will know? After a while, the pasture ecosystem starts to be degraded because there are too many goats grazing; the grass is eaten down to the ground and the soil begins to erode. But in the meantime, the person with the fifteen goats has become rich from selling goat's milk and cheese.

The point is that the individual with the fifteen goats had every opportunity and motivation to profit from the exploitation of the commonly owned resource but no particular reason to pay for controlling or repairing the damage caused in the process. It's not a very flattering portrayal of human nature, but unfortunately it is often true: when access to or control over a resource is shared, and is not closely monitored or tightly controlled, individuals do not always act in the best interest of the resource as a whole. This is particularly true when an individual's share of the profits is allowed to exceed his or her share of the resulting damage to the resource. In the real world, there are many common property resources that have enormous environmental importance, and our ability to police the use of these resources is limited.

Can you think of how the tragedy of the commons is expressed in the management of the atmosphere and the ocean?

Are there enough resources available to raise the living standards of all people to the levels they desire? The answer to this question is not clear. It is difficult to know what Earth's carrying capacity may turn out to be and whether we have reached or even surpassed it. The human population continues to grow, and we continue to find, process, and make use of resources in a wide variety of ways. Perhaps the only firm conclusion we can reach is that while Earth's carrying capacity is very large, it is not infinite. The available resources can be extended through careful use and conservation practices, new technologies, and new discoveries. In the long run, however, there must eventually be some limits to growth. (*Limits to Growth* [Earth Island Limited, 1972] is the title of a book by Dennis and Donella Meadows. It was one of the first studies that used computer modeling to make projections about remaining supplies of resources.) It seems inevitable that our descendants will have to manage Earth resources very differently from how we and our parents and grandparents have managed them.

SELF-TEST

These questions are designed to help you assess how well you have learned the concepts presented in chapter 7. The answers are given at the end. If you get any of the questions wrong, be sure to troubleshoot by going back into that part of the chapter to find the correct answer.

1. The doubling time for a population growing at an annual rate of 5 percent would be ______________.
   a. 1,400 years
   b. 140 years
   c. 14 years
   d. 1.4 years

2. The average number of children born to each woman during her reproductive lifetime is the ______________.
   a. birth rate
   b. biologic productivity
   c. biotic potential
   d. fertility rate

3. Pressures that tend to lead to increases in birth rates in a human population are called ______________.
   a. fertility rates
   b. the demographic transition
   c. neo-Malthusians
   d. pronatalist influences

4. The study of human population is called ______________.

5. The mathematic equation that describes exponential population growth is ______________.

6. ______________ growth shows a constant *rate* of increase over time, whereas ______________ growth shows a constant *amount* of increase over time.

7. A population in which the birth rate is at replacement levels (one child born to each person) will have achieved zero population growth. (True or False)

8. A straight-line graph sloping upward is characteristic of exponential growth. (True or False)

9. What is a limiting factor? Give some examples.

10. What is sustainable development?

11. What is the tragedy of the commons?

12. What is the present human population of the world, its current rate of growth, and its current doubling time?

13. Why did the human population grow dramatically as a result of the Agricultural Revolution?

ANSWERS

1. c 2. d 3. d

4. demography 5. $dN/dt = rN$

6. Geometric (or exponential); arithmetic

7. False (Not necessarily; even if the birth rate is at the replacement level, there may still be considerable population growth as a result of immigration, as well as from the momentum of a large proportion of the population being under the age of fifteen—that is, prereproductive age.)

8. False

9. Anything that acts to limit or control population growth. Examples from the text include poor access to food in a test tube crowded with bacteria; lack of light on the ground floor of a densely vegetated rain forest; lack of water in a desert; poor availability of bamboo forest habitat for pandas; abundance of lynx (predator) for snowshoe hares.

10. Development that meets the resource needs of the present generation without compromising the ability of future generations to meet their needs.

11. The tragedy of the commons is that it is often in an individual's best interest to profit from shared or commonly owned resources without regard for the damage caused as a result. However, this action may not be in the best interest of the resource as a whole or of the group that owns the resource.

12. About 6.2 billion; 1.4 percent per year; 50 years

13. Agriculture and the domestication of animals effectively increased the carrying capacity of a given area of land, allowing it to support a greater number of people and making food resources more readily available. A sedentary lifestyle became possible, and both population and population density increased in response.

KEY WORDS

arithmetic growth
biotic potential
birth rate
carrying capacity
common property resource
death (mortality) rate
demographic transition
demography
doubling time
ecologic footprint
environmental resistance
exponential growth
fertility rate
flow
geometric growth
global commons
growth rate
inexhaustible resource
infant mortality
J-curve
life expectancy
limiting factor
neo-Malthusian
nonrenewable resource
population density
population pyramid
pronatalist
renewable (replenishable) resource
stock
sustainability
sustainable development
tragedy of the commons
zero population growth (ZPG)

8 Living Resources I: Forests, Wildlife, and Fisheries

Trees are poems that the Earth writes upon the sky.

—Kahlil Gibran

Objectives

In this chapter you will learn about:

- forest resources;
- wildlife and wilderness;
- fisheries resources; and
- the exploitation, conservation, and management of living resources.

Forest Resources

The world's forest resources encompass a great variety of forest types, as discussed in chapter 5. Among these are temperate and tropical closed forests, with continuous canopies and a high proportion of trees relative to other vegetation. Open forests or **woodlands** have less densely spaced trees interspersed with grasses and other vegetation; the savanna is an example. About 95 percent of the world's forests are natural forests; the remaining 5 percent are forest **plantations**—managed forests where trees have been planted in rows, usually for the eventual purpose of harvesting.

This agroforestry worker in Honduras is extracting resin from a pine tree without harming the tree.

Natural forests are not intentionally cultivated, but neither are they typically free of human intervention. Natural forests host a wider variety of plant life than plantations. The diverse ecologic communities that constitute a natural forest develop and change over time; this is called a **succession**. In a **primary succession**, a plant community takes hold and begins to grow in an area of bare rock or soil, such as a volcanic terrain, a sand dune, or a glacial deposit. If a primary succession is disturbed by a natural event, such as a forest fire or a landslide, or by human intervention, such as logging or clearing for agriculture, the community that subsequently establishes itself and develops on the disturbed site is a **secondary succession**.

In a primary succession, the first plants to successfully colonize bare ground comprise the **pioneer community**. These can be lichens, mosses, grasses, or other hardy plants; commonly, they are plants that propagate via wind-borne seeds. The pioneer community helps create soil where none existed previously. As years go by, the pioneer community is replaced by an increasing diversity of shrubs and trees of varying ages. Each stage in the succession is characterized by a particular plant community. As this community changes over time, the habitat it offers and the animal life it hosts also change.

A primary forest that has endured for hundreds or thousands of years and has never undergone extensive human intervention is called an **old-growth forest**.

Old-growth forests are characterized by very large, old trees, high species and habitat diversity, and a range of vegetation ages. They are irreplaceable ecosystems, but they are also highly valued for their timber, which makes them controversial from an environmental perspective.

Forests are economically important because of the products they provide. Timber and fuelwood are chief among these, but there is also a great variety of nonwood forest products, including latex, nuts, citrus fruits, bananas, oils, coffee, and bush meat. The world's great forests host many species, as yet undiscovered, which may be useful in the development of new pharmaceuticals. Forests also provide recreational opportunities, and they are of great sociocultural significance for aboriginal peoples whose communities and livelihood depend on them.

The environmental services provided by forests are also of great value (Figure 8.1). Trees stabilize the soil by binding it and holding it in place with root systems. They keep the soil replenished with organic material by returning nutrients to the forest floor in the form of dead leaves and branches, or **litter**. Trees play an important role in the hydrologic cycle, drawing water from the soil through their roots and returning it to the atmosphere by the process of transpiration. They intercept water that falls as precipitation, returning much of it (as much as 75 percent in rain forests) to the atmosphere by evaporation and transpiration from leaf surfaces. Trees also hold humidity under their canopies and prevent high winds from reaching the soil surface.

On a global scale, forests harbor extensive reserves of biodiversity. As much as half of all plant and animal species live in tropical rain forests. Forests regulate the global climate by serving as an enormous reservoir for carbon. Approximately 600

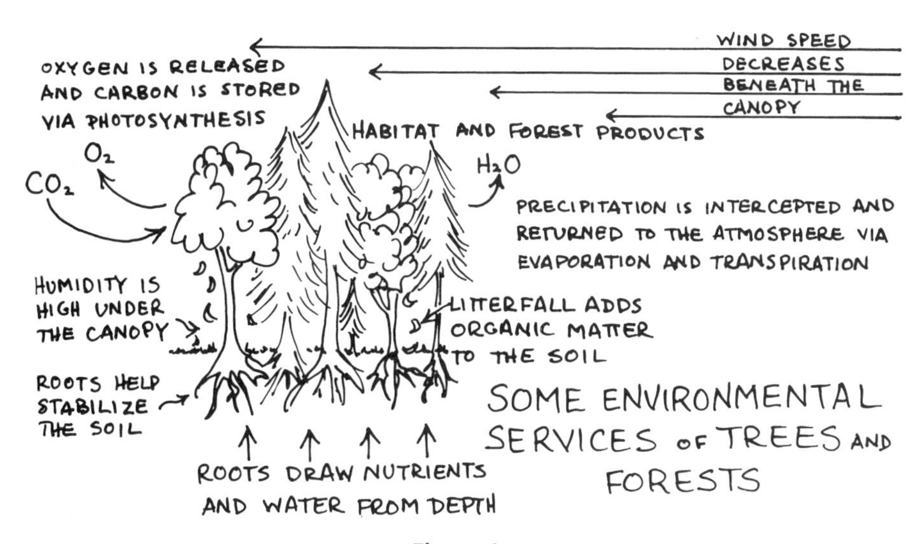

Figure 8.1.

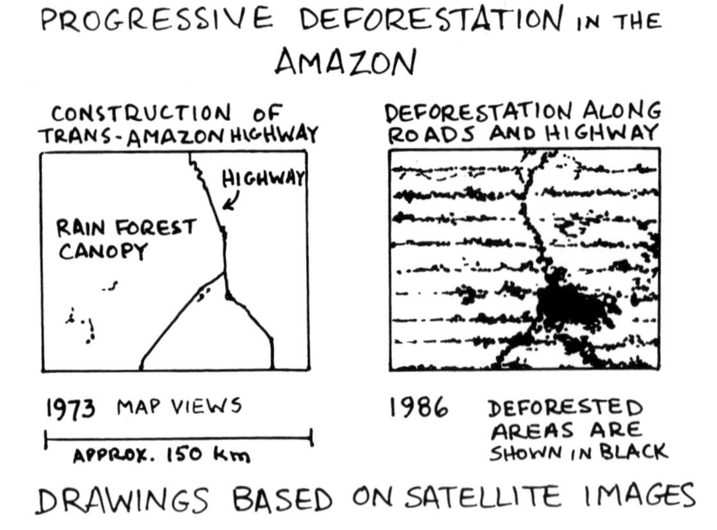

Figure 8.2.

billion metric tons of carbon reside in aboveground biomass in forests, with another 1.5 trillion metric tons as litter and soil organic matter. This is almost three times as much carbon as in the atmospheric reservoir. If this carbon were released, Earth's climate system would undergo an enormous upheaval.

It is challenging to keep track of natural and anthropogenic changes in forest resources. This task is undertaken by the Food and Agricultural Organization (FAO) of the United Nations through its Global Forest Assessment. The assessment process combines a variety of ground-based information sources with data from **remote sensing**—the collection of information from a distance. In forest monitoring, satellite images are analyzed to determine the amount, type, and health of forest cover (Figure 8.2). By comparing images from year to year, scientists can determine the extent of changes in the amount and quality of forested land. The 2001 assessment revealed a global forested area of 3.87 billion ha. Over the past decade, the average amount of forest added each year has been approximately 5.2 million ha, compared to 14.6 million ha of forest area lost each year. Thus, the average net annual change in forested area over the past decade has been a loss of approximately 9.4 million ha.

What is an old-growth forest? ______________

Answer: A primary forest that has endured for many years without extensive human intervention; characterized by large, old trees, high biodiversity, and a range of vegetation ages.

Logging, Forest Management, and Agroforestry

There are a number of ways to harvest trees from a forest. **Clear-cutting** involves the removal of all trees from an area. This is the most economically efficient way to harvest wood, but it is ecologically destructive if it covers an extensive area. Removing trees damages the undergrowth and eliminates habitat, killing or driving away much of the animal life in the area. More subtle but equally harmful are the impacts of clear-cutting on soil and the cycling of water and nutrients. Soils that are exposed to the effects of sun and wind quickly dry out, becoming susceptible to erosion. Experiments carried out by the U.S. Forest Service and others have shown that clear-cutting greatly accelerates the loss of soil nutrients, which are washed away by runoff. Clear-cutting also disrupts the hydrologic cycle. Depending on the specific circumstances, a forest that has been clear-cut may be able to regenerate only after many years, or not at all.

In contrast, forests can recover and in some cases even benefit from natural disturbances such as forest fires and landslides. **Selective cutting**—the removal of certain trees or selected small areas of trees, leaving the rest of the forest intact—can be designed to mimic the characteristics of clearings that result from natural disturbances. Timber harvesting that minimizes harm to the forest ecosystem in this and other ways is called **reduced-impact logging**. Another approach in forest management is the harvesting of timber from plantations rather than from natural forests. The planting and management of forest plantations for the purpose of timber harvesting is called **silviculture**. Modern silviculture focuses on maintaining a **sustainable yield**—taking only as many trees as can be replaced by new, harvestable growth each year. When replacement trees are planted in a cleared area, it is called **reforestation**. When trees are planted in an effort to establish a forest where none has existed in the recent past, it is called **afforestation**.

Review the discussion of resource stocks and flows (chapter 7) and steady state (chapter 2). Can you see how these concepts relate to the idea of sustainable yields in the management of forests and other renewable resources?

Another important development in forest management is the increasing use of **agroforestry**, in which trees and crops are planted together. Agroforestry has many benefits. The trees provide habitat for natural enemies of pests that might otherwise damage the crops. Trees typically have much deeper root systems than agricultural crops; the deep roots stabilize the soil and allow the trees to draw water from deep in the ground. Trees also contribute to the organic content of the soil through litterfall. If the trees grow so tall that they begin to shade the crops, they can be selectively harvested, providing fuelwood, timber, fruits, nuts, leaves, and other clippings to feed domesticated animals.

What is reduced-impact logging? ______________

Answer: Logging that minimizes harm to the forest ecosystem through practices such as selective cutting and mimicking natural disturbances.

The Causes of Deforestation

According to recent assessments by the FAO, forest cover in developed countries is stable or slightly increasing, while **deforestation**—the loss of forest cover—continues at a rapid pace in many of the poorer countries of the developing world. This does *not* mean that logging does not occur in wealthy countries. In fact, extensive logging and clearing of land has occurred historically throughout the developed world, particularly during the latter part of the nineteenth century and the early twentieth century, and it continues in many areas. Only about 5 percent of original forested land remains intact in North America. However, the logging that occurs today in developed countries typically is balanced by the planting of new trees. Reforestation can protect forested area in terms of hectares of coverage, but it does not resolve the environmental controversies surrounding logging in irreplaceable old-growth forests.

Unfortunately, many of the newer approaches to logging and forest management, such as reduced-impact logging, reforestation, and sustainable yield forestry, have not been widely implemented in the developing world, where deforestation is rampant. Of particular concern are tropical rain forests. The three remaining large expanses of tropical rain forest today are in South America (mainly Brazil), Central Africa (mainly Democratic Republic of Congo), and Southeast Asia (mainly Indonesia). In Brazil alone, the loss of forested area averages 5 million ha per year. Estimates suggest that if present rates of deforestation continue, the world's tropical rain forests will be at risk of virtual extinction by the end of this century.

Deforestation is especially problematic in tropical rain forests not only because of the loss of biodiversity and habitat but also because tropical soils are fragile and highly susceptible to drying out, compaction, and the loss of nutrients. Although rain forests are lush and thick with vegetation, most of the nutrients and organic matter reside in the vegetation, not in the soil (Figure 8.3). Deforestation removes

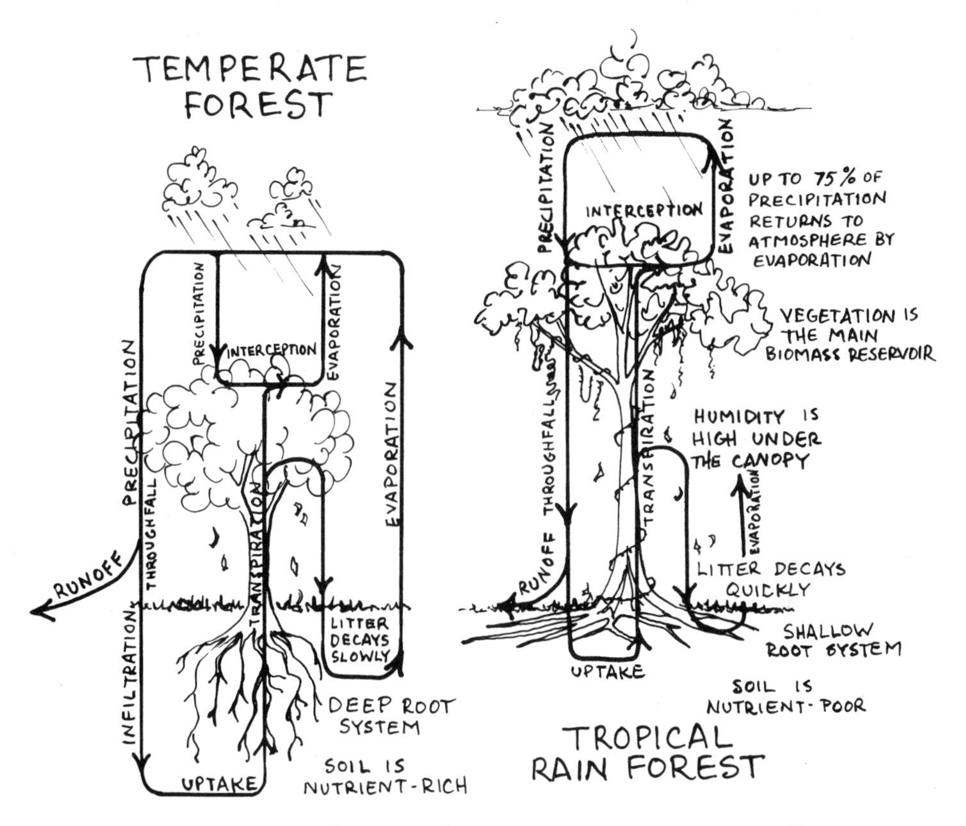

Figure 8.3.

most of the organic matter from the system, leaving the nutrient-poor soils exposed to erosion. In temperate forests, the deeper root systems stabilize the soil. Litter decays more slowly in temperate forests, which means that organic matter has a chance to accumulate at the surface. This provides nutrients and holds moisture in the top layer of the soil.

If you can take a walk in a northern forest, notice that the floor is covered with a thick, springy layer of pine needles and other organic litter. The floor of a tropical rain forest is crowded with *living* plants, but the soil underneath is typically highly weathered and poor in organic matter.

Deforestation in the developing world has many causes. Aside from logging, the most important direct cause of deforestation is land clearing for agriculture and ranching. Traditionally, farmers in many parts of the developing world use a slash-and-burn approach, deliberately setting fires to clear the land. Because tropical soils are nutrient-poor, the cleared areas typically sustain farming for only a year or two before being depleted in nutrients. Farmers must then move on to clear a new plot of land, leaving the original clearing irretrievably damaged. Remotely sensed images taken over a period of years demonstrate the progressive effects of the clearing of forested land for agriculture and ranching. Road building contributes to deforestation by providing access into the forest, whereby other types of development can move in and take hold (see Figure 8.2). Hydroelectric dams are also significant contributors to deforestation because of the extensive flooding behind the dam when a reservoir is created.

Cutting trees for fuelwood is another important cause of deforestation in developing countries, where wood and charcoal still account for a large proportion of household energy use. The use of wood for fuel actually outstrips the use of wood for construction in the developing world. Fuelwood use is especially problematic in areas of open forest and savanna where trees are sparsely distributed, such as the Sahel region of sub-Saharan Africa. Although the population of the area is relatively low, the demand for fuelwood exceeds the carrying capacity of this fragile environment. Charcoal is the fuel of choice in many urban centers in Africa and Asia because it is more convenient, hotter-burning, and less bulky than wood. However, charcoal is particularly damaging because its production is very wood-intensive: it takes more wood to obtain energy from charcoal than to obtain the same amount of energy from fuelwood.

Some causes of deforestation are indirect. For example, forests in northeastern North America, Europe, and Southeast Asia are threatened by air pollution, particularly acid precipitation. Acid precipitation weakens the trees, making them more susceptible to pests and disease. Other types of environmental change, both natural and anthropogenic, such as drought, severe weather conditions, and climatic change, can also weaken trees and allow pest populations to take hold.

It is important to also consider the underlying economic causes of deforestation. Poverty and the debt loads of developing nations—currently more than $2 trillion in total—are leading causes of deforestation. Wood is the fuel of last resort for the poorest people in many parts of the world; a tree can be cut for fuel even if someone has no money with which to purchase kerosene or oil. Brazil, which carries the highest debt in the developing world, is typical of countries that are compelled to sell off rich natural resources to service the interest payments on their national debts. Until the cycle of debt is broken, and until nations are able to recognize and capitalize upon the asset value of their natural resources without depleting and destroying them, deforestation will continue in the developing world.

Many environmental organizations are working locally and internationally to save the world's forests. One of the most famous is Chipko Andolan, literally the "tree-hugging" movement. Chipko was started by a group of women in northern India in the 1970s. The women were dismayed by what they perceived as wanton destruction of the forests by logging, which threatened their livelihood and way of life. They courageously intervened by placing their bodies in the way of the logging equipment. Their efforts came to the attention of Indira Gandhi, then the prime minister of India, who eventually paved the way for the protection of 12,000 km^2 of forested land in northern India. Chipko has since grown into an environmental movement that has had a significant impact on forest conservation in India. It has also come to symbolize the importance of nonviolent intervention and grassroots community-level environmental action.

Another well-known grassroots environmental organization is the Green Belt Movement, based in Kenya. Its founder, Wangari Maathai, was awarded the Nobel Prize for Peace in 2004—the first time that the Nobel Committee has included environment as a consideration and the first time that the Peace Prize has gone to a black woman from a developing country.

What are the main causes of deforestation today? ______________

Answer: Logging; conversion of land for agriculture and ranching; pollution, especially acid precipitation; severe weather, drought, and other types of environmental change; fuelwood use; road-building, hydroelectric dams, and other development; poverty; and national debt.

Wildlife and Wilderness

Wilderness is an area where natural forces are more significant than human intervention and where people do not live on a permanent basis. The protection of wilderness areas began with the national parks movement in the United States and the creation of Yosemite Park by law in 1864 and Yellowstone Park in 1872. Since then, virtually every country of the world has set aside some land (and water) areas to be protected in one way or another, either as state, provincial, or national parks and forests, international preserves, or **wildlife refuges**—sanctuaries for the animals that inhabit wilderness areas. Still, there are many questions and controversies about wilderness preservation. Why should we protect wilderness and wildlife? Is it possible to protect them but still allow other activities to occur in protected areas? Is there a necessary or optimal size and shape for a protected area? Are parks and refuges the only way to preserve wildlife? Finally, in a world with many economic and political pressures, how can we ensure that wilderness and wildlife will continue to be protected? Let's look more closely at these questions.

The question of why wilderness and wildlife should be protected speaks to the value of ecosystems and biodiversity (chapter 6). There are many anthropocentric and utilitarian reasons to protect wilderness and wildlife. In parks, forests, and wildlife refuges, however, there are conflicts between preservation and human use. Do we set aside wilderness areas for protection because of the inherent value of wilderness and wildlife? Or do we protect wilderness so that people can visit these special areas to be rejuvenated physically and mentally, to experience the peacefulness of nature, and to learn about wildlife and the world around us? Are preservation and human use mutually exclusive? Some parks are so heavily visited that it can be challenging to find a quiet place in which to experience the wilderness. Almost by definition, wilderness should be remote and difficult to access. But does wilderness that is rarely visited justify its own existence? Perhaps the value of wilderness parks, aside from their broader ecosystem value, is that they are places that have not been tamed. They are great laboratories of successful natural communities, and they provide a lesson for a world of limited resources, promoting intensive experience rather than intensive use. (The discussion in this paragraph is based on *Mountains Without Handrails: Reflections on the National Parks*, UMP, 1980, by Joseph Sax—an interesting book that explores the roots of this controversy.)

Most parks and forests are designed to fulfill more than one purpose; this is a **multiple-use** approach. Recreational use, logging, mineral exploration, oil drilling, hunting, and even agriculture and ranching are permitted in many protected areas. Some of these uses are incompatible. For example, logging does not make sense in an area where the preservation of natural forest is of prime concern. Tourism and ecotourism are economic uses of land set aside for the purpose of protection and conservation. However, certain types of tourist use, such as off-road vehicles, dune buggies, mountain bikes, and even heavy trail use, can be damaging to the environment. Hunting and ranching may be inconsistent with wildlife and habitat preservation. Even outside of the boundaries of a park such as Yellowstone, controversies rage about the preservation of wolves, as they defy park boundaries and prey on the sheep of nearby ranchers. Enormous controversy also surrounds the use of wilderness areas for mineral and oil exploitation because of the potential for devastating environmental impacts. In considering multiple uses, it is important to distinguish between **conservation**, the management and protection of natural heritage and resources, and **preservation**, the maintenance of natural wilderness areas in near-pristine condition.

Habitat fragmentation occurs when the size of an area of natural habitat is reduced, or when a natural area is cut off and isolated by a highway or other developments. Fragmentation can irretrievably damage the quality of habitat even if it does not dramatically reduce the spatial extent of the habitat (Figure 8.4). For example, some species require habitat in the interior of a forested area. Near the edge of a forest, the characteristics of the ecosystem grade into those of adjacent areas; this is called the **edge effect**. If a forested area is cut or otherwise disturbed, new edge is created, which damages interior habitat. **Migratory species** require habitat that allows them to move freely from one natural area to another. Some migratory species, such as elephants, travel very long distances. This often puts elephants into direct conflict with human land development needs. Migratory wildlife can coexist with development if natural **corridors** are preserved that connect habitat areas together, allowing wildlife to migrate freely. The classic work on this subject is by Robert MacArthur and Edward Wilson, *The Theory of Island Biogeography* (Princeton University Press, 1967).

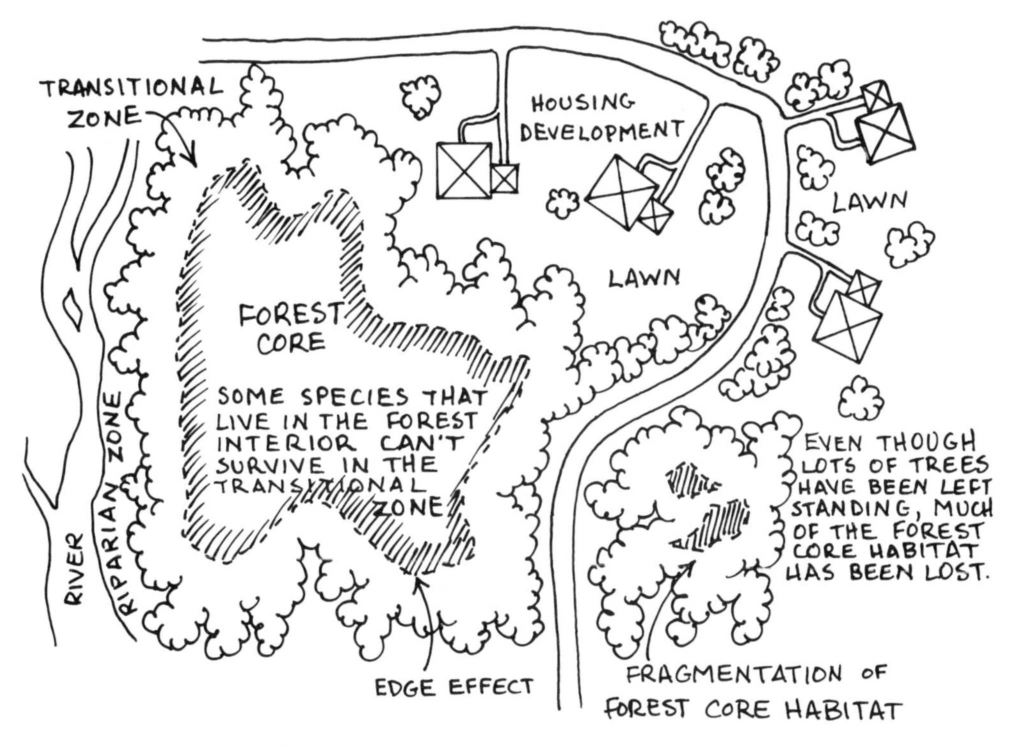

Figure 8.4.

Much of what is known about habitat fragmentation comes from **island biogeography**, the study of biologic diversity on islands. The basic premise of island biogeography is that species diversity in an isolated area is a balance between the loss of species through extinction and the arrival of new species through colonization (immigration). An island that is small or far from land has less chance of being colonized by new species. A large island has greater habitat diversity and can support more species; one that is close to land has a better chance of being colonized. Thus, a large island that is close to land should have greater species diversity than a small island far from land. This has been supported by many observations of species diversity on actual islands. An extension of this concept considers small fragments of land-based habitat to be like islands, in the sense that they are limited in size and isolated from similar habitat areas (Figure 8.5). As with islands, the size and connectedness of terrestrial habitat fragments influences their ability to support species and habitat diversity.

The preservation of biodiversity in forests, parks, and wildlife refuges is **in situ conservation** (*in situ*, meaning "in place"). When the natural population of an endangered species drops too low or its habitat has become too severely damaged or

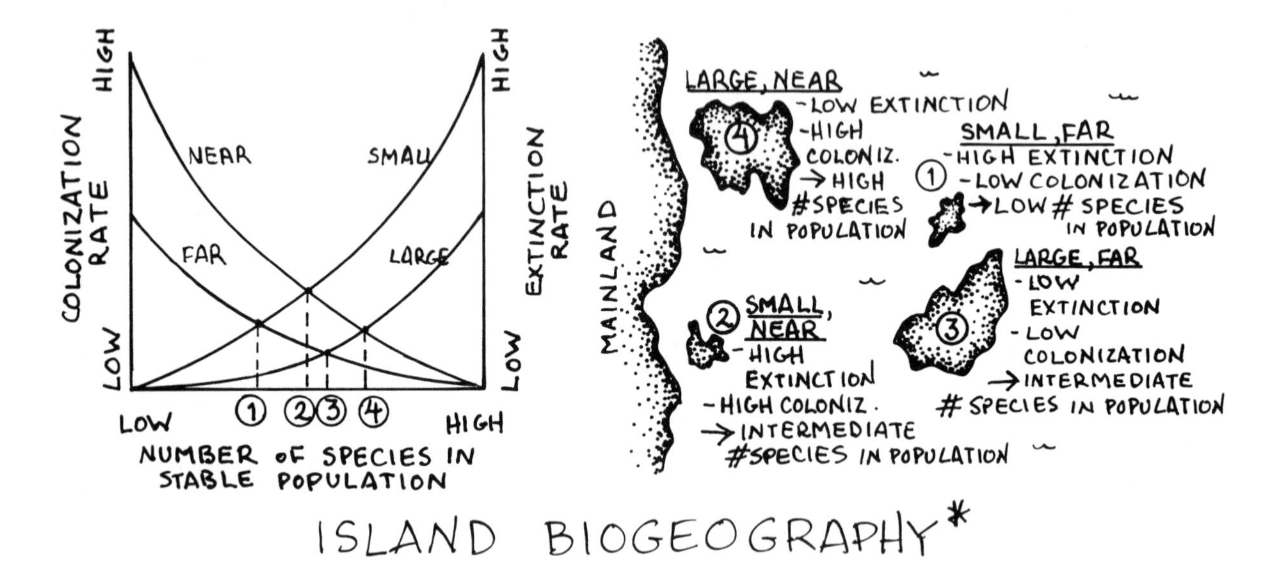

Figure 8.5.

fragmented, it may be impossible for the species to successfully endure on its own. In such cases, the only way to preserve the species may be through **ex situ conservation** (*ex situ*, meaning "out of place"), such as a captive breeding program at a zoo or an aquarium. Breeding in captivity is obviously not as desirable as helping a species to survive in the wild. Some species are difficult to breed in captivity. Others bred this way can never be released back into the wild because they lack crucial survival skills or because their natural habitat has been irretrievably damaged. If the existing population of a species is very small, its genetic diversity may be so low that the species will never be self-sustaining outside of the zoo or aquarium.

Botanic gardens and seed collections play a crucial role in the ex situ conservation of plant biodiversity. For many years, farmers and botanists interested in preserving genetic diversity in plants have maintained living and preserved sample collections. There are currently more than a hundred **seed banks** in the world, where more than 3 million seed samples have been dried and preserved. Many of these are wild varieties of plants that were long ago domesticated as crops and have subsequently lost some of their genetic diversity.

Natural habitat and wilderness are under increasing pressure from economic and political forces that promote industrial development, urbanization, and the exploitation of natural lands and resources. Today only about 6.4 percent of the world's land area outside of Antarctica is protected (7.2 percent in developed countries, 5.8 percent in developing countries). Protection of natural lands can occur through international agreements to establish biosphere reserves and natural heritage sites, or under national jurisdiction in the form of nature reserves, wilderness areas, national parks, natural monuments, habitat or species management areas, and protected landscapes. Protected seascapes and marine ecosystems are less well documented than terrestrial areas, and there are fewer of them. Protective legislation for natural areas and species and the extent of enforcement vary widely from country to country.

One international mechanism through which natural ecosystems can be protected is UNESCO's Man and Biosphere program, which allows for the identification and legal protection of biosphere reserves. Another is the Convention on Wetlands of International Importance (1971), in which each country that becomes a party to the convention agrees to protect at least one significant wetland. The World Heritage Convention (1972) is an international agreement that identifies sites of universal value because of their outstanding natural or cultural features. A number of international agreements are designed to protect plants and animals. Among these is the Convention on International Trade in Endangered Species of Flora and Fauna (CITES, 1975), which regulates the hunting and trade of endangered and threatened species. Other international agreements are aimed at the protection of specific species such as whales. Most countries have built some form of protection of natural lands and species into their national legislation. Examples in the United States include the Wilderness Act (1964), Endangered Species Act (1973), and Wild and Scenic Rivers Act (1968).

Economic incentives and planning tools can also contribute to the preservation of species and habitat. For example, the World Conservation Strategy (1980, modified in 1991) serves as a template for national strategic plans for the conservation of biodiversity. Some countries have successfully translated the preservation of wilderness and wildlife into economic benefits. For example, government programs encourage farmers in Zimbabwe to switch from cattle ranching to **game farming**—that is, ranching of wildlife such as elephants, lions, and giraffes. The financial return from ecotourism and big game hunting is greater than the farmers would normally realize through cattle ranching, and game farming is less harmful to the environment. Another important economic tool is the **debt-for-nature swap**, in which an organization, usually an environmental group, agrees to take on a portion of a developing country's debt in exchange for some kind of protection for the natural environment, usually through the establishment of a park or reserve. Although there are problems with enforcement, debt-for-nature swaps represent an innovative approach to the preservation of habitat and biodiversity in developing countries.

How much of the world's land area is currently protected under national or international laws?_______________

Answer: About 6.4 percent.

Fisheries and Aquaculture

Living resources in aquatic environments are under increasing pressure. Preserving species and habitat diversity and avoiding overexploitation of these resources present some unique challenges. Aquatic environments, particularly marine environments, are less well known and less well protected by law than terrestrial environments even though some marine environments, notably coral reefs, may be as biologically rich as the most diverse terrestrial ecosystems. National jurisdiction over marine fisheries—perhaps the best means of controlling and monitoring their use—is subject to constant challenges and potential misuse by competing nations. The open ocean beyond national territorial waters is much more difficult to regulate.

The FAO, which monitors the state of forest resources, also undertakes a periodic review of the world's fisheries. This is even more challenging than forest assessment because it relies on individual fishing vessels and countries to report their catch in a truthful, accurate, and timely manner. It is difficult to monitor and assess the status of a resource that is so large, so widely dispersed, and hidden from direct view. Therefore, uncertainty and inadequacy of information are of major concern in assessing the state of the world's fisheries.

Total world fish production is currently about 125 million tonnes per year. About 75 percent goes to human consumption; most of the rest is made into oils or fishmeal, which is often for animal consumption. Marine **capture fisheries**, where fish are caught in the wild rather than raised in captivity, account for the majority of world production (Figure 8.6). Capture fishery production increased dramatically from the 1950s to the 1970s mainly as a result of greatly improved fishing gear, expanded fleets, and newly exploited fisheries. The rate of increase slowed in the 1970s and 1980s and leveled off altogether in the 1990s, probably because many fish stocks are nearing their productive limits. There was a significant decrease in capture fishery production in 1998, which may be partly attributable to the effects of the severe El Niño event of 1997–1998. Production recovered somewhat in 1999, reaching 92 million tonnes. Inland (freshwater) capture fisheries, for which there is much less information, account for about 8 million tonnes, or roughly 9 percent of the total capture fishery

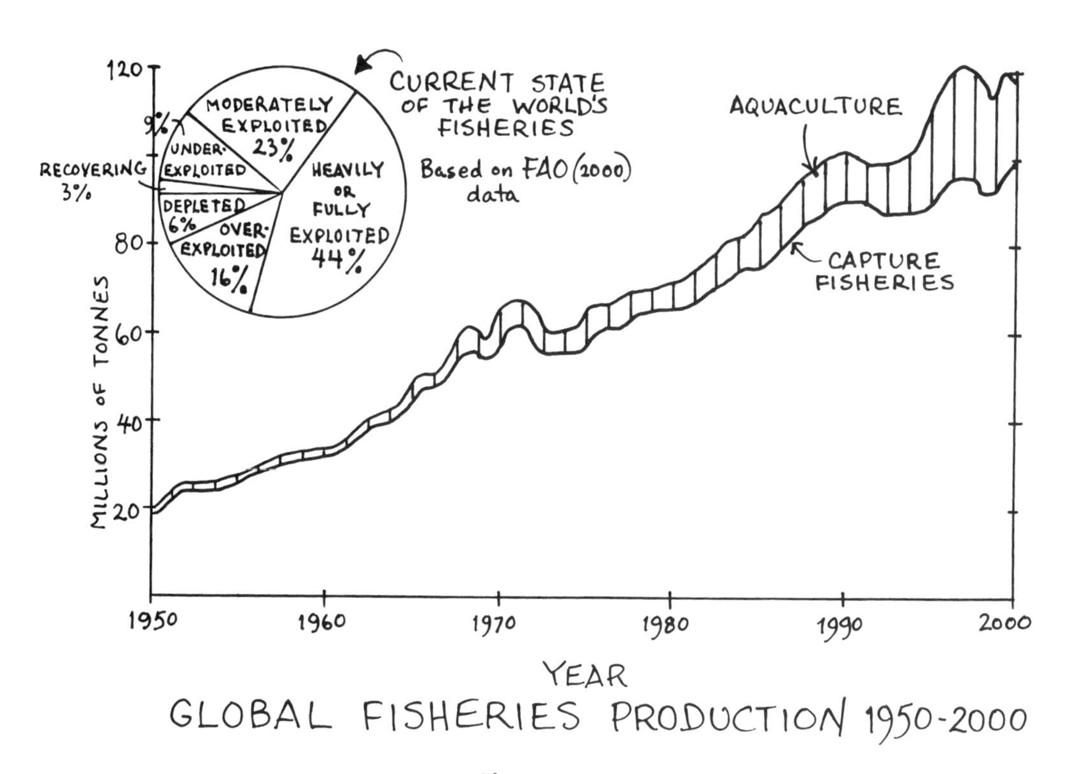

Figure 8.6.

production. More than 90 percent of inland capture fishery production occurs in developing countries.

Production from marine and inland **aquaculture**, the raising of fish, shellfish, crustaceans, and aquatic plants in captivity, has increased dramatically over the past twenty years. Current production is more than 32 million tonnes, up from less than 5 million tonnes in the 1980s. Aquaculture is seen as offering the potential to enhance incomes in many poor and rural areas in developing countries. Most of the growth has occurred in Asia, and China currently accounts for about 70 percent of the world's aquaculture production.

The current status of fish stocks and the potential for overexploitation are of great concern. As with all renewable resources, the balance of stocks and flows and maintenance of sustainable yields is of primary importance in fisheries management. However, the FAO reported in 2000 that 47 to 50 percent of the world's major fish stocks are fully exploited—that is, producing catches that are close to their maximum sustainable yield. Another 15 to 18 percent of stocks are overexploited and at risk of permanent damage if steps are not taken to limit overfishing; and 9 to 10 percent of stocks are depleted or recovering from depletion. Thus, more than 70 percent of the world's fish stocks are fully exploited, overexploited, or depleted. This is mainly due to overfishing, although environmental change can also have a significant impact on fish stocks.

An example of a depleted stock is the Atlantic cod off the eastern coast of Canada, which were fished to **commercial extinction**, meaning that so few cod remained that it was no longer economically viable to fish. A moratorium was placed on Atlantic cod fishing in Canada in 1992; the United States followed in 1994 by closing the Georges Bank cod fishery off the New England coast. The total spawning biomass of cod at the time of the moratoria was depleted by 75 to 98 percent, and some stocks have continued to decline since then. It is not known when—if ever—the fishery will recover.

Aside from the depletion of stocks, the environmental impacts of fishing are of significant concern. These concerns relate mostly to the use of environmentally damaging gear. For example, *drift nets*, which are illegal, are enormous nets that are dragged behind a boat, entangling fish and other marine organisms such as dolphins. *Purse seines* are floating nets that sometimes ensnare seabirds. *Trawls* are huge baglike nets that are dragged along the ocean bottom; they can cause considerable damage to sensitive seafloor habitats such as coral reefs. A *longline* is a fishing line more than 100 km in length with thousands of hooks. This type of gear is responsible for huge volumes of **by-catch**, which are nontarget fish and other marine organisms that are caught and killed by mistake. Up to 25 percent of marine organisms that are inadvertently caught by fishers are discarded as by-catch.

To minimize environmental impacts, fishing gear and practices are more closely monitored and regulated than ever before, as are the size and capacity of fleets. Other advances, such as the use of sonar and computerized satellite tracking systems, have made it easier to find and capture fish, contributing to concerns about the depletion of stocks. Even some small-scale or **artisanal fisheries** can negatively affect the environment through damaging practices such as the use of explosives to kill fish. There are also environmental problems with inland, freshwater fisheries. The FAO reports that as many as 20 percent of freshwater fish species may be extinct, endangered, or threatened mainly as a result of overfishing and habitat destruction.

Review the discussion of the tragedy of the commons in chapter 7. In what way is the overexploitation of fish stocks an example of this phenomenon?

Aquaculture, or fish farming, is increasing in importance as the world's natural fisheries come under greater stress. Much aquaculture production focuses on products that are of high economic value, notably shrimp and salmon. The specific approach depends on the type of organism being farmed. For example, salmon are started (spawned) in captivity, then released to the wild, where they mature. Because of their natural breeding habits, the salmon eventually return to their original spawning ground to breed; this predictability of behavior allows them to be harvested. Other aquatic organisms, such as shrimp, are raised in enclosures in brackish-water coastal areas. Saltwater aquaculture is used mainly for aquatic plants, such as kelp, and mollusks, such as oysters. Inland or freshwater aquaculture is used mainly for finned fish, such as carp, and for recreational fishing purposes. Aquaculture has a relatively high environmental cost. For example, if cultured organisms escape, they can contaminate the natural gene pool or spread diseases into wild populations. In Southeast Asia, coastal aquaculture farms have contributed significantly to the loss of coastal wetlands and mangrove forests.

Fish is an important source of animal protein in low-income, food-deficient countries. About 1 billion people in the world depend on fish as their main source of animal protein. In developing countries, fish accounts for almost 20 percent of animal protein consumption compared to less than 6 percent in wealthier countries, although the amount of fish consumed per capita in wealthy nations is actually higher than in developing countries. As stocks become depleted, higher prices make it more difficult for people in poorer nations to have access to this very important food source. Fishing is also of great socioeconomic importance; about 36 million people depend on fishing for their livelihood. The majority are employed in small-scale, artisanal fisheries even though the largest proportion of the economic return from fishing comes from the large, long-distance fleets maintained by countries like the United States, Russia, Japan, and China.

What is the current status of the world's major fish stocks? ______________

Answer: Fully exploited: 47 to 50 percent; overexploited: 15 to 18 percent; depleted or recovering from depletion: 9 to 10 percent; the remaining 25 to 30 percent are underexploited.

The basic concepts involved in the management of renewable resources seem simple: balance stocks and flows, monitor the state of the ecosystem, and maintain a sustainable yield. In practice, however, these are difficult concepts to implement, and renewable resources are subject to many types of uncertainties and pressures. As a global community, our record in the management of renewable resources has been less than commendable. We must improve if we hope to protect these precious living resources.

SELF-TEST

These questions are designed to help you assess how well you have learned the concepts presented in chapter 8. The answers are given at the end. If you get any of the questions wrong, be sure to troubleshoot by going back into that part of the chapter to find the correct answer.

1. A primary forest that has endured for a long time without significant human intervention is called a(n) ______________.
   a. ecosystem
   b. pioneer community
   c. old-growth forest
   d. wildlife refuge
2. In the management of renewable resources like fisheries and forests, it is important to harvest no more than can be replaced by harvestable growth each year. This is called ______________.
   a. sustainable yield
   b. multiple-use approach
   c. cultivation
   d. preservation
3. The majority of world fish production comes from ______________.
   a. inland (freshwater) capture fisheries
   b. marine capture fisheries
   c. freshwater aquaculture
   d. saltwater and brackish-water aquaculture

4. Leaves, branches, and other organic material that falls to the ground from trees is called ________________.

5. The maintenance of ________________ can help migratory species deal with habitat fragmentation and coexist with human development.

6. Agroforestry is the planting of trees for the purpose of harvesting them as crops. (True or False)

7. Parks, forests, and wildlife reserves are places where in situ conservation occurs, whereas zoos, botanic gardens, and aquaria are places where ex situ conservation occurs. (True or False)

8. Since a moratorium was placed on the Atlantic cod fishery off the east coast of Canada in 1992, the fish population has reestablished itself, and a recovery is anticipated in the near future. (True or False)

9. What is the difference between a primary succession and a secondary succession?

10. How does island biogeography contribute to our understanding of land-based habitat fragmentation?

11. How does clear-cutting affect soil?

12. What are the main environmental impacts of fishing?

13. What is a debt-for-nature swap?

ANSWERS

1. c   2. a   3. b   4. litter   5. natural corridors

6. False (This can be partially true, but there is much more involved in agroforestry than the harvesting of trees.)

7. True   8. False

9. Primary succession: a succession of ecologic communities—that is, vegetation and habitat that changes over time after having established itself in an area of bare ground where no soil existed. Secondary succession: a succession that establishes itself in an area that had previously been vegetated but was disturbed.

10. Island biogeography tells us that species diversity on an island is a balance between the loss of species through extinction and the arrival of new species through colonization, which in turn suggests that a large island close to land should have greater species

diversity than a small island far from land. Similarly, small fragments of land-based habitat are islands, in the sense that they are limited in size and isolated from similar habitat areas. Thus, the size and connectedness of land-based habitat fragments influences their ability to support species and habitat diversity.

11. Clear-cutting exposes the soil, leaving it open to drying and compaction. It also facilitates the stripping of organic content from the soil by runoff. These impacts, in turn, make the soil more susceptible to erosion.

12. Overfishing and depletion of stocks; damage to bottom ecosystems by trawling; by-catch of nontarget fish, marine mammal, and seabird species by drift nets, purse seines, longlines, and trawls.

13. An organization, usually an environmental group, agrees to take over a portion of the debt of a developing country in exchange for some form of protection of the natural environment, usually the establishment of a park or wildlife refuge.

KEY WORDS

afforestation
agroforestry
aquaculture
artisanal fishery
by-catch
capture fishery
clear-cutting
commercial extinction
conservation
corridor
debt-for-nature swap
deforestation
edge effect
ex situ conservation
game farming
habitat fragmentation
in situ conservation
island biogeography
litter
migratory species
multiple use
old-growth forest
pioneer community
plantation
preservation
primary succession
reduced-impact logging
reforestation
remote sensing
secondary succession
seed bank
selective cutting
silviculture
succession
sustainable yield
wilderness
wildlife refuge
woodland

9 Living Resources II: Soils and Agriculture

What, O Earth, I dig out of thee, quickly shall that grow again.

—Prithvi Sukta, Atharva Veda

Objectives

In this chapter you will learn about:

- the formation and characteristics of soils and the process of erosion;
- traditional and modern agricultural techniques and their environmental impacts;
- the Green Revolution and genetically modified foods; and
- sustainable agriculture and global food security.

Soils and Soil Nutrients

Soil is weathered rock material that has been altered by the presence of living and decaying organisms such that it can support rooted plant life. With its partly mineral, partly organic composition, soil forms a link between the lithosphere and the biosphere. Like liquid water, soil is one of the features that make Earth a unique planet. The availability of **arable** soils—soils that are suited for growing agricultural crops—is crucial for the maintenance of our global food supply. Therefore, we begin this chapter with a look at the science of soils.

If you examine soil with a microscope or a magnifying glass, you will see that it contains a variety of materials, including tiny fragments of rocks and minerals. The smallest are **colloids**, which are extremely fine particles (less than 0.0002 mm) that are easily suspended in water. Next are **clay** particles, which are mineral fragments less than 0.002 mm in diameter. **Silt** particles are slightly larger (0.002 to 0.005 mm), and **sand** particles are the largest and most easily visible (0.05 to 2 mm). Larger rock fragments may also be present. Soil also contains **humus**, which is partially decayed organic matter, as well as a variety of small living organisms such as worms, spiders, and mites, and microorganisms such as bacteria and fungi.

The relative proportions of these constituents determine the characteristics of the soil. For example, loam is an organic-rich soil that is particularly well suited for growing plants. In addition to humus, loam consists of about 40 percent sand, 40 percent silt, and 20 percent clay particles. These constituents give the soil just the right texture to be able to hold moisture and supply **nutrients**—the chemicals required for plant growth—to root systems. A soil that is richer in one or another of these constituents will have different physical, chemical, and biologic properties.

It can take many thousands of years for soil to form from bare rock. The formation rate depends on a variety of factors. Most important is the rate of weathering, which in turn is influenced by the composition of the rock. Some minerals weather readily, whereas others such as quartz are highly resistant. High temperatures and abundant rainfall promote rapid **chemical weathering** by enhancing the rate of chemical breakdown of the minerals (Figure 9.1). Weathering is more intense and

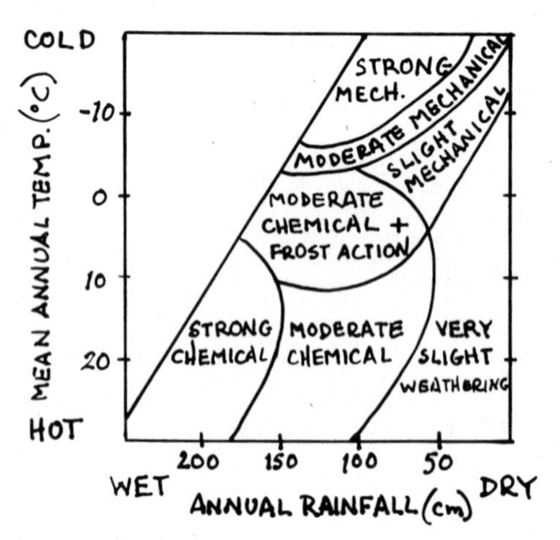

Figure 9.1.

extends to greater depths in warm, wet, tropical climates than in cold, dry, arctic and alpine climates. In cold, dry regions and on high mountain peaks, weathering is shallow and proceeds slowly by **mechanical weathering** processes such as cracking and breaking. Rocks on steep slopes also weather more rapidly than on gentle slopes (Table 9.1).

As the bedrock weathers, soil gradually develops downward from the surface. A fully developed soil consists of a succession of roughly horizontal layers, or **soil horizons**. Each horizon has distinct characteristics of color, texture, chemistry, organic content, and water content. The entire sequence of horizons from the surface to the underlying bedrock constitutes the **soil profile**, which varies considerably from one location to another. The uppermost horizon in many soil profiles is an accumulation of decaying organic matter commonly underlain by a dark, humus-rich horizon; these horizons together constitute **topsoil**. Any of a variety of mineral-rich horizons may underlie the topsoil. This is determined by the chemistry of the soil, the climate, and other physical, chemical, and biologic conditions in which the soil formed. The deepest horizon typically contains large fragments of the original bedrock from which the soil formed (Figure 9.2).

Because soil formation is influenced by many variables, it is not surprising that the classifications used by soil scientists are very complicated. The classification scheme used in the United States and many other countries is a hierarchic one headed by eleven soil orders (appendix 3). Each order is distinguished by specific characteristics such as the presence or absence of well-developed horizons, accumulations of certain minerals, distinctive color, and acidity. The orders are divided into suborders and then in increasing detail into groups, subgroups, families, series, and types. At least 17,000 types of soils are defined, and each of them has a name!

Soil fertility is the ability of soil to provide nutrients, such as phosphorus, nitrogen, and potassium, to growing plants. Humus is a crucial component in soil fertility because it retains some of the chemical nutrients released by decaying organisms

Table 9.1 Factors That Influence Weathering

	Rate of Weathering	
	Slow	**Fast**
Mineral resistance	High (e.g., quartz sandstone)	Low (e.g., limestone)
Fractures	Few	Many
Steepness of slope	Gentle	Steep
Vegetation	Dense	Sparse
Temperature	Cold	Hot
Rainfall	Low (<40 cm/yr)	High (>130 cm/yr)
Burrowing animals	Rare	Abundant

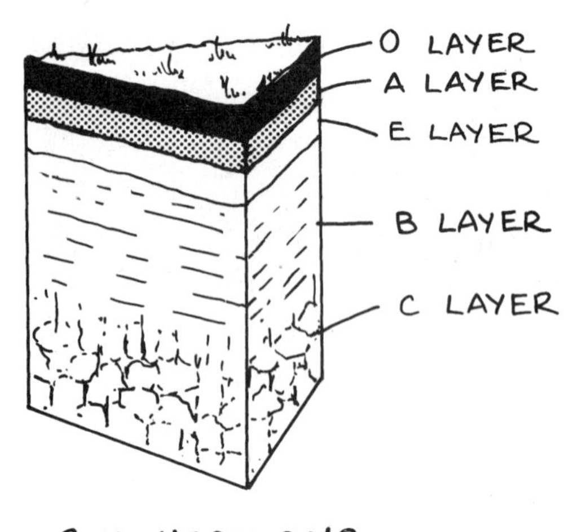

Figure 9.2.

and the weathering of minerals. Water is essential, too, because plants can only utilize nutrients that are in aqueous, or dissolved, form. **Soil water** (or **soil moisture**) fills or partially fills the **pore spaces** between the mineral grains in the soil (Figure 9.3). A soil (or other material) that has a high proportion of pore space relative to solid material is said to have high **porosity.**

Soil water occurs as a thin film that adheres to the surfaces of mineral particles. Most of the components dissolved in the water are cations, with a positive electric charge.* Some soil water components that are important for plant growth are K^+ (potassium), Ca^{2+} (calcium), Na^+ (sodium), NH_4^+ (ammonia), Mg^{2+} (magnesium), and H^+ (hydrogen). In contrast, clay particles in the soil carry a negative electric charge; they are anions. The negatively charged clay particles attract the positively charged cations, holding them on their surfaces until the nutrients are needed by the plant. Many contaminants in groundwater and soils are also positively charged, including heavy metals such as mercury, cadmium, and lead. Such contaminants may accumulate in the soil by adhering to the surfaces of clay particles; this will be discussed further in chapter 13.

---

*An atom that has an excess positive or negative electrical charge caused by the loss, addition, or sharing of an electron is called an **ion**. When the charge is positive, the atom has given up electrons, and we call it a **cation**. When the charge is negative, the atom has gained electrons, and we call it an **anion**.

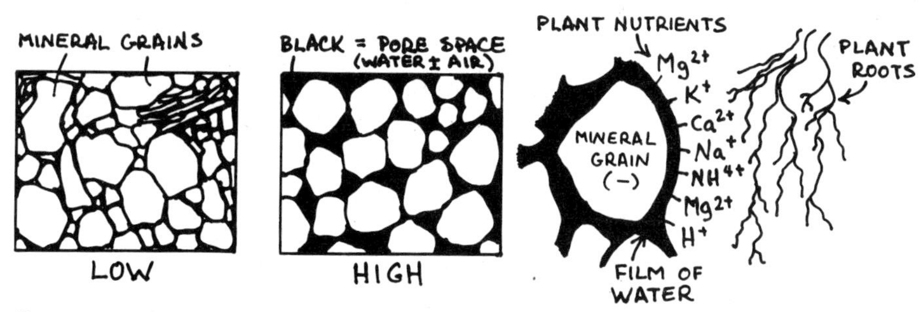

Figure 9.3.

Hydrogen (H^+) ions are particularly important because their concentration determines the **pH**—that is, the **acidity** or **alkalinity**—of the soil. The pH scale runs from 0, very strongly acidic, to 14, very strongly alkaline; pure water has a pH of 7, neutral. The pH of most soils falls in the range of 4 to 8. Soil pH is important because it affects the **solubility** of mineral nutrients—that is, the ease with which they dissolve in water. For example, aluminum, which is potentially toxic to both plants and animals, becomes more water soluble at a low pH. Other mineral nutrients that are essential for plant growth become less soluble and therefore less available at a higher pH.

For optimal growth, plants require nutrients in specific proportions. If one type of plant is grown in the same soil over a long period, the soil will become depleted of the specific nutrients required by that plant. If most nutrients are available in abundance but one essential nutrient is not, the lack of this one nutrient becomes a **limiting factor**—it limits the growth potential of the plant. (Review the concept of limiting factors in chapter 7.) For these reasons, farmers supplement the nutrient contents of agricultural soils with plant **fertilizers**—mixtures of mineral nutrients (mainly nitrates, phosphates, and potassium in various proportions) designed to supplement soil nutrients that may be limited in availability.

What is topsoil? ______________

Answer: The organic and humus-rich soil horizons.

Traditional and Modern Agriculture

Thousands of years ago, humans were nomadic **hunter-gatherers**. They traveled from place to place in small groups, extracting the foods they required from the environment around them. The availability of food resources varied with the season and location. About 12,000 years ago, for reasons that are not entirely understood, people began to settle into a sedentary lifestyle that involved production rather than just extraction of food resources. As people selectively chose to cultivate and breed certain types of plants and animals, **domestication** occurred, and agriculture was born.

The hunter-gatherer lifestyle still practiced by some nomadic groups today requires a large land area to support a small human population. Agriculture entails a much more intensive use of environmental resources. Therefore, the development of agriculture effectively increased the carrying capacity of the land, allowing human populations to increase accordingly. In the Industrial Revolution, which began in England in the eighteenth century, the invention of the steam engine and resulting development of mechanized farming tools again spurred a huge increase in agricultural production. This revolution was the beginning of modern agriculture.

Farmers in many small villages in the developing world still practice traditional **subsistence agriculture**, growing only enough food to feed their families. In **swidden** or **shifting agriculture**, the farmer moves from one plot of land to the next every few years in search of fertile soils (Figure 9.4). The original field is left **fallow** (unplanted) in the meantime, allowing the soil to replenish its nutrients naturally. Some traditional farmers employ a **slash-and-burn** approach, using fires to clear new agricultural fields. In tropical areas where soils are fragile and nutrient-poor, particularly if fallow periods are short, slash-and-burn agriculture can be devastating to both soils and forests.

Traditional agriculture typically involves the planting of diverse crop types in one field. In contrast, modern agriculture relies on the planting of large tracts of a single

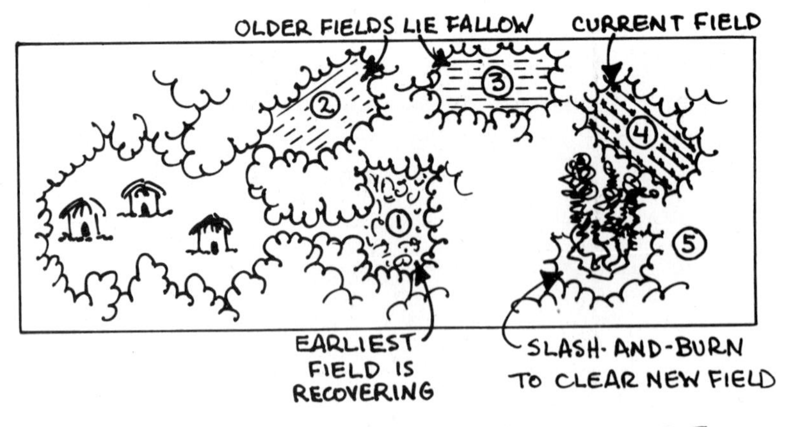

Figure 9.4.

Women in Cameroon carry the cotton harvest in traditional baskets.

crop, or **monoculture**. Overall, only about 5,000 plant species are cultivated, and 95 percent of our nutrition comes from just 30 of these. This has led to *genetic uniformity*—a decrease in the genetic diversity of crop species. Genetic uniformity is problematic because it renders a species less able to respond to pests, diseases, and changes in the physical environment. Seed banks (chapter 8) are an effort to protect the vanishing genetic diversity of crops. Modern agriculture also tends toward shorter fallow periods, intensive irrigation, and the use of heavy farm machinery, all of which can be damaging to soil quality. Industrial agriculture is extremely energy intensive as well. The energy inputs come from **agrochemicals**, such as fertilizers and pesticides, and from the fossil fuels required to run large pieces of farm equipment. Traditional agriculture generally relies on human or animal energy rather than chemical energy.

What is swidden agriculture, and who practices this type of farming?

Answer: Swidden is shifting agriculture, in which farmers move from one field to the next every few years in search of more fertile soil. It is practiced by traditional farmers in many developing countries.

The Green Revolution and the Environmental Impacts of Modern Agriculture

After the Industrial Revolution, the large-scale, mechanized techniques of modern farming began to take hold, crowding out smaller landowners in most parts of the world. Over the past forty or fifty years, in particular, modern farming methods have led to an enormous increase in agricultural productivity. For example, wheat and corn production almost tripled in the decades following the 1950s. Rice and other grains have more than doubled, and similar increases have occurred for fish and livestock. This tremendous increase in productivity is referred to as the **Green Revolution**. The Green Revolution was brought about by greatly increased use of agrochemicals, greatly expanded irrigation, and the development of high-yield, disease-resistant seed types (Figure 9.5), in addition to enormous increases in the amount of land under cultivation.

There may be limits to the tremendous increases in productivity realized as a result of the Green Revolution. In the mid-1980s, productivity in some agricultural sectors began to level off or increase at a slower rate. When combined with global population growth, this meant that productivity per capita had begun to stagnate or even decline. It was also observed that crop yields per hectare had begun to decline. Why?

The practices that led to the Green Revolution have had some negative impacts on environmental quality, which in turn have affected agricultural productivity. For example, the increased use of chemical pesticides has led to the emergence of pesticide-resistant insect species. This suggests that the effectiveness of some

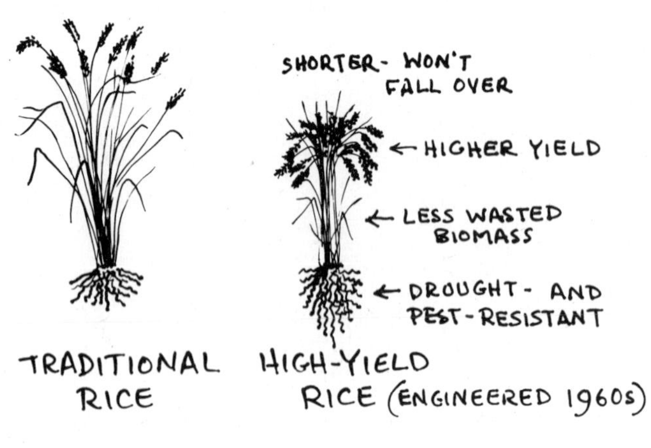

Figure 9.5.

pesticides has been diminished, leading farmers to use more and more chemicals with less satisfactory results. The increased use of fertilizers that has been required by the short fallow times and the monoculture typical of modern agriculture have led to surface and groundwater pollution from nitrate- and phosphate-bearing runoff.

Agrochemicals are also used to increase productivity in domesticated farm animals (livestock). For example, cattle are routinely administered hormones to enhance their rate of growth. This has led to concerns about the levels of growth hormones in beef. Other livestock, including pigs, chickens, and cows, are given antibiotics to prevent bacterial infections. This, too, is of concern because of the potential impacts on humans who consume the livestock. Even more worrisome, however, is that the indiscriminate use of antibiotics has led to the emergence of antibiotic-resistant bacterial strains.

Greatly expanded irrigation was another hallmark of the Green Revolution; this, too, has had some drawbacks. When soils in arid and semiarid environments are overirrigated, **salinization** (or **salination**) can occur. The top layer of the soil becomes soaked, or **waterlogged**. As this wet layer dries in the hot sun, salts are drawn up from deeper in the soil by capillary action, leaving a crust of dried salt on the surface. Waterlogging and associated salinization can cause **land degradation**, which permanently damages agricultural land. Extensive irrigation projects have also led to the overexploitation and depletion of groundwater resources, hindering agricultural productivity in some areas. Agricultural researchers in arid regions have begun to develop new, high-efficiency irrigation techniques in response to this problem. An example is **drip irrigation**, which minimizes water waste by delivering exactly the right amount of water directly to the root of the plant at just the right time.

High-yield and pesticide- and disease-resistant crop varieties also contributed greatly to the Green Revolution. Traditionally, new crop varieties have been developed by agricultural researchers through **selective breeding**—choosing crop varieties with specific desirable features for propagation. Early in the Green Revolution, selective breeding enhanced the yields of many crops, including corn, rice, and wheat, by maximizing the amount of biomass contained in the productive part of the grain and minimizing unwanted stems and leaves (see Figure 9.5).

The methods of development of new crop types changed dramatically with advancements in **biotechnology**. Through **genetic engineering** (or **bioengineering**), specific genes are removed from an organism and implanted in an individual of another species. This process creates a **transgenic** plant or animal in which the selected genetic trait may be expressed. With genetic engineering, scientists have developed crop species with inherent resistance to specific pests and diseases.

For example, potatoes have been designed to contain a natural toxin that renders them resistant to attacks by potato beetles. One of the earliest genetically engineered crops still marketed today is Bt corn, which is engineered to produce a continuous supply of a toxin derived from the Bt bacterium. The Bt toxin protects the corn from various pests without the need for pesticide spraying. Genetic engineering has also been used to enhance productivity and disease resistance in livestock. For example, cattle can be genetically engineered to resist viral infections, and trout can be implanted with a gene that triples their rate of growth.

Many controversies surround the genetic engineering of crops and livestock. Much of the discussion has focused on the use of these **genetically modified organisms** (**GMOs**) in food products for human consumption. Scientists still don't understand all of the potential impacts of genetically modified foods on human health. Activists call attention to the fact that GMOs are routinely used in many processed food products, but this may not be revealed in the product labeling. Farmers who practice organic farming—farming with minimal or no use of agrochemicals—are concerned about the possibility of genetic contamination of their crops. Genetic contamination of wild plants in natural ecosystems is also a concern. And farmers who have used Bt-engineered crops for a number of years have noted decreasing efficiency in pest control, perhaps indicating that pest species have become resistant to the engineered toxins.

Agricultural productivity per capita has slowed in recent years as a result of limitations in the overall amount of permanent cropland available. Most of the best farmland in the world is already being used for agriculture; with further population growth, each plot of land must feed more people. This tends to cause an increase in the intensity of agricultural use of the land. Furthermore, much land formerly devoted to food crops has been switched to cash crops, such as coffee and cotton, or to **fodder** (animal food). The pressures of these and other types of development, including urbanization, have led some farmers to rely on marginal lands—lands that are not well suited for agriculture because they are too dry, infertile, or steeply sloping. The per-hectare productivity of these marginal lands is lower than for more fertile lands.

What are agrochemicals and what are they used for? ______________

Answer: Common types of agrochemicals include fertilizers (to enhance availability of nutrients for plant growth); pesticides (to kill off pests that threaten crops); antibiotics (to prevent bacterial infections in livestock); and growth hormones (to enhance growth rates in livestock).

Erosion and Loss of Agricultural Soil

The Green Revolution led to greatly increased agricultural productivity but with an environmental cost. Now the environmental impacts and the limitations of the physical environment are taking a toll on that productivity. In the long run, at least until the science of **hydroponics** (growing plants without soil in nutrient-enriched solutions) has been perfected, the lack of arable soil may be the most serious limitation to agricultural productivity worldwide. The estimated global rate of soil loss through erosion now exceeds 25 billion tons each year. At this rate of loss, the world's productive soils are being depleted by 7 percent each decade. This clearly represents an unsustainable pattern of utilization of a critically important resource.

Soil loss doesn't mean that the soil just vanishes; it refers to the process of erosion, in which topsoil is carried away from agricultural lands. Eroded topsoil generally winds up in a body of water, where it may clog a stream channel or contribute contaminants that have adhered to the surfaces of clay particles in the soil. Thus, soil loss is a serious problem on farms and rangelands, but it also causes problems in the streams, rivers, and lakes where the soil is deposited. Contaminants such as agrochemicals, and farm waste, including manure and organic materials derived from field stubble and dead animals, also end up in water bodies and can have a serious impact on water quality. The deaths of seven people in Walkerton, Ontario, in 2000 from *Escherichia coli* infections have been attributed to the contamination of drinking water by pig manure in runoff from nearby farms.

Soil scientists often talk about a "tolerable" annual rate of soil loss. This refers to losses that are balanced or offset by the natural regeneration of the soil. Obviously, of greatest concern is soil loss that is *in excess* of the rate at which topsoil is being regenerated. In the United States, for example, the amount of arable soil lost to erosion at the beginning of the 1990s exceeded the amount of newly formed soil by almost 2 billion tons per year. The rate of loss since then has been reduced substantially by the implementation of aggressive soil erosion control programs. However, soil loss in the other major food-producing countries—China and Russia—is at least as rapid, while in India the soil erosion rate is estimated to be more than twice as high.

In what ways is the concept of a tolerable annual rate of soil loss similar to sustainable yield in a forest or fishery (chapter 8)? In what ways are these concepts different?

Erosion that leads to soil loss most commonly happens as a result of wind or water streaming over bare soil. Water begins to cause erosion as soon as rain hits the ground. It happens by impact as raindrops dislodge small particles and by sheets of water flowing downslope during heavy rains. This **overland flow**, or **sheetflow**, quickly organizes into channels. If depressions or furrows are present in the soil from plowed rows, tire tracks, animal pathways, or natural lines of drainage, these will concentrate the surface runoff into **rills**, which are small, narrow channels that cut into the top layer of the soil. As erosion progresses, the running water will begin to develop **gullies**, which are distinct, narrow stream channels that are larger and deeper than rills, with steep walls.

The ability of water to erode and carry away soil particles is related to the way the water moves through the channel (Figure 9.6). The higher the velocity and the more **turbulent**, or chaotic, the flow, the greater is the ability of the water to pick up particles and carry them away. In contrast, quieter **laminar** flow has less capacity to pick up and transport particles. The **load** of a stream is the material being carried by the water. This includes particles that bounce and roll along the bottom by a process called **saltation**; finer particles that are carried by the water in **suspension**; and dissolved substances from rock weathering, fertilizers, or contaminants. Gullies and rills are effective at transporting soil, but they are not true streams because they typically carry water only during and immediately after rainstorms or during periods of snowmelt.

Wind is also an important agent of erosion (Figure 9.7; note that this diagram could be used equally well to illustrate erosion by water). **Eolian** (wind) **erosion** is most effective in arid and semiarid regions. The most important eolian process in agricultural areas is **deflation**, which occurs when the wind picks up and removes loose particles of sand and dust. Deflation takes place mainly where there is little or

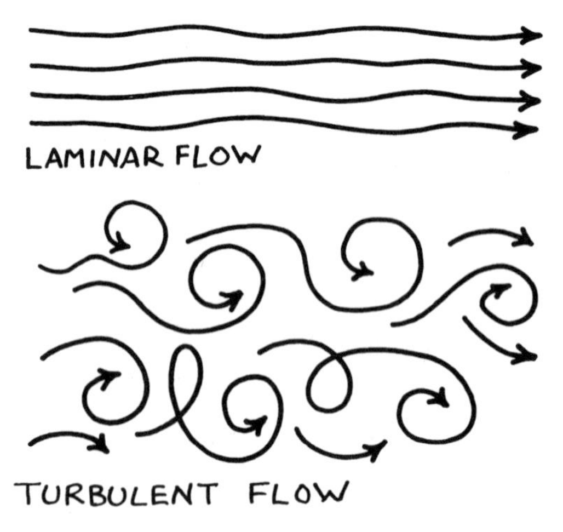

Figure 9.6.

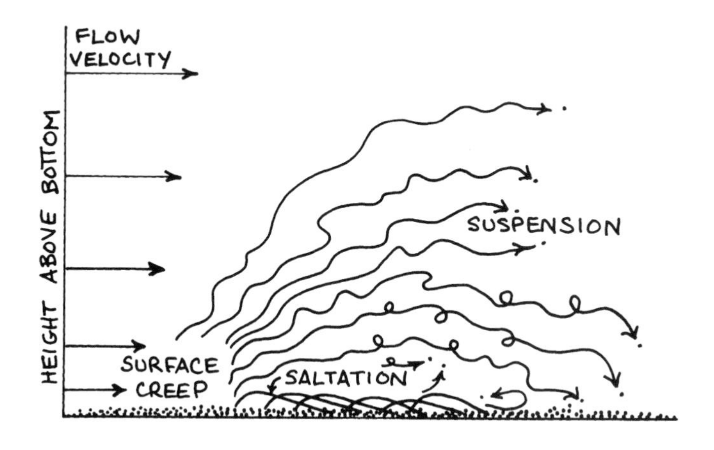

Figure 9.7.

no vegetation and where the loose particles are fine enough to be picked up by the wind. It is especially severe in deserts but can occur elsewhere during times of drought when no moisture is present to hold soil particles together. During the Dust Bowl era in the 1930s in North America, deflation occurred on a grand scale throughout the Great Plains, in the agricultural heartland of the United States and Canada.

Human activities contribute significantly to the loss of arable soil, primarily through the removal of vegetation. This can happen naturally—for example, through landslides—or as a result of activities such as clear-cut logging, slash-and-burn agriculture, or overly intensive grazing by animals. Deep tillage causes problems by overturning the soil, exposing it, and allowing it to dry out. Fertile soil is also lost each year through contamination and through the use of prime agricultural land for urban and suburban development.

Although soil erosion and the resulting land degradation are severely affecting many countries, some relatively low-tech, inexpensive control measures can substantially reduce them. The most serious soil erosion problems occur on steep hill slopes. Steep agricultural slopes can be protected through terracing, in which flat, steplike surfaces are constructed on the hill slope. Terraces provide level ground for planting, and they interfere with the downslope momentum of flowing water. Terracing is often combined with the use of **micro-catchments**, which are small basins constructed around the stem or trunk of a single tree or a small planted area. The micro-catchments hold water close to the plant's roots, binding the soil, conserving water, and preventing downslope erosional flow. Contour plowing, in which the plowed rows are designed to follow the topographic contours of the hill slope, works on the same premise as terracing, but it is used on shallower slopes. Reducing the depth of tillage can also help prevent compaction, drying, and loss of nutrients from soils.

Crop rotation and intercropping can help diminish soil loss. As discussed, land that is planted year after year with the same crop will eventually become stripped of the nutrients used by that particular crop. Soil loss from land planted continuously in corn can be reduced by as much as 85 percent when corn, wheat, and clover are rotated. Soil planted continuously with corn becomes depleted in the nutrients required by corn. The land is also more susceptible to erosion when exposed between rows of corn than with a more continuous cover of wheat or clover. Increasing fallow periods and leaving field stubble and other organic litter on the field after the harvest can also restore nutrients and protect the soil.

Agroforestry—interplanting trees and crops in close proximity—is another useful approach that can help conserve soil and increase productivity. When trees and crops are planted together, the tree roots bind the soil and pump nutrients and water from greater depths than those normally reached by crop roots (Figure 9.8). Trees also provide shelter, acting as a windbreak to protect the land from wind erosion, and they contribute organic litter to the soil. Leaves and branches can be sustainably harvested from the trees and fed to small livestock, which then contribute organic material to the soil via their manure. (Agroforestry was also discussed in chapter 8.)

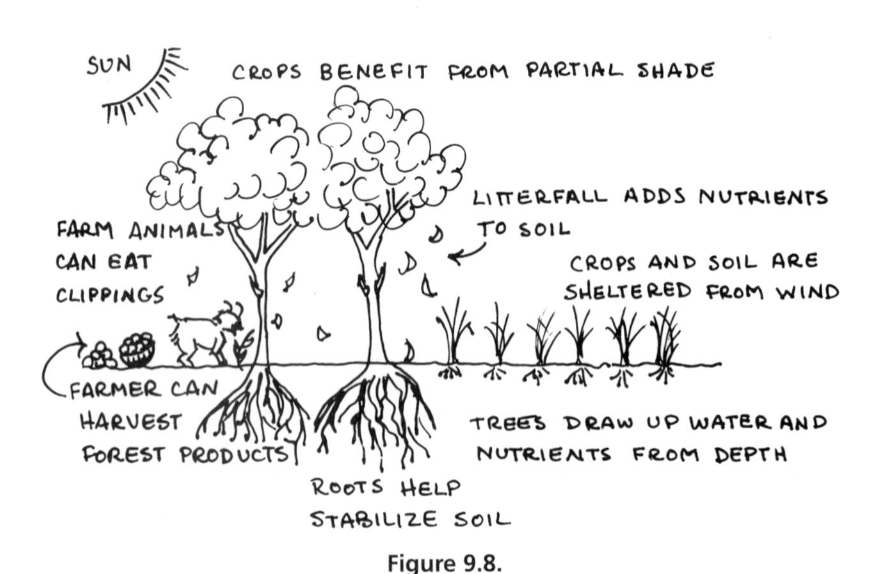

Figure 9.8.

Summarize the effective, low-tech approaches to soil conservation and how they work. ______________

Answer: Terraces and contour plowing interfere with downslope momentum of flowing water on hillslopes; micro-catchments hold water and prevent downslope flow; reduced-depth tillage prevents loss of moisture and nutrients; rotating crops, interplanting, increasing fallow periods, and leaving field stubble restore or maintain nutrients and protect soil; agroforestry protects soil because the tree roots bind the soil, pump nutrients and water from depth, and act as a windbreak.

Sustainable Agriculture and the World Food Supply

The Green Revolution has supported enormous population growth and has led to reduced hunger, improved nutrition, and food self-sufficiency in many parts of the world. At the beginning of the 2000s, the worldwide average of food consumed per day was 2,760 calories per capita—a level that is more than nutritionally adequate. However, food production, buying power, consumption, and dietary diversity are not distributed evenly over the world's population; there are large surpluses and deficits at regional, national, and local levels. Specifically, a nutritional gap of about 965 calories per capita per day exists between industrialized and developing countries. In sub-Saharan Africa in particular, the situation is actually worse now than it was in 1970.

The unequal global distribution of food and resulting food insufficiency in certain areas has arisen from a combination of social, economic, environmental, and political factors. These include inadequate distributional mechanisms, agricultural trade barriers and government subsidies, environmental problems such as drought and land degradation, agricultural practices that are ill-advised or improperly applied, and inequitable access to land and other agricultural resources, expertise, training, funding, and production technologies. With global population projected to increase by 50 percent or more by the middle of the twenty-first century, pressures to increase agricultural productivity and improve food distribution will be even greater.

Why should we be concerned about food shortages in the developing world? Since the 1970s, various international gatherings have stated and restated the goals that no one should go to bed hungry and that no human being's physical or mental potential should be stunted by malnutrition. These goals remain unmet; there are still as many as 800 million people in the world who are chronically hungry. As ethical beings we have a responsibility to be concerned for the welfare of others.

However, if we require a more selfish reason to be concerned, it is worth noting that national security is increasingly defined in terms of food security, income security, and environmental quality, rather than in military terms. The lack of these forms of security and the lack of access to land and basic resources are making more and more people into environmental refugees with nowhere to turn but the industrialized world.

The concept of **food security** is particularly important. There are five main factors that define food security: (1) *availability*, implying that there is adequate food production to meet needs; (2) *accessibility*, implying that food actually reaches the people who need it; (3) *affordability*, implying that food is within people's financial means; (4) *adequacy*, implying that the food will meet people's nutritional needs; and (5) *acceptability*, implying that the available food is in a form that is culturally suitable. Global food security is sometimes described in terms of **carryover grain stocks**—essentially, the amount of grain left in the world's storage bins as one growing season ends and the next begins. To ensure global food security, it is considered that there should be carryover grain stocks sufficient to feed the world's people for seventy-two days; there are currently stocks for about sixty days.

Some would argue that "acceptability" should not be part of the definition of food security. If you are chronically hungry, shouldn't you just eat whatever is offered to you? But what if the only available foods were grubs and termites? These foods are acceptable in some cultures but highly unpalatable to most North Americans. And what about foods that are unacceptable for religious reasons, such as pork for Jews or Muslims. Under what circumstances should people have to eat food they consider culturally unacceptable?

A complicating factor is the shift from locally grown grains to more processed grains and meat as dietary staples. This change, which correlates with an increase in income, is referred to as the **diet transition**. The diet transition is relevant in a discussion of global food security because far more energy and land are required to feed a person with meat than with grains. This is because there is a significant loss of energy associated with eating at a higher trophic level on the food chain. It takes 3 to 10 kg of grain to produce 1 kg of meat; currently, 38 percent of the world's total grain production is fed to livestock rather than to people. As meat consumption increases, there is pressure to develop new rangelands. In the meat-producing countries of South America, for example, this is a significant cause of tropical deforestation. Thus, when people in wealthy nations consume meat that was produced in tropical regions, they may unknowingly contribute to deforestation of rain forests; this has been dubbed the "**hamburger connection**."

What are the five factors that determine food security? _______________

Answer: Availability, accessibility, affordability, adequacy, and acceptability.

In spite of the many concerns discussed in this chapter, modern agriculture is probably capable of meeting the challenge of feeding a population of 8 to 10 billion by the middle of the twenty-first century. What is less clear is whether this can be achieved using methods that are environmentally and socially acceptable and sustainable. **Sustainable agriculture** refers to agricultural production that is sufficient and secure without compromising the environmental integrity of the land. Certainly sustainable agriculture implies the utilization of effective land management and soil conservation techniques—those discussed in this chapter and many others. Wise and restrained use of agrochemicals will also be a cornerstone of sustainable agriculture in the future. For example, it is becoming increasingly clear that there are both environmental and economic benefits for the use of **integrated pest management (IPM)**. IPM involves minimal, strategic use of pesticides while maximizing natural and biologic pest control. This is done by encouraging natural predators and by intercropping species that are known to inhibit diseases and pest infestations.

Some people also consider that sustainable agriculture must involve strengthening support for small-scale and subsistence farmers, and encouraging more community-based agricultural and food distribution policies. Many countries in the developing world, in particular, are now choosing to recognize small-scale village-level and subsistence farmers, and even urban farmers, who were formerly ignored in many schemes of agricultural development.

SELF-TEST

These questions are designed to help you assess how well you have learned the concepts presented in chapter 9. The answers are given at the end. If you get any of the questions wrong, be sure to troubleshoot by going back into that part of the chapter to find the correct answer.

1. The smallest mineral particles in soils are _______________.
   a. humus
   b. clay
   c. silt
   d. sand

2. Humans developed agriculture and began to adopt a sedentary lifestyle around ______________ years ago.
   a. 1 million
   b. 3,000
   c. 12,000
   d. 2 million

3. The world's current carryover grain stock is approximately ______________ days; for global food security, approximately ______________ days of carryover grain stock is required.
   a. 450; 260
   b. 250; 360
   c. 100; 175
   d. 60; 72

4. A plant or animal that has been genetically modified by the implantation of a gene taken from another species is called ______________.

5. A sequence of soil horizons from the surface down to the bedrock is called a(n) ______________.

6. The technique of growing plants in an aqueous, nutrient-enriched solution without soil is called ______________.

7. Eolian erosion is most effective in arid and semiarid regions. (True or False)

8. A hunter-gatherer lifestyle is based on a less-intensive utilization of environmental resources than a sedentary, agriculture-based lifestyle; therefore, the hunter-gatherer lifestyle requires less land to support a given population. (True or False)

9. Discuss the factors that influence the rate of rock weathering.

10. Why is soil moisture important in maintaining soil fertility?

11. What are the four main practices that led to the dramatic increases in agricultural productivity that characterize the Green Revolution?

12. How can irrigation in arid and semiarid areas cause land degradation?

13. What is the diet transition and why is it important in a discussion about global food security?

ANSWERS

1. b

2. c

3. d

4. transgenic (or GMO)

5. soil profile

6. hydroponics

7. True

8. False

9. Rock composition (some minerals weather more readily than other, more resistant minerals); temperature (high temperatures promote chemical weathering, whereas mechanical weathering is more common in cold, dry climates); rainfall (high precipitation promotes chemical weathering); steepness of slope (steeper slopes promote faster weathering).

10. Plants can only access nutrients that are in aqueous (dissolved) form. Chemical nutrients in the soil are held in solution in soil water that adheres to mineral particles until the nutrients are needed by the plant.

11. Increased use of agrochemicals; increase in irrigation; development of high-yield and pest- and disease-resistant crop types; increase in the amount of cultivated land.

12. Through waterlogging and salinization. When land in an arid or semiarid region is overirrigated, the top layer can become waterlogged. As the water evaporates in the hot sun, more water is drawn up by capillary action, bringing dissolved mineral salts. The salts are left on the surface as the water evaporates, leaving a crust of salts that can render the land unusable.

13. The diet transition is the change associated with increasing income from the reliance on locally grown grains to more processed grains and meat as dietary staples. It is important for global food security because it takes far more land and energy inputs to feed a person adequately with meat than with grains.

KEY WORDS

acidity
agrochemical
agroforestry
alkalinity
anion
arable
bioengineering
biotechnology
carryover grain stock
cation
chemical weathering
clay
colloid
deflation
domestication
diet transition
drip irrigation
eolian erosion
fallow
fertilizer
fodder
food security
genetic engineering
genetically modified organism (GMO)
Green Revolution
gully
"hamburger connection"
humus
hunter-gatherer
hydroponics
integrated pest management (IPM)
ion
laminar (flow)
land degradation
limiting factor
load
mechanical weathering
micro-catchment
monoculture
nutrient
overland flow (sheetflow)
pH
pore space
porosity
rill
salinization (salination)
saltation
sand
selective breeding
silt
slash-and-burn
soil
soil fertility
soil horizon
soil profile
soil water (soil moisture)
solubility
subsistence agriculture
suspension
sustainable agriculture
swidden (shifting) agriculture
topsoil
transgenic
turbulent (flow)
waterlogging

10 Mineral Resources

What we are pleased to call the riches of a mine are riches relative to a distinction which nature does not recognize.

—John Playfair

Objectives

In this chapter you will learn about:

- the dependence of human society on mineral resources;
- the uses of mineral resources;
- the formation of mineral deposits; and
- the environmental impacts of mining.

Mineral Resources and Modern Society

Each of us directly and indirectly uses a very large amount of material derived from nonrenewable mineral resources. Our modern world could not operate without mineral resources. Without them, we could not build planes, cars, televisions, or computers. We could not distribute electric power or build tractors to till the fields and produce food. We use mineral resources to make clothes, build shelter, fertilize crops, provide transportation, and communicate electronically. Without mineral resources, modern agriculture and industry would collapse.

Look around. Even if you are reading this book under a tree in a meadow, you are surrounded by products made from mineral resources. Your watch has metal parts, and the face is glass (a silica product) or plastic. Your pen is metal or plastic (a petroleum product). Your jewelry, your glasses, and even your clothes (if you are wearing synthetic fabrics, which are petroleum products) are made from mineral resources. The food in your lunch was grown in soil, tilled by oil-powered metal machinery, and helped by mineral fertilizers. Even the paper in this book has mineral additives to give it texture and color. And virtually every part of the car, bicycle, or bus that transported you to the meadow was made from mineral resources.

We don't know when or where mining started, but long before recorded history, trade routes for the exchange of mineral resources, notably salt, crisscrossed the globe. Metals were first used about 17,000 years ago. Copper and gold both occur naturally as pure or **native metals**, and these were the first metals used. These elements are rare, so eventually other sources were needed. About 6,000 years ago, our ancestors learned how to extract copper from certain minerals by **smelting**, an ore processing technique that involves heating ores to such extreme temperatures that the metals they contain melt and separate from the other mineral content of the rock.

By 5,000 years ago, our ancestors had learned how to smelt lead, tin, zinc, silver, and other metals, and how to mix metals to make **alloys** such as bronze (copper + tin) and pewter (tin + lead + copper). The smelting of iron is more difficult than the smelting of copper and was only achieved about 3,300 years ago. By the time the Roman Empire came into existence, about 2,500 years ago, humans had come to depend on a very wide range of mineral resources, including not just metals but processed materials such as cement, plaster, glass, porcelain, and other pottery ceramics.

The list of materials we mine, process, and use has grown enormously since that time. Today we have uses for almost all the naturally occurring chemical elements, and more than two hundred different kinds of minerals are mined. In environmental science and management there are three principal concerns associated with mining. We seek to understand the resource itself—how mineral deposits are formed, where they occur, how to locate them, and how to exploit them. We are also concerned with the effective management of mineral resources, because as nonrenewable resources, they are finite. Finally, we need to learn as much as possible about the environmental impacts of mining and how these impacts can be minimized.

What is an alloy? ________________

Answer: A mixture of metals.

Finding and Exploiting Mineral Resources

Nearly every kind of rock and mineral can be used for something, although those that are most valuable tend to be rare. It is convenient to group mineral resources according to how they are used (Table 10.1). **Metallic minerals** are mined specifically for the metals that can be extracted by smelting. Examples are the minerals sphalerite (zinc sulfide, ZnS), from which zinc is recovered, and galena (lead sulfide, PbS), from which lead is recovered. **Nonmetallic minerals** are mined for their properties as minerals, not for the metals they contain. Examples are salt, clay, and gemstones.

A **mineral deposit** is a local concentration or enrichment of a given mineral. Deposits are sought that will yield the highest quality and quantity of materials at the lowest cost. The costs of extraction vary widely depending on the location of the deposit, how concentrated it is, how deeply buried it is, how big it is, what kinds of technologies must be used to extract and process the material, and how far the

Table 10.1 Mineral Resources

Metals	**Examples**
Abundant metals	Iron, aluminum, magnesium, manganese, titanium, silicon
Scarce and rare metals	Copper, lead, zinc, nickel, chromium, gold, silver, tin, tungsten,mercury, molybdenum, uranium, platinum, and many others
Nonmetals	**Examples**
Used for chemicals	Sodium chloride (halite), sodium carbonate, borax, calcium fluoride (fluorite)
Used for fertilizers	Calcium phosphate (apatite), potassium chloride, sulfur, calcium carbonate (limestone), sodium nitrate
Used for building	Gypsum (for plaster), limestone (for cement), clay (for brick and tile), asbestos, sand, gravel, crushed rock, shale (for brickmaking), decorative stone
Used for jewelry	Diamond, corundum (ruby and sapphire), garnet, amethyst, beryl (emerald), and many others
Used for glass and ceramics	Clays, feldspar, quartz (silica sand)
Used for abrasives	Diamond, garnet, corundum, pumice, quartz

processed material must be transported to get to a market, as well as the costs of labor, mine management, and environmental protection.

To distinguish between profitable and unprofitable mineral deposits, we use the word **ore**, a deposit from which one or more minerals can be extracted profitably. Whether a given mineral deposit is an ore is determined by how much it costs to extract the mineral and how much the market is prepared to pay for it. The level of concentration, or **grade**, of the ore is an important factor. In general, the more highly concentrated an ore is, the higher the grade and the more valuable the deposit. Ores that are less concentrated—that is, low-grade—contain a higher proportion of nonvaluable minerals called **gangue** (pronounced "gang").

Mineral resources are nonrenewable and therefore exhaustible by mining. Furthermore, economically exploitable occurrences of minerals are distinctly localized within Earth's crust. This uneven distribution is the main reason that no nation is self-sufficient in all types of minerals. It is difficult to assess the available quantity of a given material and even more difficult to anticipate whether new deposits will be discovered. A country that can meet its needs for a given mineral substance today may face a future in which it will become an importing nation. For example, a century ago England was a great mining nation, producing and exporting such materials as tin, copper, tungsten, lead, and iron. Today most of England's known mineral deposits have been exhausted. This pattern—intensive mining followed by depletion of the resource, declining production and exports, and increasing dependence on imports—can be applied in a local, regional, or global context to estimate the remaining lifetime of a given mine or mineral resource.

Sustainable development is commonly defined as meeting the resource needs of the present generation without compromising the ability of future generations to meet their needs. Given that definition, do you think sustainable development of nonrenewable mineral resources is possible?

Over the past few decades, there has been a slow but steady shift in the emphasis of mineral exploration and production away from the industrialized nations and toward the developing nations. In the industrialized world the geologic locations that are most favorable for conventional mineral exploration have already been prospected, assessed, and in some cases mined and depleted. This doesn't mean that there are no more mineral deposits to be found, but geologists must look harder, utilize innovative exploration techniques, and look for mineral deposits in unconventional locations.

To assess how much of a particular resource is left, we use the concept of a **reserve**—the portion of the resource that has been identified and is economically

Table 10.2 Resources and Reserves

	Identified	Undiscovered	
		Known Districts	Undiscovered Districts or Forms
Economic	Reserves	Hypothetical Resources	Speculative Resources
Marginally Economic	Marginal Reserves		
Subeconomic	Subeconomic Resources		

Reserve
Resource

extractable using current technologies (Table 10.2). "Identified" refers to deposits that have been found and studied and for which scientists have relatively detailed information concerning the grade, location, and extent of the deposits. Deposits that are less well known are speculative or hypothetical resources, not reserves. In addition, it must be economically feasible to mine the material using known technologies at its present market value.

What is the difference between a mineral deposit and an ore? ______________

Answer: Mineral deposit: a local concentration of one or more minerals. Ore: a mineral deposit that contains valuable minerals, meaning minerals that can be extracted profitably.

How Mineral Deposits Are Formed

For a mineral deposit to form, a geologic process or combination of processes must produce a localized enrichment of one or more minerals. Minerals can become concentrated as a result of:

- hot fluids flowing through fractures and spaces in rocks;
- metamorphic processes;
- magmatic processes;
- evaporation or precipitation from lake water or seawater;
- the action of waves or currents in flowing surface water; or
- weathering processes.

Many famous mines contain ores that formed when minerals were deposited from **hydrothermal solutions**—hot, aqueous (watery), metal-saturated fluids (from the Greek words *hydro*, meaning "water," and *therme*, meaning "heat"). As hydrothermal solutions move through cracks and spaces in rocks, they deposit their dissolved constituents, forming veinlike concentrations of ore minerals (Figure 10.1). Hydrothermal deposits are the primary sources of many metals, including copper, lead, zinc, mercury, tin, molybdenum, tungsten, gold, and silver.

The geologic process of rock **metamorphism**—the alteration and recrystallization of rocks as a result of exposure to high temperatures and pressures—can act to concentrate minerals. In response to the changes that occur during metamorphism, minerals may become segregated and concentrated into distinct bands or layers. A wide variety of valuable nonmetallic mineral resources are concentrated in metamorphic rocks, including mica, asbestos, graphite, marble, and some gem-

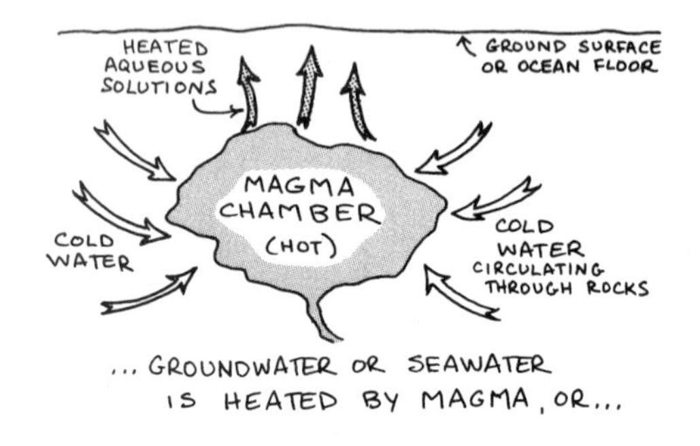

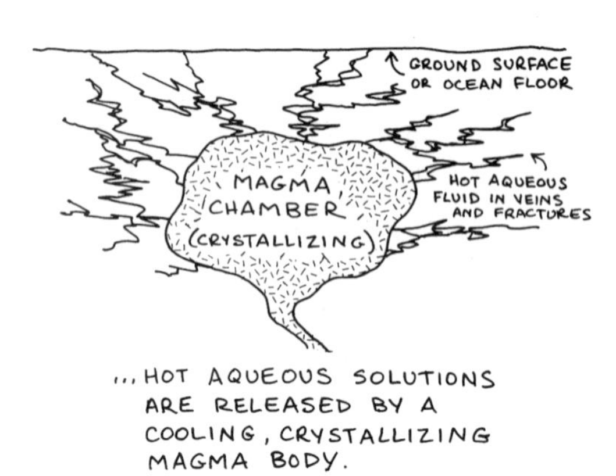

Figure 10.1.

stones. Rock metamorphism can be closely associated with hydrothermal processes; if fluids are circulating while metamorphism occurs. When metamorphic fluids dissolve and transport materials from a rock, the process is called **metasomatism**.

There are several ways by which ore minerals may become concentrated as a result of igneous or magmatic processes—that is, geologic processes involving magma or lava (molten rock). For example, when a certain type of magma solidifies into rock, one of the first minerals to crystallize is chromite. Chromite is the main ore mineral from which we derive chromium, an important constituent of steel. Chromite crystals, which are denser than the magma that contains them, may settle and accumulate at the bottom of the magma chamber. This process can produce pure layers of chromite. Another important kind of magmatic mineral deposit is **kimberlite**, a long, thin, pipelike body of rock made from magma that originates deep in the mantle (Figure 10.2). Kimberlite magma rises explosively, transporting broken fragments of mantle rock (called *xenoliths*) to the surface. One of the mineral

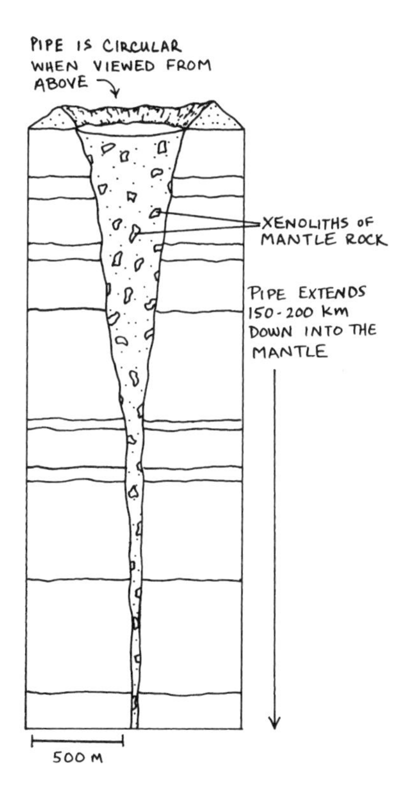

Figure 10.2.

constituents of the rocks in kimberlite deposits is diamond, a high-pressure mineral that forms only in the mantle, at depths greater than 150 km (about 90 mi). The only way diamonds can reach the surface is through kimberlite pipes.

Mineral deposits can form when dissolved substances precipitate from lake water or seawater. One cause is evaporation, in which layers of salts are left behind, forming **evaporite deposits**. Sodium carbonate, sodium sulfate, and borax come from deposits formed by the evaporation of lake water. Marine evaporites, from the evaporation of seawater, produce gypsum (used to make plaster), halite (salt), and a variety of potassium salts used in fertilizers. Biochemical reactions in seawater can also cause precipitation. Such reactions lead to the precipitation of the mineral apatite, the main source of phosphate fertilizers. In sedimentary rocks older than about 2 billion years, there are unusual iron-rich rocks called banded-iron formations, which are ancient biochemical precipitates from seawater. Historically, banded-iron formations have been important sources of iron ore.

Heavy mineral grains can be concentrated as a result of the sifting or winnowing action of flowing water (Figure 10.3). The flowing water may be in a stream or

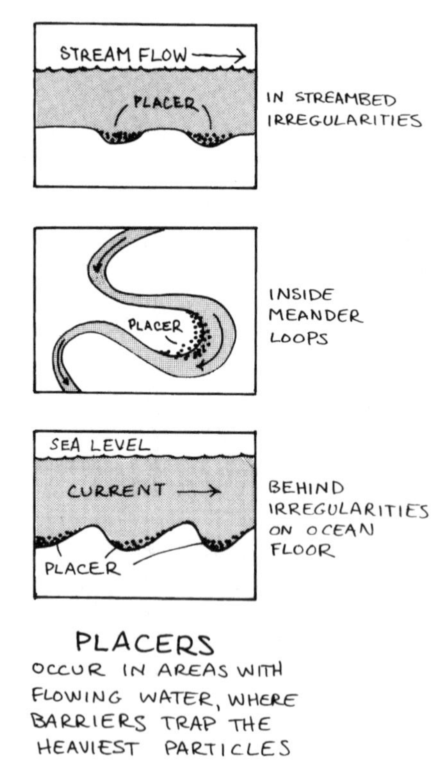

Figure 10.3.

in waves along a shoreline. Flowing water carries a sediment load. When the current slows down for one reason or another—if the stream goes around a bend or flows over a large boulder, for example—the stream will lose energy and drop the heaviest particles from its load. As waves sift through the sand along the shoreline, the heaviest minerals become concentrated in layers. A deposit that forms in one of these ways is called a **placer deposit**. Heavy minerals that typically become concentrated in placer deposits are gold, platinum, diamonds, chromite, and minerals that contain zirconium or titanium.

Chemical weathering in tropical climates can lead to the concentration of minerals by removing the most soluble material. The dissolved material is carried downward by infiltrating rainwater. When the water encounters a zone of sharply different temperature or composition, it may deposit all of its dissolved materials in one concentrated layer (Figure 10.4). These **residual deposits** form mainly in the tropics, where chemical weathering is most effective because of high rainfall and high temperatures. A common example of a residual deposit is laterite, a hard, highly weathered material that is mined for iron and nickel. Laterite is by far the most common kind of residual deposit, but the most important for human exploitation is bauxite, the source of aluminum ore. Where bauxite is found in present-day temperate conditions, such as France, China, Hungary, and Arkansas, we can infer that the climate was tropical at the time it formed.

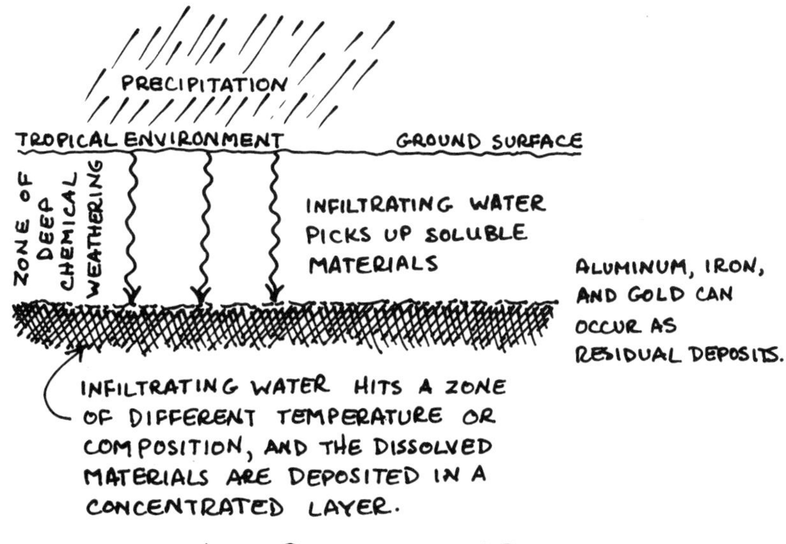

Figure 10.4.

Where do diamonds come from? ______________

Answer: The mantle, about 150 km deep in Earth; they are brought to the surface by kimberlite magmas and may be concentrated as heavy minerals in placer deposits.

Environmental Impacts: The Mining Process

Mining is of critical importance to industrialized society. It is also a controversial industry that suffers from a negative public image partly due to many years of neglectful practices. The mining industry has undergone an enormous change in attitudes and practices since the 1970s; it is now a tightly regulated industry in terms of environmental impacts. Many large North American mining companies are committed to employing environmentally sound practices in their operations in other countries even if those countries have less rigorous environmental regulations.

An issue that remains particularly controversial is whether mining should be permitted in parks, forests, and wilderness areas. In favor are those who believe that we cannot afford to forego development of these important mineral (and energy) resources. In some cases it may be possible to mine and then restore the land to its former state for recreational or other purposes. This is called **sequential land use**. In very remote areas, however, this may not be economically feasible. Furthermore, in fragile environments such as the Arctic, unique ecosystems, or areas inhabited by threatened or endangered species, the environmental damage caused by mining can be permanent. For these reasons and others, environmentalists tend to be vehement opponents of mining and oil drilling in parks and wilderness areas.

The production of metals and minerals involves many stages, from finding the ore to extraction, processing, and refining. Wastes, emissions, and environmental impacts occur at every stage in the process, although some parts of the mining process entail more impacts than others. In general, the **prospecting** or **exploration** phase, during which an area is assessed for ore potential, is short-lived and involves little or no permanent environmental damage. Dirt roads or tracks may be constructed, and there may be drilling. The noise of helicopters used by exploration geologists has been an issue in some wilderness areas because of the potential impacts on animal migration.

The potential for real environmental disruption begins when commercial production is undertaken at a mine site. The way the mineral or metal occurs in nature determines the type of mining operation needed and the amount of waste produced in the extraction, separation, and concentration of the ore. In the case of sand and gravel extraction, very little material is discarded. Because sand and gravel are not worth very much in terms of their market value, such deposits tend not to be economically viable unless they are quite high-grade, meaning very little waste rock is present. In other cases, huge quantities of rock must be removed and discarded to obtain a relatively small amount of a very valuable ore. For example, a typical copper grade of 2 percent will produce just 20 kg of pure metal from each tonne of ore processed. It is worthwhile to extract a lot of rock for a small quantity of metal only because the market value for the material is high.

Once the ore has been extracted, it is crushed and concentrated; this is the **milling** phase. For many nonmetallic minerals and construction materials (such as sand, limestone, and gravel), basic processing is all that is needed before the substance is transported and used. Other nonmetallic minerals (such as asbestos, talc, and potash) and metallic ores require further processing of the concentrate to separate the ore from the gangue. This may involve the use of chemical reagents such as kerosene, cyanide solutions, or mercury. The material that is discarded after the initial processing, usually at the mine site, is called **tailings**. After the concentrate leaves the mine site, metallic minerals, in particular, must be further processed at refineries using high-temperature smelting and/or chemical processes. The tailings piles, other waste products, and air pollution from these operations constitute some of the most significant environmental impacts of mining.

Finally, the mine site itself must be managed in the **postoperational** phase after the deposit has been exhausted and mining has ceased. When the mine is closed, steps must be taken to ensure that discarded materials are properly contained and monitored on an ongoing basis. Once a mine is closed permanently, laws require that the mine site be cleaned up and, if appropriate, reclaimed for other uses. Shutting down a mine site and preparing it for long-term postoperational management, including cleaning up any contamination at the site, is called **decommissioning**. Mine site decommissioning is a strictly regulated and monitored procedure in most industrialized countries. In some cases, the restoration of the land to its former status is a legal requirement that companies must agree to before they are even granted a permit to begin mining. The restoration process, which includes removal of buildings and other works, decontamination, regrading of the topography, and replanting of vegetation, is called mine site or land **reclamation**.

Serious problems remain, however, with contamination from old, abandoned mine sites. Strict environmental regulations are relatively recent, even in North America, and mines abandoned before the 1970s typically did not receive the kind of attention now given to waste control and reclamation, either during production or decommissioning. Some environmental problems associated with abandoned mines include water pollution, underground fires, subsidence of the ground surface, and collapse of mine structures. The environmental impacts of abandoned mines can be especially difficult to handle if there is no one with legal or financial responsibility for the site.

In addition, there can be extensive social and economic impacts in the postoperational phase of mining. When a mine is decommissioned, miners and other long-term mine employees lose their jobs. In some cases, the entire structure of the town may have developed with the mine as its central economic feature; without it, the town may risk economic collapse. Many mining companies now plan for the eventuality of the mine's closure by setting aside a fund for future employment retraining for mine employees.

What are tailings? ______________

Answer: Waste material—including gangue and chemical residues—from the early stages of ore processing, usually at the mine site.

Impacts on Land, Air, Water, and the Biosphere

The impacts of mining on land (Table 10.3) depend on the type of mine. **Open-pit mines** are huge, open holes; these are obviously very disruptive to land. **Underground** (or **subsurface**) **mines**, with small openings but extensive underground works, are less disruptive at the surface, although subsidence of the land overlying the underground works can be a problem. In **strip mines**, the vegetation and top layer of regolith are removed so that the underlying ore can be extracted. Strip min-

Table 10.3 The Potential Environmental Impacts of Mining

Stage of Mining	Potential Impacts on Land	Potential Impacts on Air	Potential Impacts on Water	Potential Impacts on Health and Biosphere
Exploration	Usually low	Usually low	Usually low	Usually low
Mining/milling	High	Low–medium	Medium–high	Medium–high
Smelting/processing	Low–medium	High	Low–medium	Medium–high
Post-operational	Depends on management	Usually low	Can be high	Usually low

The environmental impacts of mining always depend heavily on the specific geologic and ecologic setting and on the mine management process. Environmental impacts can be minimized if the mining process is managed properly.

ing can be very disruptive, although in some cases it is possible to restore the land after mining has been completed. Other impacts on land include the construction of roads and buildings, which can be particularly damaging in wilderness areas, and the storage of rock waste, tailings, and other discarded materials.

Smelting and refining processes emit pollutants into the air unless the substances are captured by air pollution control equipment. The pollutants from these operations include particulates (smoke and fine particles), nitrogen oxides and sulfur oxides (which have a role in the production of acid precipitation), vaporized metals, and volatile organic compounds; all of these will be discussed in chapter 14. Airborne substances may be deposited locally or transported over long distances. Another problem can be blowing dust from waste rock piles and tailings. Sometimes there are potentially hazardous materials in the tailings, such as chromium or asbestos. Mining companies today generally seal their tailings and ore stockpiles either by spraying on a latex cover, maintaining a water cover, or revegetating the area.

Liquid wastes containing a wide range of contaminants can be generated during the milling and processing stages of mining. Flowing liquid wastes, or **effluents**, may carry toxic metals, such as arsenic, cadmium, iron, lead, and others, and chemical reagents, such as cyanide, kerosene, mercury, organic flotation agents, activated carbon, and sulfuric acid. **Acid mine drainage**—possibly the single

This is the site of the abandoned Crystal gold mine in the Boulder River Basin, southwestern Montana. Acidified discharge from the mine is draining into the Uncle Sam Gulch on the left side of the photo.

most challenging environmental problem associated with mining today—results when water interacts chemically with sulfide minerals in waste rock and tailings piles, creating sulfuric acid. Mine drainage associated with some types of mineral deposits is saline or alkaline (caustic). In some cases, as in the mining of uranium, mine wastes contain radioactive materials. Mining companies deal with liquid waste by designing the operation to function, as much as possible, as a closed system. Wastes are generally stored and treated on site until they are clean enough to be released back into the environment.

All of the environmental impacts described in this chapter can have negative effects on plants and animals, including humans. Of particular concern for local ecosystems is acid mine drainage, and air pollution from smelting can be particularly worrisome for human health. Some health hazards are directly related to the mining process, although health regulations for miners are now strictly enforced. The health impacts of mining include black lung disease and coal dust explosions, which are associated with coal mining; mine collapse; and various cancers, such as lung cancer and mesothelioma, which are mainly associated with asbestos and uranium mining.

Why are the environmental impacts of the postoperational phase of mining so variable? ______________

Answer: The postoperational impacts of mining depend on the type of ore deposit and the type of mine; the effectiveness of the decommissioning process and the management plan that has been put in place; and how effectively the monitoring, maintenance, and restoration of the mine site are carried out.

SELF-TEST

These questions are designed to help you assess how well you have learned the concepts presented in chapter 10. The answers are given at the end. If you get any of the questions wrong, be sure to troubleshoot by going back into that part of the chapter to find the correct answer.

1. The nonvaluable minerals associated with ores are called ______________.
   a. tailings
   b. gangue
   c. residuals
   d. reserves

2. Which one of the following is not a metallic mineral resource?
   a. iron c. copper
   b. gold d. diamond

3. The phase of the mining process in which the ore is crushed and concentrated is called ________.
   a. refining c. milling
   b. prospecting d. smelting

4. A high-temperature process for refining an ore concentrate is called ________.

5. Ore deposits that form as a result of the winnowing action of flowing water or waves are called ________.

6. The prospecting or exploration phase of mining commonly entails intense environmental damage. (True or False)

7. A kimberlite is a type of evaporite deposit. (True or False)

8. Mining companies sometimes seal waste rock and tailings piles with a latex spray to prevent harmful dust from blowing off the piles. (True or False)

9. Should mining be allowed in national parks, forests, and wilderness areas?

10. What is sequential land use?

11. What causes acid mine drainage?

12. What is the difference between a resource and a reserve?

13. What are some of the social impacts associated with the decommissioning of a mine?

ANSWERS

1. b 2. d 3. c 4. smelting

5. placers 6. False 7. False 8. True

9. For you to decide; weigh the issues.

10. When land is used for a purpose such as mining and subsequently reclaimed so that it can be used for another purpose such as recreational use.
11. When sulfur-bearing minerals in rock waste and tailings piles react with water, sulfuric (and other) acids are formed.
12. A resource includes the entire body of material; a reserve is that portion of the resource that is identified and economically extractable using present technologies.
13. Miners and other mine workers may lose their jobs; the entire town's welfare may be at stake if its central economic structure was the mine.

KEY WORDS

acid mine drainage
alloy
decommissioning
effluent
evaporite deposit
exploration
gangue
grade
hydrothermal solution
kimberlite
metallic mineral
metamorphism
metasomatism
milling
mineral deposit
native metal
nonmetallic mineral
open-pit mine
ore
placer deposit
postoperational (mine management)
prospecting
reclamation (land or mine site)
reserve
residual deposit
sequential land use
smelting
strip mine
tailing
underground (subsurface) mine

11 Energy Resources

If sunbeams were weapons of war, we would have had solar energy long ago.

—George Porter

Objectives

In this chapter you will learn about:

- Earth's energy cycle;
- fossil fuels and how their use affects the environment;
- nuclear energy and how its use affects the environment; and
- the pros and cons of alternative energy sources.

Energy Basics

Earth's energy comes from three sources: solar radiation; geothermal energy from natural radioactivity and other heat sources inside the planet; and tidal energy from Earth's gravitational interactions with the Sun and the Moon (Figure 11.1). Energy circulates through the many pathways and reservoirs of Earth's **energy cycle**, driving processes such as photosynthesis and atmospheric circulation. The energy flowing in the Earth system, all of which could theoretically be used, is $174,000 \times 10^{12}$

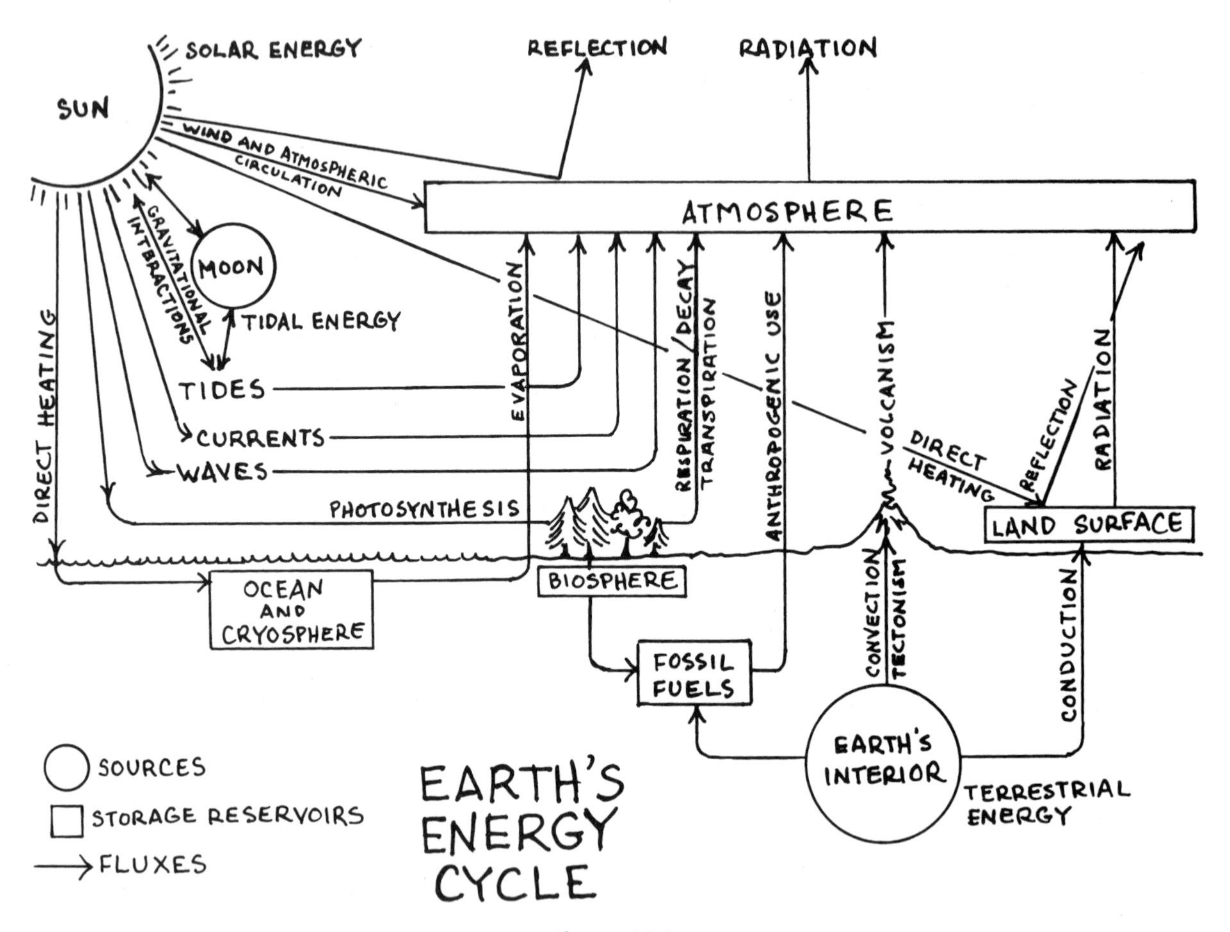

Figure 11.1.

watts (174,000 terawatts*). The world's more than 6 billion people use power at an average of 1,600 watts per capita, for a total of 9.6×10^{12} watts (9.6 terawatts). From these figures, it should be clear that we are not on the verge of running out of energy in an absolute sense. However, we do need to find sources of energy that will meet society's needs in a socially, environmentally, and economically acceptable way. Commercially produced energy is used in every aspect of our lives, from food pro-

*A watt is a unit of power rather than energy; it refers to an amount of energy used per unit of time. Specifically, 1 watt = 1 joule per second (J/s). This can also be expressed in calories per minute (1 watt = 14.34 cal/min) or in British thermal units per hour (1 watt = 3.4129 Btu/hr). The prefix "tera-" means 10^{12}, or 1,000,000,000,000. For information about units, conversions, and large numbers, see appendix 1.

duction to transportation, housing, manufacturing, and recreation, and North Americans are the world's biggest users of energy per capita.

Some energy sources are renewable or inexhaustible; others are nonrenewable. (See chapter 7 for a review of basic resource types.) Examples of renewable energy sources are fuelwood, wind, and water power. Energy sources driven by the Sun and gravity are inexhaustible; as long as there is a Sun–Moon–Earth system with an atmosphere and ocean, there will be winds, tides, and solar energy. Earth's own internal energy is also, in principle, inexhaustible. Energy sources that are nonrenewable on a humanly accessible time scale include coal, oil, and natural gas. Nuclear energy is truly nonrenewable—once an atom has been split, there is no way to put it back together.

Everywhere in the world, even in the least developed countries, nonrenewable sources supply at least half of the energy used. The main source of commercial energy today is **fossil fuel**, which includes coal, oil, and natural gas. Throughout history, energy consumption and the character of the world's energy mix have changed (Figure 11.2). During the Industrial Revolution, coal replaced wood as the main energy source in industrializing societies. Coal was burned to heat water, creating steam to drive the newly invented steam engines. After World War II, oil rose to dominance in the world's energy mix. The shift from coal to oil was driven by new technologies, primarily the internal combustion engine, which required fuel in liquid form. Recently, the use of natural gas has increased, driven by price as well as environmental concerns.

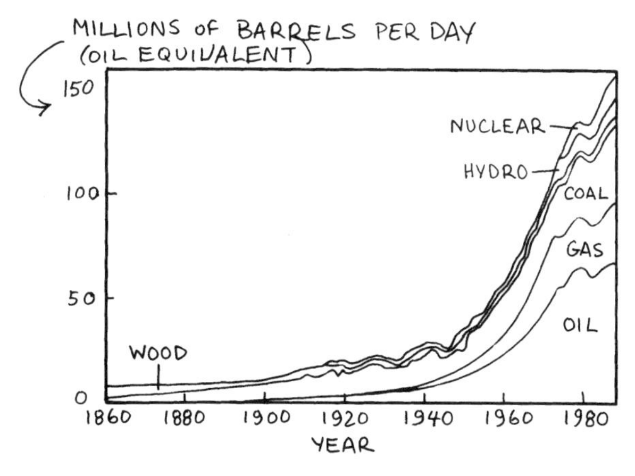

Figure 11.2.

What were the two new technologies that drove the shift from wood to coal and from coal to oil in the world's energy mix? ______________

Answer: The steam engine and the internal combustion engine.

Fossil Fuels

Fossil fuels consist of altered organic matter from the remains of plants or animals trapped in a sediment or sedimentary rock. Various changes occur during and subsequent to the burial of the organic remains. The kind of sediment, the kind of organic matter, and the kinds of postburial changes determine which type of fossil fuel will form.

Fossil fuels ultimately derive their energy from the Sun. Plants use the Sun's energy through photosynthesis to combine water (H_2O) and carbon dioxide (CO_2) into oxygen and organic carbohydrates. When organic matter decays, the solar energy stored in the organic matter is released. Any organic matter that escapes decay and is buried in sediment is a potential long-term storage reservoir for solar energy. The chemical composition of this organic matter is dominated by the elements hydrogen and carbon, so an alternate name for fossil fuels is **hydrocarbons**.

Organic matter on land mainly comes from trees, bushes, and grasses. In water-saturated places, such as swamps and bogs, organic remains accumulate to form **peat**, an unconsolidated deposit with a high carbon content (Figure 11.3). Peat is

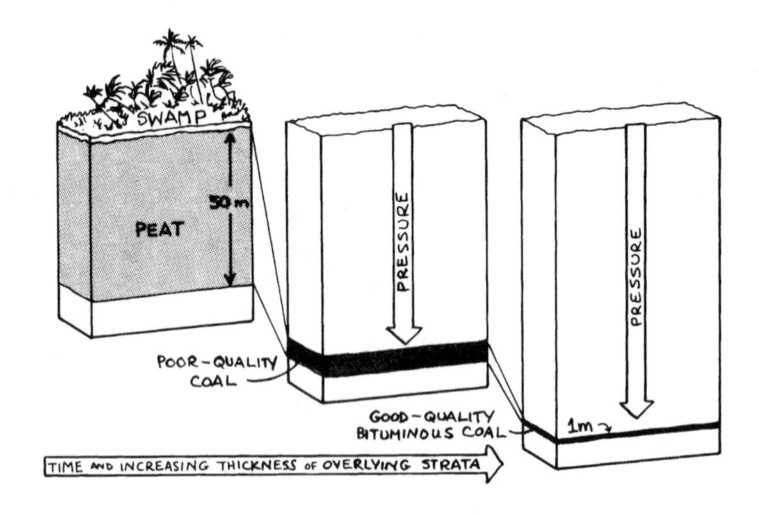

Figure 11.3.

the initial formation stage of **coal**, a black, combustible, carbon-rich rock. Over millions of years, the peat is compressed by overlying sediments. Water is squeezed out, and gaseous (volatile) compounds such as carbon dioxide (CO_2) and methane (CH_4) escape. By compaction and gas escape, the thickness of the layer of peat is reduced by 90 percent, and it is converted into a layer or *seam* of coal. Peat and coal have been forming more or less continuously since land plants first appeared 400 million years ago. Today peat formation is occurring in wetlands such as the Okefenokee Swamp and the Great Dismal Swamp.

In the ocean, microscopic phytoplankton and bacteria are the main sources of organic matter trapped in sediment. When these marine microorganisms die, their remains settle to the bottom and collect in the fine seafloor mud, where they start to decay. The decay process quickly uses up any oxygen that is present. The remaining organic material is preserved and covered with more layers of mud and decaying organisms. Over time, with continued burial, the muddy sediment with the partially decayed organic material is subjected to heat and pressure, eventually being transformed into sedimentary rock. During burial and the conversion of seafloor sediments into rock, the organic compounds are chemically transformed into **petroleum** (from the Latin words *petra*, meaning "rock," and *oleum*, meaning "oil")—naturally occurring gaseous, liquid, and semisolid substances that consist chiefly of hydrocarbon compounds (Figure 11.4). The burial and chemical transformation process that leads to the formation of petroleum is called **maturation**.

Oil, the liquid form of petroleum, came into use in 1847 when oil from natural seeps was used as a lubricant. The first oil wells were hand dug in Oil Springs, Ontario, after a Canadian chemist discovered how to produce kerosene, a distilled

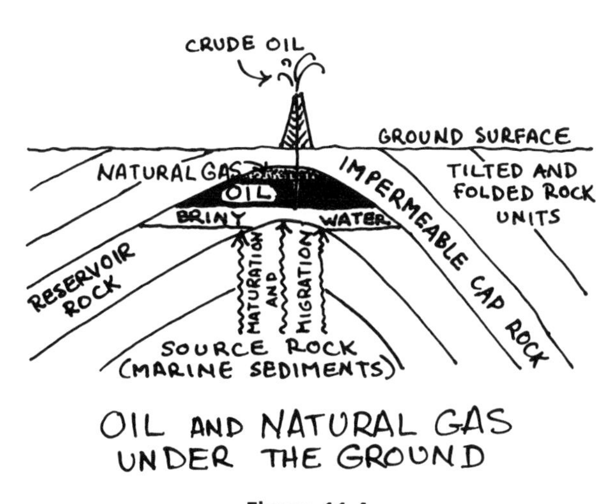

Figure 11.4.

oil that could be burned in lamps. In 1859, the first commercial well was drilled in Titusville, Pennsylvania. **Natural gas**, a naturally occurring hydrocarbon that is gaseous at surface temperature and pressure, was discovered in 1821 in Fredonia, New York, when a water well produced bubbles of a mysterious gas that burned. Before long, natural gas was being piped for the purpose of lighting lamps.

Petroleum is called **crude oil** when it emerges from the ground. From this state it must be distilled and refined. The refining process involves separating the heavy, medium, and light fractions of the oil, and breaking down or "cracking" the long-chain hydrocarbons to form compounds with lower boiling points such as gasoline and kerosene. The hydrocarbon components of petroleum are also used to make fertilizers, lubricants, asphalt, and an array of synthetic materials such as plastic, as well as fabrics such as nylon and rayon.

What is petroleum? ______________

Answer: Naturally occurring gaseous, liquid, and semisolid hydrocarbons, including oil and natural gas.

Fossil Fuels and the Environment

Fossil fuel use causes environmental impacts at every stage, from extraction, refining, and transportation to the emissions that result from its combustion. Coal is the "dirtiest"—that is, the most environmentally damaging—of the fossil fuels. It is often strip mined (chapter 10), causing extensive land degradation. Underground coal mines can be particularly hazardous. Coal dust is highly combustible, and many of history's great mine disasters, including underground fires, explosions, and collapses, have occurred in coal mines. If the rock waste from coal mining is exposed to rain, stream water, or groundwater, it will react to form acid mine drainage, which is extremely damaging to the aquatic environment. Extensive areas of farmland have been lost as a result of the ground collapsing into abandoned underground coal mines. Coal miners historically have suffered from chronic illnesses such as black lung disease.

Natural gas and oil are fluids, so they are extracted from the ground by drilling and transported by pipelines, trucks, and oil tankers. One of the biggest environmental controversies over the past few decades has been the debate about whether to allow oil drilling in environmentally sensitive areas, such as offshore areas, national parks, and wildlife refuges. This debate is intensified by the fact that many domestic oil and natural gas deposits are located in wilderness areas—for example, Alaska's Prudhoe Bay. Aside from land disruption and the potential for spills, drilling can have local environmental consequences such as the release of briny or heated fluids (used to aid in extracting the hydrocarbons from the ground).

Oil spills are the most visible impact of both drilling and the transport of oil by pipeline or tankers. Crude oil consists of hundreds of chemical components, many of them toxic, and spills can be devastating to the local environment. One of the best-known spills, although not the largest, happened in 1989 when the tanker *Exxon Valdez* ran aground off the coast of Prince William Sound, Alaska. Approximately 10.9 million gallons (260,000 barrels) of oil were spilled, spreading to cover thousands of square kilometers of open water. Hundreds of kilometers of Alaska's shoreline were contaminated with crude oil, killing tens of thousands of birds, fish, and mammals. The largest oil spill in history was intentional rather than accidental. It occurred in 1991 during the Persian Gulf War, when hundreds of oil wells were bombed by the Iraqi army in Kuwait. Over 250 million gallons (6 million barrels) of oil—more than twenty times as much as from the *Exxon Valdez*—spilled into the Persian Gulf and onto the desert.

The *Exxon Valdez* spill and other major spills and cleanups have revealed gaps in scientific knowledge about the natural processes by which oil is dispersed in oceanic and coastal environments, and the technologies that are most effective in cleaning up spilled oil, especially in cold environments. The *Exxon Valdez* spill also raised many issues concerning the legal and financial responsibility for the spill and the cleanup, as well as the technologic and procedural aspects of transport. The litigation associated with this spill is ongoing; it has been estimated that overall the spill will cost Exxon more than $10 billion.

As visible as they are, however, oil spills are *not* the principal environmental problem associated with fossil fuel use. The combustion of fossil fuels releases atmospheric emissions, including CO_2, SO_2, NO_x, and particulate matter, all of which cause extensive harm to the environment. Airborne particulates that result from the burning of fossil fuels are extremely tiny particles—soot, essentially—that cause respiratory problems and have other human health impacts. Carbon dioxide (CO_2) contributes to the anthropogenic greenhouse effect implicated in global warming. Sulfur dioxide (SO_2) and nitrogen oxides (NO_x) react with water in the atmosphere to form acids, which then fall to the ground as either dry acidic deposition or acid precipitation. Coal is particularly high in both CO_2 and SO_2 emissions. Coal-fired power plants are often installed with technologies that remove some of the sulfur from postcombustion emissions.

Increasing concerns about the environmental impacts of fossil fuel use will inevitably lead to greater interest in alternative sources of energy. Further concerns arise from issues of national security and the need to import oil from politically unstable regions. In the meantime, as dwindling supplies of fossil fuels drive prices up, more attention will be paid to enhancing **energy efficiency**, the amount of energy consumed per unit of productive output. The energy efficiency of technologies used in industrialized countries has been increasing (less energy is being used to accomplish the same tasks) over the past decade or so.

Fossil fuels are nonrenewable and therefore finite. What happens when we reach the limit? We can hope that the demand for energy will level off as a result of increased energy efficiency and conservation. Realistically, however, it is unlikely that the demand for fossil fuels will decrease substantially or that another fuel source will replace oil in the near future. For the next few decades, the known deposits of natural gas and oil will likely be sufficient to meet demand. As supplies dwindle, however, prices will increase. Only coal—the most abundant fossil fuel but also the most environmentally damaging—may be in sufficient supply to meet the world's demands beyond the middle of the twenty-first century. Therefore, it makes sense to look for alternative renewable or inexhaustible sources of energy to fill a greater proportion of our energy needs.

Summarize the strategic, economic, and environmental concerns associated with the use of fossil fuels. ________________

Answer: Strategic concerns: heavy reliance on oil imported from politically unstable regions. Economic concerns: rising prices driven by a variety of market forces including dwindling supply. Environmental concerns: impacts of extraction including strip mining, mine collapse, black lung disease, acid mine drainage, and various effluents associated with oil drilling; transport including oil spills; and combustion (particulate matter, which causes respiratory problems; CO_2, which is implicated in global climatic change; and SO_2, which causes acid precipitation).

Unconventional Hydrocarbons and New Hydrocarbon Technologies

Hydrocarbon resources other than oil, natural gas, and coal are called unconventional hydrocarbons, or synthetic fuels, or **synfuels** (even if they are naturally occurring). In this section we will look at synfuels, as well as new hydrocarbon processing technologies. It is tempting to call synfuels and new hydrocarbon technologies alternative energy sources. However, they are not truly alternatives in the sense that they represent an extension rather than a departure from our existing dependency on fossil fuels.

Tar sands are deposits of dense, thick, asphaltlike oil that cannot be pumped easily. Tar is found in a variety of sedimentary rocks and unconsolidated sediments (not just sand). Tar sands may be petroleum deposits in which the volatile components have migrated away, leaving behind a residual, tarry material. Alternatively, they may be immature deposits in which the chemical alterations that form liquid and gaseous hydrocarbons have not yet been completed. The largest known occurrence is in Alberta, Canada, where the Athabasca Tar Sand covers an area of 5,000 km^2 (almost 2,000 mi^2) and reaches a thickness of 60 m (almost 200 ft). The Athabasca deposit may contain as much as 600 billion barrels of oil from tar.

When organic material is buried, compacted, and cemented in very fine-grained sedimentary rocks, a waxlike compound called **kerogen** may be formed if burial temperatures are not high enough to initiate the chemical breakdowns that lead to the formation of oil and natural gas. If kerogen is heated, it breaks down into liquid and gaseous hydrocarbons similar to those in oil and gas. All fine-grained sedimentary rocks, such as shale, typically contain some kerogen. To be considered an energy resource, however, the kerogen in an **oil shale** must yield more energy than is required to mine and heat it. The world's largest deposit of oil shale is located in the United States in the Green River Oil Shale deposits of Colorado, Wyoming, and Utah. These deposits could ultimately yield about 2 trillion barrels of oil.

Gas hydrates are another unconventional hydrocarbon resource that has yet to be widely exploited. These deposits of methane (CH_4) are frozen into ice in permafrost and in seafloor sediments. The amount of methane trapped in gas hydrates may be greater than all known deposits of oil and natural gas. To date, however, it is still economically unfeasible to exploit methane gas hydrates. There is some concern that global climatic change may lead to the warming of ocean water, which in turn would release gas hydrates that are currently frozen in seafloor sediments. This could result in a positive feedback cycle, leading to more greenhouse warming because methane is a highly effective greenhouse gas.

In addition to these naturally occurring synfuels, new hydrocarbon processing technologies are being developed that may extend existing hydrocarbon resources or limit the negative environmental impacts of their use. Among these is fluidized-bed combustion, in which coal is burned more efficiently but at lower temperatures than in a conventional power plant. The more efficient burning produces more heat and less CO_2. If carried out in a pressurized environment, the technology can also remove SO_2 and NO_x from the emissions.

Other technologies have been developed to turn solid coal into a liquid fuel that is similar to oil and less polluting than coal; this is called **coal liquefaction**. Technologies that use oil could be retrofitted to use liquefied coal relatively easily. Other technologies can produce methane gas from coal. An advantage of **coal gasification** is that impurities, notably sulfur compounds, can be removed in the precombustion stage, eliminating the need for scrubbers.

Where are the world's largest known deposits of tar sands and oil shales?

Answer: The Athabasca Tar Sand deposits in Alberta, Canada, and the Green River Oil Shales in Colorado, Wyoming, and Utah.

Alternatives to Fossil Fuels

Consider the power sources in your home or office. Chances are that you have a pipeline through which natural gas is delivered to the building, or perhaps a truck comes periodically to fill your oil tank. These fuels are often used in homes and offices to run appliances such as furnaces, air conditioners, and stoves. Other appliances use batteries, a form of electrochemical energy. You also have a line that joins your home or office building to an extensive grid system through which electricity is distributed. Typically, a utility company owns and maintains the power plants that generate the electricity, and another company distributes it to homes, offices, and factories through the **power grid**. The power plants that generate the electricity are coal-fired, natural gas, hydroelectric, or nuclear generators.

A growing trend among environmentalists is to find ways of living "off the grid"—that is, to generate some or all of their required electricity on-site using alternative energy sources rather than importing electricity from the grid. In some areas, homeowners receive financial credit for returning any excess electricity they generate to the grid. It has also become possible through deregulation for some homeowners to decide upon the source of the electricity they purchase; thus, the homeowner may decide to purchase **green power**, which is electricity generated by a nonpolluting technology such as wind energy.

Find out whether the main source of power where you live comes from hydroelectric, nuclear, natural gas, or coal-fired power plants. Do some research to find out if it is possible for you to purchase green power through your electricity provider. Is green power more expensive than electricity from other sources? Why?

A technology that can greatly enhance energy efficiency is **cogeneration**. In cogeneration (or "cogen"), a conventional fuel is used to create steam, which turns a turbine to generate electricity. Normally, the steam is allowed to cool off before being recycled—the heat in the steam is wasted. In cogeneration, after the electricity has been produced, the heat from the steam is used. Cogen is being used successfully and economically, especially for local applications such as heating office buildings.

New hydrocarbon technologies and applications may extend the lifetime or enhance the environmental acceptability of some fossil fuel energy sources. However, there are other energy sources that are true alternatives to fossil fuels. We will begin by considering the various ways in which we can directly utilize energy from the Sun. Then we will examine indirect sources of solar energy, including biomass, wind, and wave energy. We will look at sources of energy that are derived from gravity (tidal and hydroelectric), chemical sources, and Earth's own internal heat. Nuclear energy comes with its own attendant environmental problems; we will consider it separately in the final section.

What is cogeneration? ______________

Answer: A technology in which a conventional fuel is used to produce electricity and the steam from this process is used instead of being wasted.

Alternative Energy: Solar, Wind, Hydrogen, and Biomass

Solar energy reaches Earth from the Sun at a rate more than 10,000 times greater than human power use from all sources combined (see Figure 11.1). Direct solar energy is best suited to supply heat for such applications as home and water heating; this is called **passive solar heating**. In other applications, solar energy is collected, usually by solar panels located on a rooftop, and the heat is stored and distributed by fans or pumps; this is **active solar heating**. The most significant problem associated with the use of solar energy is that insolation varies from place to place and from time to time—it is highest during the summer in the middle of a noncloudy day near the equator. Because peak energy requirements do not always meet these specifications, it is necessary to store the solar-generated power for later use or install a backup power source.

Solar energy can be converted into electricity through **solar thermal electric generation**. Mirrors or lenses concentrate the sunlight, heating a circulating fluid.

The heat from this fluid is used to create steam, which turns a turbine and produces electricity. Solar energy can also be directly converted into electricity through the use of **photovoltaic cells**—thin wafers or films that are treated chemically so that they absorb solar energy, emitting a stream of electrons in response. So far, the cost of photovoltaic cells is high and their efficiency too low for most uses; also, the electricity they produce must be used immediately. Consequently, they are mainly used in small appliances such as solar-powered calculators, watches, and radios. Photovoltaic technology is constantly improving, however, and the cost is decreasing.

Another way to convert solar energy into usable power is to use the electricity from a photovoltaic cell to split water into its component parts (hydrogen and oxygen), a process called electrolysis. Hydrogen released by electrolysis or derived from other sources can be used as a fuel. **Hydrogen fuel** can be stored or transported long distances via pipelines, which is an advantage over other alternative power sources. One of the potential uses of hydrogen is in **fuel cells**, which are similar in some respects to huge batteries. Normal batteries produce power from electrochemical reactions between the chemicals stored inside them. Fuel cells also run on electrochemical reactions, but the chemical fuel (such as hydrogen, although other fuels can be used) is imported into the fuel cell from an external source. Many experts feel that our future energy economy will be largely based on hydrogen and fuel cells (Figure 11.5).

What is a photovoltaic cell? ______________

Answer: A thin wafer of material that is specially treated so that it absorbs solar energy and emits a stream of electrons—that is, electricity—as a result.

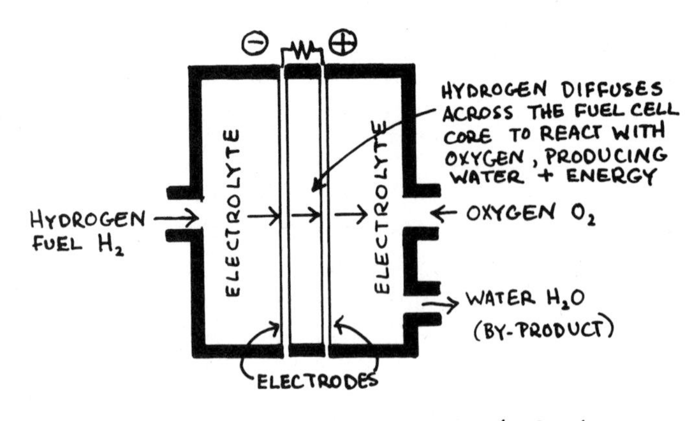

Figure 11.5.

Biomass energy is energy derived from Earth's plant life. In the form of fuelwood, biomass was the dominant source of energy until the end of the nineteenth century, when it was displaced by coal. Biomass fuels, primarily wood, peat, animal dung, and agricultural wastes, are widely used throughout the world, especially in developing countries. Today there are still more than 1 billion people who use fuelwood for cooking and heating.

Organic materials such as animal dung can be burned or converted into methane for power generation. This converted material is called **biogas**. It has become an important fuel in developing countries, particularly China and India, where it is used for cooking and lighting purposes. Methane gas from the breakdown of organic garbage is also collected from some landfill sites. Biomass can be converted into a variety of liquid fuels, notably the **alcohol fuels** methanol and ethanol. This is most commonly done using crop residues. For example, Brazil has an extensive program to produce alcohol fuel from sugarcane residues, which would otherwise be agricultural waste.

Biomass fuels are advantageous because they are widely available. This is particularly important in developing countries where other sources of energy may be financially inaccessible. Biomass energy also makes use of materials that might otherwise go to waste such as agricultural cuttings. A disadvantage, however, is that the removal of organic wastes diverts valuable nutrients from being recycled into the soil. The use of some types of biomass energy, notably fuelwood, has been associated with deforestation, soil depletion, and resulting land degradation.

Wind energy is another indirect expression of solar energy because energy from the Sun is what causes the wind to blow. For thousands of years, wind has been used as a source of power. Today huge windmill "farms" are being erected in suitably windy places. In Denmark, where tax incentives encourage the development of alternative energy sources, about 6,000 wind turbines supply electricity throughout the country. Although there are some problems with windmill technologies, it is likely that windmills will soon be cost-competitive with coal-burning power plants in some localities. Unfortunately, steady surface winds can provide only about 10 percent of the amount of energy now used by humans. Therefore, wind power may be locally important but probably will not become globally significant.

Waves are created by winds; therefore, **wave energy** is another indirect expression of solar energy. Waves contain an enormous amount of energy, but so far no one has discovered how to tap them as a source of power on a large scale. Locally, wave power stations produce electricity using hollow, tubelike chambers. As a wave rises and crests, it pushes air up into the tube. In turn, the air spins a turbine, which generates electricity. Wave power stations have been plagued by problems related to exposure to weather and the battering power of waves.

What materials can be used as biomass fuels? ________________

Answer: Almost any organic material, notably wood, animal dung, peat, agricultural wastes, organic materials from municipal garbage, and the biogas and alcohol fuels that are derived from them.

Alternative Energy: Gravitational and Geothermal Energy

Hydroelectric power is generated from the energy of a stream of water flowing downhill; thus, it is primarily gravitational energy. Hydroelectric power is the only form of water-derived power that currently fulfills a significant portion of the world's energy needs. To convert the power of flowing water into electricity, it is necessary to access a waterfall or build a dam. The flowing water is used to run turbines, which convert the energy into electricity. Hydroelectric power is very important for countries with large rivers and suitable dam sites such as Canada. However, the total recoverable energy from the water flowing in all of the world's streams has

This is the Manantali Hydroelectric Dam on the Bafing River in Mali, West Africa. The controversial dam required the relocation of a number of villages.

been estimated to be equivalent to the energy obtained by burning 15 billion barrels of oil per year. Thus, even if *all* of the potential hydroelectric power in the world were developed, it could not satisfy today's energy needs (currently about 30 billion barrels of oil equivalent per year).

Hydroelectric power is a "clean" source of energy because it has no damaging atmospheric emissions. However, there are many negative environmental side effects associated with large hydroelectric dams, including deforestation and the loss of natural habitat. The transformation of a dynamic stream into a large body of standing water causes problems, too, such as creating breeding grounds for mosquitoes and other disease-bearing insects. Downstream areas are starved for sediments and mineral nutrients once delivered by rushing streams. Reservoirs fill with silt, limiting the productive lifetime of the dam. In some countries there have been significant problems associated with the loss of farmland and cultural or historic artifacts, and the displacement of people—entire towns, in some cases—to make way for the filling of major reservoirs.

Tidal energy is another water- and gravity-related power source. The energy in tides comes from the rotation of Earth and its gravitational interaction with the Moon. The usefulness of tidal energy in any particular location depends on the configuration of the coastline. Most advantageous are coastlines with long, narrow bays and a large tidal range (the difference between water levels at low tide and high tide). A dam is constructed across the mouth of the narrow bay. With the gates open, water flows in during high tide. The gates are then closed and the water is trapped behind the dam. The water is released at low tide, driving a turbine as it rushes out. The use of tidal energy is not new; a mill in Britain dating from 1170 still runs on tidal power. Like wind power, tidal power is insufficient to satisfy more than a small fraction of human energy needs and thus can only be locally important.

Geothermal energy, Earth's internal energy, is used commercially in a number of countries, including New Zealand, Italy, Iceland, and the United States. People in Iceland use water warmed by hot volcanic rocks to heat their houses, grow plants in hothouses, and swim in naturally heated pools. Icelanders also use volcanically produced steam to generate most of their electricity. The most easily exploited geothermal deposits are hydrothermal reservoirs—underground systems of circulating hot water and/or steam in fractured or porous rocks near the surface. Most hydrothermal reservoirs are close to the margins of tectonic plates where recent volcanic activity has occurred and hot rocks or magma are close to the surface. To be used efficiently for geothermal power, hydrothermal reservoirs must be 200°C (almost 400°F) or hotter, and this temperature must be reached within 3 km (less than 2 mi) of the surface. The Geysers in California is the largest producer of geothermal power in the world. In principle, geothermal energy is inexhaustible. However, intensive exploitation of a particular reservoir can lead to local depletion to the point where the reservoir is no longer useful.

What are the negative environmental impacts associated with large hydroelectric dams? ______________

Answer: Deforestation and loss of natural habitat; creation of breeding grounds for mosquitoes and other disease-bearing insects; sediment starvation of downstream areas; reservoir siltation behind the dam; loss of farmland and cultural and historic artifacts; and displacement of people.

Nuclear Energy and the Environment

Nuclear energy comes from the heat energy produced during the induced transformation of a chemical element into other chemical elements. It can be generated in two ways: by inducing a heavy atom to split into lighter atoms, or by causing two light atoms to combine to make a heavier atom.

Splitting heavy atoms into lighter atoms is called **fission**. Fission is induced by bombarding fissionable atoms with **neutrons**, which are electronically neutral particles from the nucleus of an atom (Figure 11.6). When a fissionable atom is hit by

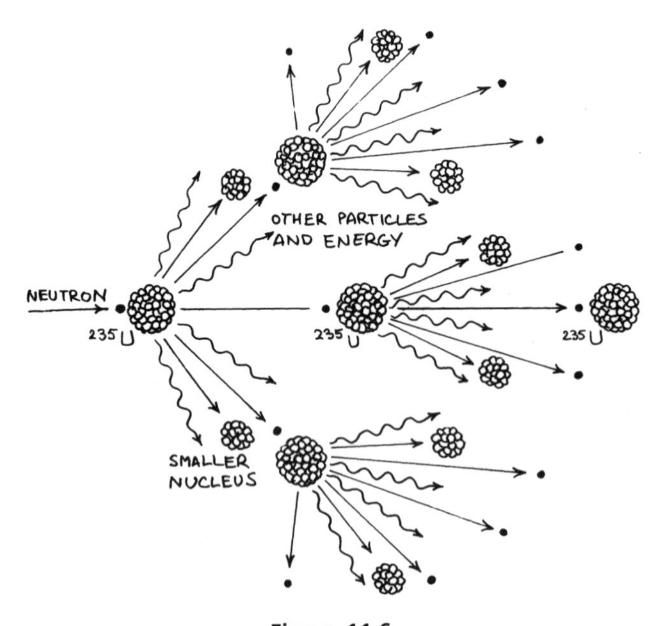

Figure 11.6.

a neutron, it splits into two different atoms, both lighter than the original. Some matter is "lost" during the splitting process. However, because of the Law of Conservation of Matter and Energy (which tells us that matter and energy can be neither created nor destroyed, only transformed), this matter is not actually lost; it is transformed into energy. Thus, when the atom splits, it creates two new (lighter) atoms and releases the leftover energy. The split atom also ejects some neutrons from its own nucleus. These neutrons can be used to induce more atoms to split, creating a chain reaction. The entire process is carried out inside a **nuclear reactor**. The rate of neutron bombardment is controlled, or moderated, usually by water. When a chain reaction is allowed to proceed without control, an atomic explosion occurs.

Uranium-235 is a naturally occurring fissionable material that is mined and used as fuel in nuclear reactors (in addition to other fissionable materials). The fissioning of just 1 gram of ^{235}U produces as much heat as the burning of 13.7 barrels of oil. The ^{235}U is processed and concentrated into fuel pellets, which are packed into a bundle of hollow tubes called **fuel rods**. The fuel rods are loaded into the core of the reactor, where the fission process is induced. The heat generated by fission is carried away by water, which also moderates the chain reaction. The heated water makes steam, which turns a turbine, producing electricity. If the heat were not removed from the fuel bundle, it would get so hot that the reactor core would melt, releasing its radioactive contents; this is called a **meltdown**, and it happened at Chernobyl in Ukraine (former Soviet Union) in 1986. Current nuclear reactor technologies are designed to minimize or eliminate the possibility of a meltdown.

What is a nuclear meltdown? _______________

Answer: A situation in which heat is not removed from the fuel bundle in a reactor core, resulting in an increase in temperature, melting of the reactor core, and the release of its radioactive contents.

Nuclear power is considered to be a clean source of energy—even by some environmentalists—because it causes no harmful atmospheric emissions. Approximately 17 percent of the world's electricity is derived from nuclear power plants. In France, more than half of all electric power comes from nuclear plants; the proportion is rising sharply in some other European countries and in Japan. The reason for the increase is that Japan and most European countries do not have adequate supplies of fossil fuels to be self-sufficient. In North America, however, the growth of the nuclear industry appears to have stalled; this is partly a result of negative public opinion stemming from nuclear catastrophes like Chernobyl and from the intractable problems associated with nuclear waste disposal.

In normal operation, nuclear power plants generate very little environmental radiation; we are actually exposed to far more radiation per year from natural sources (cosmic and terrestrial), medical sources, and even from internal sources within our own bodies than from nuclear reactors. However, nuclear fission generates highly radioactive by-products—**nuclear waste**—which must be isolated from the biosphere and the hydrosphere. This presents a technically difficult disposal problem that has not yet been resolved. Radioactive waste, or **radwaste**, consists of leftover radioactive materials or equipment from nuclear power plants, as well as from laboratories and medical procedures, uranium mining, and the decommissioning of nuclear weapons.

Radioactivity refers to energy and energetic particles (alpha, beta, and gamma radiation) that are emitted during the natural transformation of one element into another element. This process is called **radioactive decay**. Radioactivity can damage organisms, even causing death if the dose is high enough. It also causes genetic damage, which can result in damage to the organism's offspring. Radioactive elements can be toxic in minute quantities, and they can be extremely persistent, remaining radioactive for many thousands of years in some cases. Furthermore, all organisms—including humans—lack any kind of natural warning for the presence of radioactivity at any level; we can't smell it, see it, taste it, or feel it.

The decay of radioactive materials is measured in units of **half-life**, the amount of time it takes for the level of radioactivity in a material to decrease by half (that is, to 50 percent of the original level) (Figure 11.7). Thus, in two half-lives the radioactivity in the material will have decreased to 25 percent of its original level; in three

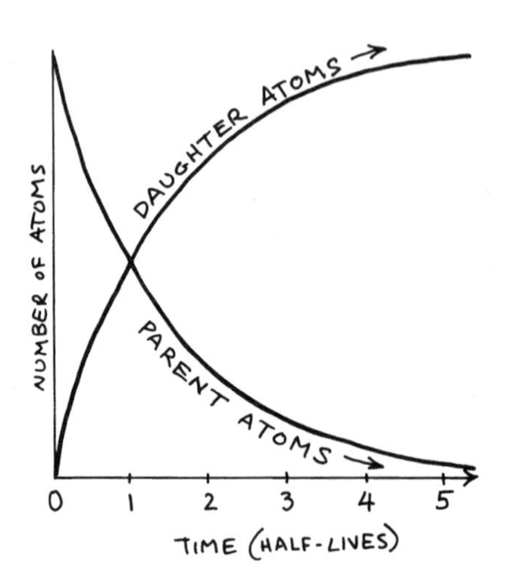

Figure 11.7.

half-lives, it will have decreased to 12.5 percent of its original level; and so on. Some radioactive materials have half-lives measured in seconds or fractions of a second. Others have half-lives measured in days, years, thousands of years, even millions of years.

Low-level radioactive wastes, which emit little radiation and have short half-lives, are contained and then released in a controlled manner once their radioactivity has dissipated. Much more problematic are the high-level radioactive wastes contained in spent fuel rods, which comprise only 1 percent of the waste generated by nuclear reactors but generate 99 percent of the radioactivity. So far, there are no permanent repositories for high-level radioactive waste. Most high-level wastes are stored temporarily in concrete bunkers or water pools (because water dissipates the heat, as well as absorbing the radioactivity). For the past two decades or more, there has been an intensive search for appropriate disposal sites and designs for permanent repositories for high-level nuclear waste.

All permanent disposal options for high-level waste involve concentration and immobilization of the waste, usually in the form of inert glass or ceramic pellets; containment in specially engineered canisters; and subsequent isolation from the biosphere and the hydrosphere. An appropriate disposal site must also be isolated from potential resources and human activities such as mining. It must be engineered so that it cannot be easily entered, damaged, or sabotaged. It must be economically feasible and capable of holding a large quantity of waste for a long time—10,000 years is commonly proposed. Ideally, disposal will not require transportation over long distances because the transport of hazardous materials is risky. Finally, the medium that surrounds the waste must be geologically stable, nonfractured (because fractures provide pathways along which escaping contaminants could travel), and provide good heat conductivity (to dissipate the heat generated by the radioactive materials) and excellent chemical absorption (Figure 11.8).

It's a tall order, but many scientists believe it is technically possible to find an appropriate site and design a permanent disposal facility for nuclear waste that will meet these criteria. Options that have been explored include sending the waste to the Sun on a rocket ship; dropping it onto the polar ice cap, where it will melt its way to the bottom of the ice sheet; or depositing canisters of nuclear waste on the ocean floor to be carried down into the mantle along with a subducting lithospheric plate. All of these seemingly wild schemes have some scientific merits, along with their obvious shortcomings. However, all serious options for the permanent disposal of high-level nuclear waste involve land-based **geologic isolation**—that is, using the properties of natural rocks to isolate and contain the material. This is combined with a **multiple barrier concept**, in which the repository is engineered to place many physical and chemical barriers in the way of any escaping contaminants.

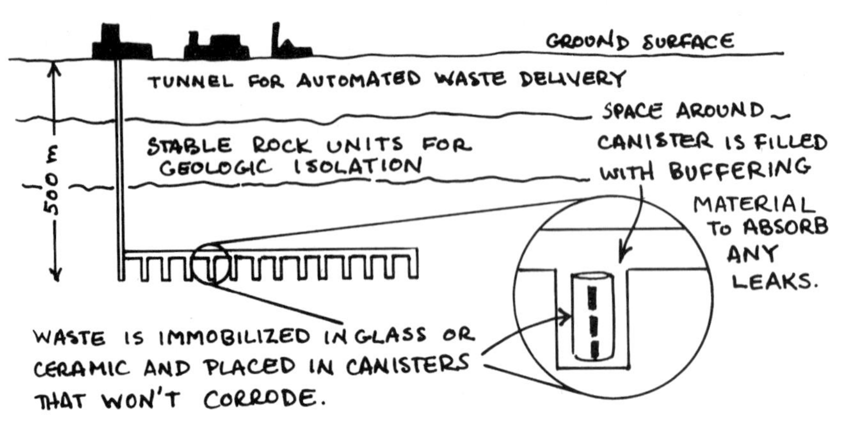

Figure 11.8.

Among the proposed sites for land-based geologic isolation are old salt mines. Salt does not fracture easily; instead, it tends to bend or flow under stress. There is also a great deal of confidence that materials contained in salt will be isolated from the hydrosphere; if water were circulating, the salt would have dissolved long ago. Deep-well injection into shale has also been proposed. This is based on scientific studies of a naturally occurring nuclear fission site in Gabon, West Africa, where shale layers have successfully contained fission by-products for at least a billion years. Canada's model involves permanent disposal of high-level waste in an underground facility deep in the stable granitic rock of the Canadian Shield. The U.S. plan calls for the permanent disposal of high-level radioactive wastes in dense volcanic rock at Yucca Mountain, Nevada. This proposal has been plagued by controversy, stemming partly from the fact that Yucca Mountain has been volcanically active in the geologic past. Transportation of radioactive wastes to the site, the presence of nearby mineral resources, and fluctuating groundwater levels have also raised questions. In spite of these concerns, the site was approved in 2002 by Congress as the first permanent repository for nuclear waste in the United States.

In principle, nuclear **fusion**—the joining together or fusing of two small atoms to create a single larger atom, with attendant release of heat energy—is another potential source of nuclear power. Nuclear fusion utilizes as its fuel a "heavy" form of hydrogen called deuterium. Earth has a virtually endless supply of deuterium in the form of a very common chemical compound—water (hydrogen dioxide, H_2O). The primary by-product of nuclear fusion would be helium, a nontoxic, chemically inert gas.

This conjures up images of a cheap, clean, virtually inexhaustible power source. So why are we not using energy provided by nuclear fusion? Fusion is the nuclear process that occurs in the cores of stars—the process responsible for the tremendous heat energy generated by the Sun. But that, in a nutshell, is the problem. For two atomic nuclei to fuse, the ambient conditions must be similar to those at the core of a star—on the order of 100 million degrees. The possibility that nuclei could be induced to fuse at room temperature (so-called cold fusion) has led many scientists to search for this holy grail of energy. In 1989, a furor arose when two researchers announced that they had achieved cold fusion in a test tube. Unfortunately, their work was not substantiated in other laboratories, and the routine use of fusion power remains a distant goal.

What is the difference between fission and fusion? _______________

Answer: Fission: splitting a heavy atom into lighter atoms. Fusion: combining two light atoms to make a heavier atom.

SELF-TEST

These questions are designed to help you assess how well you have learned the concepts presented in chapter 11. The answers are given at the end. If you get any of the questions wrong, be sure to troubleshoot by going back into that part of the chapter to find the correct answer.

1. Which one of the following is not a power source that is derived primarily from gravitational energy?
   a. wave energy
   b. tidal energy
   c. hydroelectric power
   d. All of the above are derived primarily from gravitational energy.

2. Fission is induced by bombarding fissionable atoms with _______________.
   a. electrons
   b. neutrons
   c. radioactivity
   d. radwaste

3. The world's largest known occurrence of tar sand is in _______________.
   a. the United States
   b. Canada
   c. China
   d. Iceland

4. Earth's energy comes from three sources: ______________, ______________, and ______________.
5. ______________ is a waxlike organic substance commonly found in sedimentary rocks.
6. Solar energy can be converted directly into electricity using thin wafers or films called ______________.
7. Wind energy is becoming cost-competitive with fossil fuels and may soon be meeting more than half of the world's required energy needs. (True or False)
8. There are several permanent repositories for high-level nuclear waste in North America. (True or False)
9. Wind energy is an expression of solar energy. (True or False)
10. How do wave power stations produce electricity?
11. What is the difference between fossil fuels, petroleum, and hydrocarbons?
12. Give three examples of places where geothermal energy is now exploited commercially.
13. What are some examples of modern environments in which peat is forming today?
14. Give examples of renewable, nonrenewable, and inexhaustible energy sources.

ANSWERS

1. a 2. b 3. b
4. solar radiation; geothermal energy; tidal energy (or) the Sun; Earth's internal energy; tidal or gravitational interactions among Earth, the Sun, and the Moon
5. Kerogen 6. photovoltaic cells
7. False 8. False 9. True
10. Wave power stations produce electricity using hollow, tubelike chambers. As a wave rises and crests, it pushes air up into the tube. In turn, the air spins a turbine, which generates electricity.
11. Fossil fuels are the altered remains of organic matter that has been trapped in sediment or sedimentary rock; the principal fossil fuels are coal, oil, and natural gas. Petroleum includes oil and natural gas (liquid, gaseous, and semisolid naturally occurring hydrocarbons). Hydrocarbons are compounds that consist principally of hydrogen and carbon.

12. Examples discussed in the text are New Zealand, Italy, United States (California), and Iceland; in general, geothermal deposits are located near active plate boundaries and regions of recent volcanic activity.

13. Peat is forming today in large swamps such as the Okefenokee Swamp in Florida and the Great Dismal Swamp in Virginia and North Carolina.

14. Renewable: biomass (wood, animal dung, agricultural waste, biogas, and so on). Nonrenewable: fossil fuels, nuclear energy. Inexhaustible: tidal, wind, solar, wave, geothermal, hydroelectric.

KEY WORDS

active solar heating
alcohol fuel
biogas
biomass energy
coal
coal gasification
coal liquefaction
cogeneration
crude oil
energy cycle
energy efficiency
fission
fossil fuel
fuel cell
fuel rod
fusion
gas hydrate
geologic isolation
geothermal energy
green power
half-life
hydrocarbon
hydroelectric power
hydrogen fuel
kerogen
maturation
meltdown
multiple barrier concept
natural gas
neutron
nuclear energy
nuclear reactor
nuclear waste
oil
oil shale
passive solar heating
peat
petroleum
photovoltaic cell
power grid
radioactive decay
radioactivity
radwaste
solar energy
solar thermal electric generation
synfuel
tar sand
tidal energy
wave energy
wind energy

12 Water Resources

Water, water everywhere,
Nor any drop to drink.

—Samuel Taylor Coleridge

Objectives

In this chapter you will learn about:

- surface water and groundwater;
- flooding and channel modifications;
- the exploitation and management of water resources; and
- the environmental impacts of water use.

Surface Water Processes

Surface freshwater bodies—streams, rivers, lakes, ponds, and wetlands—hold only a tiny fraction of the water in the hydrosphere, but they are our most accessible source of fresh water. People everywhere depend on surface water for drinking, agriculture, transportation, and industrial use. Freshwater ecosystems provide habitats for plants and animals, and they are important recreation sites and sources of natural beauty. As a consequence of their accessibility and utility, however, surface freshwater bodies are highly susceptible to contamination.

Surface runoff begins as sheets of water flowing overland or in small, temporary rills and gullies that form during periods of rainfall and snowmelt. If sufficient water is available, rills and gullies may grow and deepen, carving out stream channels by erosion. Streams that carry water all year, even during dry spells, are called **perennial** streams; channels that become dry between major rainstorms are called **ephemeral** streams. Ephemeral streams are fed primarily by precipitation. Perennial streams carry overland flow from precipitation, but they are also continuously replenished by groundwater discharge.

Every stream is surrounded by its **drainage basin**, the total land area from which water drains into the stream (Figure 12.1). In general, the greater a stream's annual discharge, the larger its drainage basin. For example, the drainage basin of the Mississippi River encompasses more than 40 percent of the total area of the contiguous United States. The boundary between adjacent drainage basins is called a **divide**. Divides are topographic "highs" that separate basins from one another. On continents, great mountain chains separate streams that drain toward one side of the continent from streams that drain toward the other side. The continental divide of North America lies along the length of the Rocky Mountains. Streams to the east ultimately drain into the Atlantic Ocean, while those to the west drain into the Pacific Ocean.

Lakes, ponds, and wetlands are surface water bodies with no appreciable surface gradient—that is, they hold standing water rather than flowing water. The scientific study of lakes and other inland bodies of water is called **limnology** (from the Greek word *limne*, meaning "lake" or "pond"). Lakes can appear and disappear with changes in climate. In moist climates, the water level of lakes and ponds coincides with the local water table. Seepage of groundwater into the lake, combined with

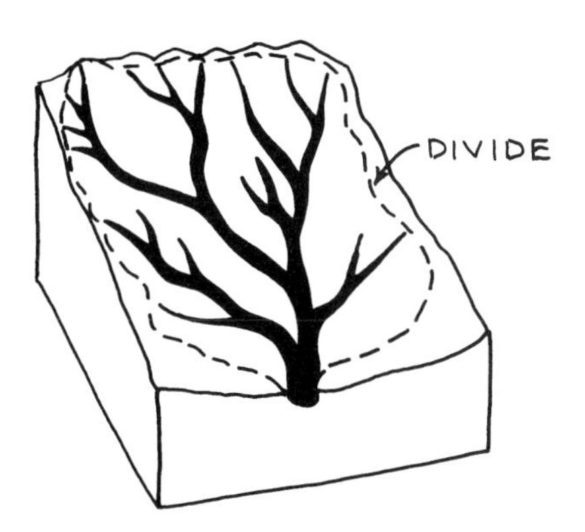

Figure 12.1.

runoff and precipitation, maintains these water surfaces throughout the year. If temperature increases or precipitation decreases (or both), evaporation may exceed input, and the lake will shrink. In regions where climatic conditions consistently favor evaporation over precipitation, lake beds may be dry or only intermittently filled with water. Streams bring dissolved solids—salts—into these ephemeral lakes. Since evaporation removes only pure water, the salts remain behind, and salinity increases. California's Mono Lake is a well-known example of an inland saltwater body; Great Salt Lake in Utah is another.

A young pond or lake that has clear water, with little organic matter or sediment, is said to be an **oligotrophic** water body. Over time, organic matter produced by plants within the lake and inorganic sediments carried into the lake by streams begin to accumulate. Eventually the lake fills in, becoming a nutrient-rich, boggy wetland with little or no standing water. A lake in the later stages of this process is called a **eutrophic** water body. The **eutrophication** process is part of the natural geologic succession of lakes and wetlands.

Wetlands are swamps, bogs, salt marshes, and other shallow depressions that are water-saturated or contain standing water for at least part of the year. Wetlands perform a number of important ecosystem functions. They are storage reservoirs in the hydrologic cycle. They are biologically productive, nutrient-rich environments that provide specialized habitat for many types of fish, aquatic birds, and other animals and plants. Wetlands also act as natural filters, trapping contaminants and removing them from water and sediments. Before their ecosystem functions and aesthetic value were clearly understood, many wetlands were drained for agricultural, industrial, or urban development. It has been estimated, for example, that more than 50 percent of the wetlands in the United States—up to 90 percent of the freshwater wetlands—have been lost to development. Wetlands are still endangered, but they are now somewhat better protected by legislation.

What is the difference between an oligotrophic lake and a eutrophic lake?

Answer: Oligotropic: a young lake with clear, nutrient-poor water. Eutrophic: an older, nutrient-rich and sediment-filled lake on its way to becoming a boggy wetland.

Flooding and Channel Modifications

All natural streams flood from time to time when the **discharge**—the flow of water per unit of time—is greater than the channel can accommodate and water spills out over the banks. The low-lying, flat land along the shoreline that becomes inundated during a flood is called the river's **floodplain**. Floodplains are often well suited for agriculture because they are constructed of soft, nutrient-rich silt and clay deposited periodically by floodwaters.

Flooding is part of the natural process of stream flow. However, floods can be disastrous for people who live along major rivers, causing both loss of life and extensive property damage. The Huang He in China, called the Yellow River because of the yellow-brown color produced by its heavy load of silt, has a long history of catastrophic floods. It is also called "China's Sorrow"; in 1931, a flood on the Huang He killed a staggering 3.7 million people. China is currently in the process of completing construction of the controversial Three Gorges Dam. One of its main goals is to control flooding on the Yangtze, another great river in China. Three Gorges Dam is the largest and costliest dam ever built—more than $25 billion, in addition to anticipated environmental, social, and cultural-historical losses.

Because floods can be so damaging, much harm can be avoided by predicting them. The average time interval between floods of the same or greater magnitude is called the **recurrence interval**. A recurrence interval of 10 years means that there is a 1-in-10 (or 10 percent) chance that a flood of that magnitude will occur in any particular year—a "10-year" flood (Figure 12.2). Short-term prediction, or **forecasting**, specifies the magnitude of the flood's **peak** (its crest, or highest discharge, expressed as water level above a reference point) and the time when it will pass a particular location. Forecasting is based on real-time monitoring of storms combined with geographically referenced information about topography, vegetation, and impermeable ground cover in the area. This information is analyzed to predict the amount of surface runoff from the storm, its velocity, and its probable course, with the goal of issuing early warnings to communities that may be affected. Flood forecasting today, like many other kinds of environmental monitoring, is handled by sophisticated software and **GIS** (**Geographic Information Systems**) that allow many layers of environmental information to be stored, compared, and analyzed quickly and accurately.

Urban development can exacerbate flooding. Urbanization greatly increases the amount of impermeable ground cover through the construction of buildings, roads, and parking lots. This can add substantially to surface runoff because water that is unable to infiltrate the impermeable ground must flow along the surface. Storm sewers allow runoff from paved areas to reach stream channels more quickly, further increasing the flow. Floods in urbanized basins have higher peaks and tend to reach their peak more quickly than floods in natural, undeveloped basins. Urban construction on soft sediments can also lead to **subsidence** (lowering of the ground surface). This has been an enormous problem in cities like Bangkok, Venice, New Orleans, and many others.

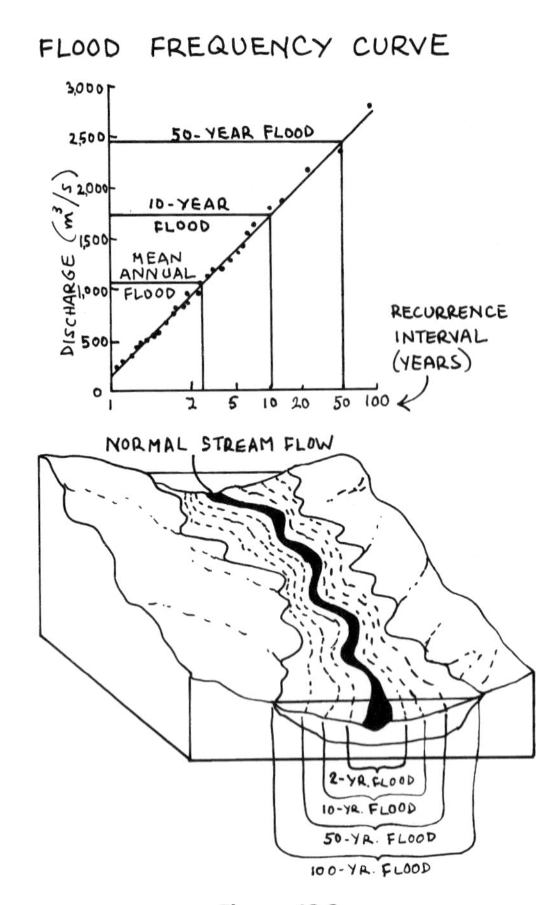

Figure 12.2.

River channels are often modified or "engineered" for flood control, or to increase access to floodplain lands, facilitate transportation, enhance drainage, or control erosion. The modifications usually consist of some combination of straightening, deepening, widening, clearing, and lining the natural channel, which is collectively referred to as **channelization** (Figure 12.3). In the context of flood control, engineering is generally designed to increase the channel's cross-sectional area, enabling the channel to handle a greater discharge of water. Diversion walls, retention ponds, levees, and dikes are other types of engineered structures aimed at flood control. Similar engineering modifications along coastlines are aimed at controlling both erosion and flooding.

River engineering, particularly channelization, is controversial because it interferes with natural ecosystems. The aesthetic value of the river can be degraded, groundwater disrupted, habitat destroyed, and water pollution aggravated. Paradoxically, channelization may control flooding in the immediate area while contributing

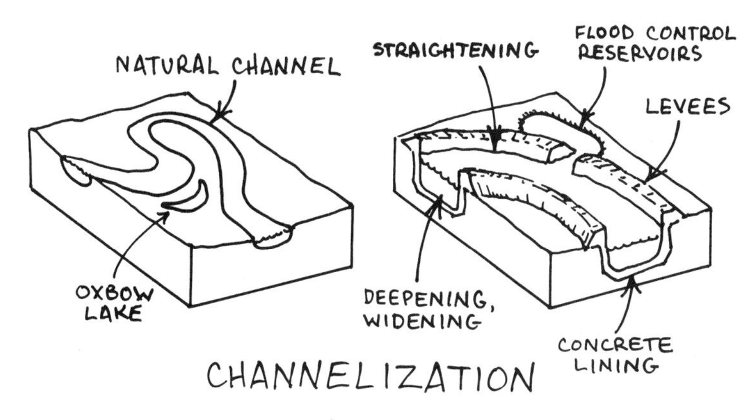

Figure 12.3.

to more intense flooding further downstream. A well-known example of poorly designed channelization is the Kissimmee River in south Florida. The project, which involved straightening, deepening, and the installation of concrete liners, began in 1962 and took nine years and $25 million to complete. The channelization did not provide the anticipated flood control, but it did cause extensive damage to aquatic habitats and water quality. In the 1990s, the state of Florida began an ambitious project to restore and rehabilitate portions of the Kissimmee River channel. In the end, the restoration will cost far more than the original channelization project.

An important part of the management of flood hazards is deciding whether damaged buildings on floodplains should be rebuilt after a major flood. Following the great Mississippi River flood of 1993, **FEMA** (the U.S. **Federal Emergency Management Agency**) modified its rules concerning floodplain development to prevent rebuilding in the most hazardous areas. The 1993 flood killed sixty people and caused approximately $10 billion in damages.

How can urbanization contribute to flooding? ______________

Answer: Construction on soft sediments can lead to subsidence and urban flooding. Impermeable ground cover adds to surface runoff. Storm sewers allow runoff from paved areas to reach river channels quickly, adding to the discharge.

Groundwater Processes

Less than 1 percent of all water in the hydrologic cycle is **groundwater**—subsurface water contained in spaces in bedrock and regolith. Even so, the volume of groundwater is about forty times greater than that of all the water in freshwater lakes or streams. Water is present everywhere beneath Earth's surface, even in hot deserts. It is stored and transmitted by water-saturated rock or regolith units called **aquifers** (from the Latin for "water carrier").

If you dig a well, the hole will ordinarily pass first through a zone in which the spaces in the soil, sediment, or rock are filled mainly with air (Figure 12.4). This is the **aerated** or **unsaturated zone**; although some water may be present, it does not completely saturate the ground. Below the aerated zone is the **saturated zone**, in which all pore spaces are filled with water. The top surface of the saturated zone is the **water table**, the upper limit of readily usable groundwater. The water table tends to imitate the shape of the land surface—it is high beneath hills and lower (deeper) beneath valleys. Where the water table intersects the land, there will be a surface water body such as a stream, a lake, or a spring. If all rainfall were to cease, the water table would slowly flatten; streams, lakes, and wells would dry up as the water table fell.

Water from rainfall or snowmelt soaks into the soil by infiltration (chapter 3) and seeps downward under the influence of gravity until it reaches the water table. This replenishment of groundwater is called **recharge**. Once in the saturated zone, groundwater flows slowly by percolation through constricted passages in sediment and rock. Unlike the swift flow of rivers, which is measured in kilometers per hour, the movement of groundwater is usually measured in centimeters per day or meters per year. Groundwater percolation is controlled by gravity, but

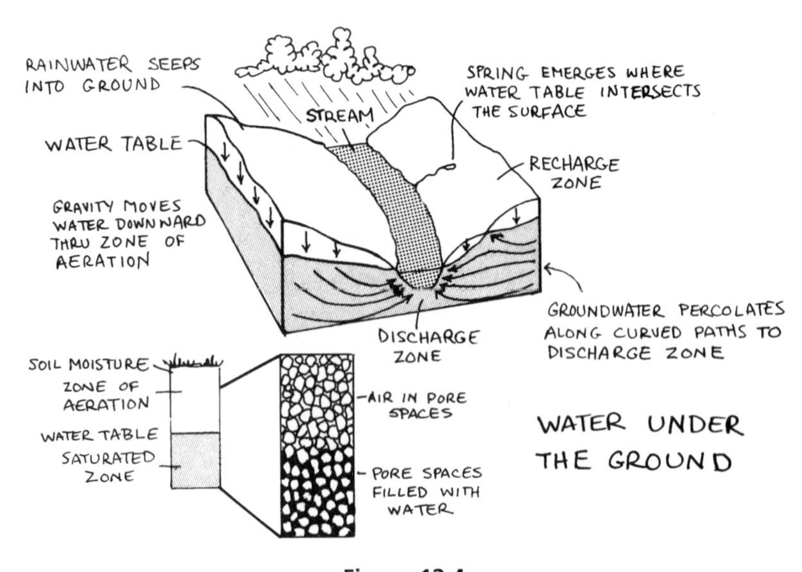

Figure 12.4.

the flow isn't always downhill in the normal sense. Rather, the water flows down the gradient of hydraulic pressure, moving from regions of high hydraulic pressure, or "head," to regions where the pressure is lower. Ultimately, groundwater flows to surface streams, lakes, or the ocean, where **discharge** occurs. (Discharge still refers to volume of flow per unit of time, as above, but the term is used slightly differently for groundwater than for surface water flow.) In a discharge zone, subsurface water becomes surface water by flowing out as a spring or by joining a surface water body.

The amount of time water takes to move through the ground from a recharge zone to a discharge zone depends on the distance and the rate of flow, which in turn depend on the physical characteristics of the rock or sediment. The porosity and permeability of the material are particularly important. **Porosity**, as defined in chapter 9, is the proportion of open spaces, or pores, in a rock or sediment. Porosity determines the amount of fluid the material can hold. The porosity of sediments and sedimentary rocks is affected by the sizes and shapes of the mineral particles; the compactness of their arrangement; and the degree to which the pores have been filled by natural mineral cement. Nonfractured igneous and metamorphic rocks, which consist of interlocking crystals, generally have lower porosities than sediments and sedimentary rocks (Table 12.1).

Permeability determines how easily a fluid can pass through a solid material; it is a measure of the interconnectedness of the pore spaces. A rock with low porosity is also likely to have low permeability. However, high porosity does not necessarily mean high permeability because the size of the pores and the extent to which they are interconnected also influence the ability of fluids to flow through the material. Gravels, sands, sandstones, and limestones generally make good aquifers because they tend to be both porous and permeable (see Table 12.1). A rock or sediment unit with low permeability that slows the transmission of groundwater, such as a shale or mudstone, is called an **aquitard**. A unit that completely blocks the flow of groundwater (and other subsurface fluids), such as a nonfractured igneous rock, is called an **aquiclude**.

In some cases, an aquifer may be bounded above and below by aquicludes. The water in this type of aquifer—a **confined aquifer**—must move within and along

Table 12.1 The Porosity and Permeability of Some Earth Materials

Rock or Sediment Type	Typical Porosity	Typical Permeability
Limestone and gravels	High	High
Sandstone and sandy sediments	High	High
Shale and clay-rich sediments	High	Low
Metamorphic and igneous rocks	Low	Low (unless fractured)

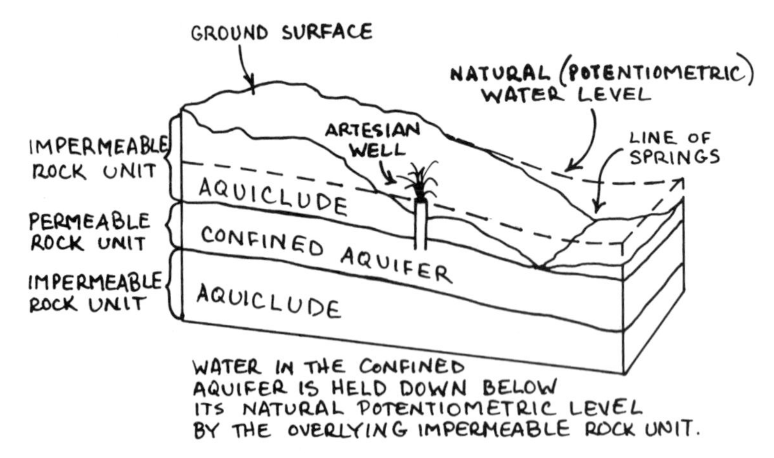

Figure 12.5.

the aquifer because it cannot pass through the unit above or below (Figure 12.5). In contrast, an **unconfined aquifer** is overlain by permeable rock or sediment units. The water in an unconfined aquifer is in contact with the atmosphere through the permeability of the overlying units. The water in a confined aquifer may be depressed by the overlying aquiclude and held below its "natural" (or potentiometric) level. If you drill into such an aquifer, the water will gush toward the surface under its own pressure. This is an **artesian well**. If the water gushes out of a natural vent rather than a drilled well, it is an artesian spring. Many people believe that artesian water has "special" properties; in fact, it is no different from any other groundwater—it just comes from a confined, artesian aquifer.

What is the difference between porosity and permeability? ______________

Answer: Porosity is a measure of the proportion of open pore space in a rock or sediment. Permeability is a measure of the interconnectedness of the pore spaces.

Fresh Water as a Resource

A reliable supply of fresh water is crucial for the survival and health of people and ecosystems and for industry, agriculture, recreation, transportation, and fisheries. Globally, crop irrigation accounts for 73 percent of the demand for water, industry for 21 percent, and domestic use for 6 percent, although the proportions vary greatly from one region to another (Figure 12.6). Global water use has more than tripled since 1950; both population growth and improved standards of living have contributed to this increase. The total amount of water withdrawn or diverted from rivers, lakes, and

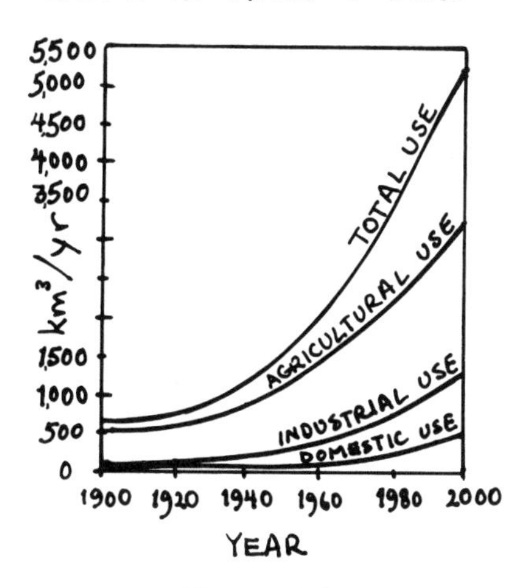

Figure 12.6.

groundwater for human use is now about 4,340 km^3/yr (1,040 mi^3/yr), which is eight times the annual flow of the Mississippi River!

As a result of population growth and development, regions with the greatest demand for water may not have an abundant and readily available supply. For this reason, surface water is often transferred from one drainage basin to another, sometimes over long distances, via canals, aqueducts, or pipelines. Aside from raising issues related to water rights, **interbasin transfer** can have many negative environmental impacts. The diversion of surface water from a stream channel can dramatically diminish the flow and increase the salinity of the water, as well as decrease the amount of sediment carried by the stream. Ecosystems and water users both upstream and downstream of the diversion may be affected; even the local climate can change.

Located on the border between Kazakhstan and Uzbekistan, the Aral Sea provides a cautionary example of the potentially disastrous effects of surface water diversion and interbasin transfer. Once the fourth largest lake in the world, the Aral Sea had a moderating influence on the climate of the region and supported a thriving fishing industry. For many decades, water has been withdrawn for agricultural and industrial purposes from the two major rivers that feed the Aral Sea: the Syr Dar'ya and the Amu Dar'ya. Today the Aral Sea is only the sixth largest lake in the world, and it is disappearing so fast that it is likely to be a desert by 2010. Because the lake is shrinking both in depth and surface area, the local rainfall is declining, average temperature is rising, and wind velocities are increasing. Most of the newly exposed lake bottom is covered with salt, which is picked up by winds and deposited

Removal of vegetation from Western Australia's farming areas has allowed groundwater to rise and salinization to set in.

on surrounding areas. Supplies of potable water have decreased, and intestinal and respiratory diseases are afflicting the local population.

Groundwater is also a major source of water for human use. Although most large public water supply systems draw from surface water sources, more than half of the people in North America get their drinking water from groundwater supplies. Aquifers are replenishable; however, if the rate of groundwater withdrawal exceeds the rate of recharge over time, the volume of stored water steadily decreases. It may take hundreds or even thousands of years for a depleted aquifer to recharge.

The results of excessive groundwater withdrawal include depression of the water table; drying up of springs, streams, and wells; compaction; and subsidence. When an aquifer suffers compaction—when its mineral grains collapse on one another because the pore water that held them apart has been removed—it is permanently damaged and may never be able to hold as much water as it did originally. Urban development also contributes to groundwater depletion not only by increasing the demand for water but also by increasing the amount of impermeable ground cover. In a recharge area that is covered by roads, parking lots, buildings, and sidewalks, the rate of groundwater replenishment is substantially reduced.

It is important to distinguish between water consumption and irretrievable consumption, or loss. Water that is "lost" is not returned to the local hydrologic cycle. Water loss mainly occurs through evaporation. Some water is also lost through leakage from underground pipes. Globally, irrigation is not only the largest consumer of water but also the largest cause of water loss. Much of the water used for irrigation is lost through evaporation and transpiration. Drip irrigation, in which a measured amount of water is delivered directly to the root of each plant, has been developed as a response to this problem. Industrial processes also consume a significant amount of water, but a large proportion of the water is eventually returned to the same surface water body. Water stored behind dams is problematic because of the very large amount of water lost to evaporation from the surface of the reservoir.

Hydrologists designate as **water-stressed** a country or a region with annual renewable water supplies of 1,000 to 2,000 m^3 per person. This refers to water that is available for *all* purposes, including industrial and agricultural use, not just personal use. For comparison, the available water supply in most parts of Canada, a water-rich country, is about 50,000 m^3 per person per year. If the available supply drops below 1,000 m^3 (about 2,700 liters, or 700 gal per person per day), the area is considered to be **water-scarce**—lack of water represents a serious constraint on agricultural production, economic development, environmental protection, and personal nutrition, health, and hygiene.

Currently at least nine countries (Libya, Qatar, United Arab Emirates, Yemen, Jordan, Israel, Saudi Arabia, Kuwait, and Bahrain) are using more than 100 percent of their internally available water. Twenty countries, with a total population of almost 250 million people, are currently designated as water-scarce, with at least ten more countries likely to be added to the category by 2025 (Table 12.2). Most of these are developing countries with rapidly increasing populations, foreshadowing

Table 12.2 Countries Currently Designated as Water-Scarce

Africa	Middle East	Other	Additional by 2010
Algeria	Bahrain	Barbados	Lebanon
Botswana	Israel	Belgium	Malawi
Burundi	Jordan	Hungary	Morocco
Cape Verde	Kuwait	Malta	Niger
Djibouti	Qatar	Netherlands	Oman
Egypt	Saudi Arabia	Singapore	Somalia
Kenya	Syria		South Africa
Libya	United Arab Emirates		Sudan
Mauritania	Yemen		
Rwanda			
Tunisia			

Per capita renewable water resources $<1{,}000 m^3/yr$

ever-increasing water problems. Even if a country is not designated as water-scarce, there may be shortages. The United States is not water-scarce; yet because of mismatched supply and demand, signs of water stress are apparent in many regions. The rates of water use in many parts of the United States far outstrip the rates at which resources are being recharged. In California, for example, there is a water surplus in the northern part of the state, but water scarcity exists in the southern part of the state as a result of the dry climate, population density, and intensive agricultural use. Much of the water used in southern California is imported from northern California or from out-of-state. It is technologically feasible for water-scarce regions to boost their supply of fresh water through desalination of ocean water, but this is a very expensive alternative.

What is interbasin transfer? ______________

Answer: Diversion of water from one drainage basin to another, sometimes over long distances, via canals or pipelines.

Water Law and Water Management

Policies concerning the allocation and regulation of water use can be very controversial; water law, in general, has a long and complicated history. Conflicts over water rights have caused or intensified many international disputes. This continues to be true in areas where water is scarce such as the Middle East. In general, water rights tend to be founded on one of two basic ideas. **Riparian rights** guarantee water access only to those who own or have legally established rights to the **riparian zone**—the land adjacent to the banks of a stream or a lake. The doctrine of **prior appropriation**, in contrast, assigns water rights to those who have established a history of water use from the source whether or not they have legal ownership of riparian lands.

Water distribution rights become much more complicated in cases where the water body crosses a political boundary, which is true of many of the world's great rivers (Table 12.3). Almost half of the world's land area lies in drainage basins shared by two or more countries. One of the most successful international collaborations on the management of trans-boundary surface water issues is the **International Joint Commission** set up by Canada and the United States to facilitate cooperation in the management of the Great Lakes.

The application of law to groundwater is even more complicated than for surface

Table 12.3 Some Examples of Internationally Shared Rivers

River (Countries)
Rhine (Switzerland, Germany, France, Luxembourg, Netherlands)
Indus (India, Pakistan)
Mekong (Vietnam, Laos, Cambodia)
Ganges (India, Bangladesh)
Nile (Egypt, Ethiopia, Sudan)
Jordan (Israel, Jordan, West Bank, Gaza)
Tigris-Euphrates (Syria, Turkey, Iraq)
Danube (Serbia, Romania, Bulgaria, Slovakia, Hungary)
Colorado (United States, Mexico)
Niger (Nigeria, Benin, Niger, Mali)
Paraná (Argentina, Uruguay, Paraguay, Brazil)

Three-fourths of the world's two hundred major watersheds are shared by at least two countries.

water because it is difficult to monitor the flow of groundwater and regulate its use. If you drill a well into an aquifer underlying your property, are you entitled to withdraw as much water as you need from that well? Should you be permitted to withdraw the water and sell it elsewhere? What happens if withdrawing the groundwater depletes the aquifer, causing your neighbor's well to run dry? Most of these issues are regulated locally or regionally (by state in the United States and by province in Canada). Similar problems arise when a private landowner's activities cause an aquifer to become contaminated. In some jurisdictions groundwater contamination is not a legal problem unless the contaminant migrates underground and crosses a boundary into a different property.

What is the purpose of the International Joint Commission? ______________

Answer: To facilitate cooperation between the United States and Canada on the management of the Great Lakes.

Water is essential—it's the lifeblood of our planet. So far, we have considered the natural cycling of water in the atmosphere and hydrosphere (chapter 3); the role of water in supporting life (chapters 4 and 8); water in ecosystems, biomes, and habitats (chapter 5); water as a solvent for contaminants in the mining context (chapter 10); and fresh water as a resource (chapter 12). In the next chapter we will look more closely at pollutants and contaminants, the impacts of pollution on water resources, and the role of water in dissolving and transporting contaminants.

SELF-TEST

These questions are designed to help you assess how well you have learned the concepts presented in chapter 12. The answers are given at the end. If you get any of the questions wrong, be sure to troubleshoot by going back into that part of the chapter to find the correct answer.

1. The top surface of the saturated zone is the ______________.
   a. water table
   b. aquifer
   c. aquiclude
   d. aerated zone

2. The average time between floods of equal or greater magnitude is the ______________.
   a. discharge
   b. recurrence interval
   c. peak
   d. forecast

3. Which one of the following is not a potential environmental impact of surface water diversion?
   a. decrease in flow and amount of sediment carried by the water
   b. increase in salinity of water
   c. changes in local climate
   d. These are all potential environmental impacts of surface water diversion.

4. A stream channel that is continuously filled with water, even in dry spells, and is fed by groundwater, is called a(n) ______________ stream. A stream channel that carries water only intermittently is called a(n) ______________ stream.

5. The study of lakes and other surface water bodies is called ______________.

6. Water consumption is also called irretrievable loss. (True or False)

7. The doctrine of prior appropriation gives water rights to those who own or have legal access to lands in the riparian zone. (True or False)

8. A country or region is said to be water-stressed if the available renewable freshwater supply drops below 10,000 m^3 per person per day. (True or False)

9. Give two definitions for the term *discharge*.

10. Describe how an inland saline lake might form. What are two examples that are mentioned in the text?

11. What is an artesian well? Is artesian spring water different from other groundwater?

12. What is the difference between an aquifer, an aquitard, and an aquiclude?
13. Why is the flow of groundwater so much slower than the flow of water in a stream? What are typical rates of flow for each?

ANSWERS

1. a
2. b
3. d
4. perennial; ephemeral
5. limnology
6. False
7. False
8. False
9. (1) For surface water: discharge is the volume of water flowing past a given point on the stream bank per unit of time. (2) For groundwater: discharge is the flow of water from the subsurface to the surface.
10. If climatic conditions consistently favor evaporation over precipitation, a lake may dry up. Since evaporation only removes fresh water, the water that remains in the lake becomes more and more saline. This can also occur if incoming freshwater streams dry up or their flow is diminished either naturally or by diversion of water from the stream channels. The examples given in the text are Mono Lake, Great Salt Lake, and the Aral Sea.
11. An artesian well is a well drilled into a confined aquifer from which water gushes under its own pressure. There is no difference between artesian spring water and other groundwater.
12. Aquifer: water-saturated porous and permeable rock or regolith unit that can hold and transmit groundwater. Aquitard: a unit of rock or regolith that slows the movement of groundwater, usually because of its low permeability. Aquiclude: an impermeable unit of rock or regolith that completely blocks the movement of groundwater.
13. Water in a stream flows quickly in an open channel, typically measured in km/hr. Groundwater must flow through tiny, constricted passages in subsurface rock or sediment units; flow rates, typically measured in cm/day or m/yr, depend on the porosity and permeability of the subsurface materials.

KEY WORDS

aerated (unsaturated) zone
aquiclude
aquifer
aquitard
artesian well
channelization
confined aquifer
discharge
divide
drainage basin
ephemeral (water body)
eutrophic (water body)
eutrophication
FEMA (Federal Emergency Management Agency)
floodplain
forecasting
GIS (Geographic Information Systems)
groundwater
interbasin transfer
International Joint Commission
limnology
oligotrophic (water body)
peak (of a flood)
perennial (water body)
permeability
porosity
prior appropriation
recharge
recurrence interval
riparian rights
riparian zone
saturated zone
subsidence
unconfined aquifer
water scarcity
water stress
water table
wetland

13 Water Pollution and Soil Pollution

Don't look back. Something or somebody might be gaining on you.

—Satchel Paige

Objectives

In this chapter you will learn about:

- drinking water quality;
- contaminants and their health effects;
- the main sources and types of water and soil pollution; and
- the most effective methods for remediation of contaminated soil and water.

Drinking Water

In North America, we are fortunate to have domestic water supplies that are generally safe and tightly regulated. Even so, the use of home filtration systems and bottled spring water has increased as people have become wary of potential contaminants in tap water. (It is worth noting, though, that spring water and artesian water are essentially the same as other groundwater, and that bottled water, in general, is not as closely monitored as tap water.) Elsewhere in the world, people are not so lucky with their drinking water. About 1.2 billion people, mainly in developing countries, do not have access to a reliable source of clean water.

Tap water is closely monitored and treated in North America, and drinking-water quality standards are high.

Do you recall the criteria under which a region is designated as water-scarce? Check chapter 12 for the answer.

The average daily water use in the United States is almost 6,000 liters (1,600 gal) per person for all purposes combined; in Canada, about 3,200 liters (850 gal) per person per day. Of this, 10 percent goes to residential use and the remaining 90 percent to industrial and agricultural use. The water that comes to our homes and offices through the tap typically is drawn from relatively clean surface water bodies or groundwater aquifers. However, water quality guidelines are defined differently for different water uses. Water for domestic use, in general, must follow drinking water standards, whereas the guidelines for other activities, such as swimming, may be less stringent.

In some areas, especially where groundwater is in use, water for domestic use may require a treatment for hardness, a high content of ions (especially calcium and magnesium) dissolved from minerals in the aquifer. Most municipalities add fluoride to the water supply (**fluoridation**), which reduces the development of dental caries (cavities). If water is saline or briny, **desalination** can be used to remove the salt from the water and render it drinkable. Desalination is energy intensive and very expensive, but it is worth the cost in some countries where water is particularly scarce.

Most importantly, water for domestic use must be filtered to remove sediments and treated by **chlorination** to kill harmful microorganisms. Sewage, with the infectious agents it may carry, can find its way into surface water bodies, groundwater wells, and ultimately to people via a wide range of environmental processes. In some water-scarce areas, previously contaminated water or treated sewage (called "gray" water) is used for purposes such as watering of parks and boulevards, aquaculture, or irrigation. It is scientifically possible to take water with almost *any* contaminant and clean it sufficiently for personal use; however, there are strong cultural taboos against the use of previously contaminated water for personal consumption.

Where does your water come from? Is it groundwater, or does it come from a surface freshwater body? Is it piped in from a great distance, or is water locally abundant?

How do scientists and regulators decide on the maximum permissible levels of salts, contaminants, and other components in drinking water? Taste, smell, and clarity are obvious criteria, but they are not sufficient indicators of the safety of water for human consumption. There are thousands of potentially harmful contaminants that may enter a water supply at any number of points in the hydrologic cycle. Ensuring safety involves determining the health effects of these substances, then setting and enforcing medically, technologically, and socially acceptable and achievable limits. In this chapter we will look at some of the great variety of water and soil contaminants; the processes by which contaminants move, interact, persist, and degrade; and their potential impacts on people, organisms, and the natural environment.

How (and why) is tap water typically treated before it is delivered for domestic use?

____

Answer: Filtered to remove sediments; chlorination to kill harmful microorganisms; fluoridation to prevent dental caries.

The Contaminants in Environmental Reservoirs

A **toxin**, or toxic substance, is a material that is harmful to organisms. An important aspect of environmental science is to understand the pathways taken by toxic substances through the environment and into organisms, including people. Some substances remain in the natural environment for a long time—that is, they have long **residence times** in environmental reservoirs. Recall (from chapter 2) that the residence time of a substance in a particular reservoir depends on a number of factors, including:

- the chemical and physical characteristics of the substance itself;
- the nature of the medium through which or by which it is being transported;
- environmental conditions; and
- the presence of other substances.

The chemical and physical characteristics of a substance are important determinants of its behavior and stability and thus its residence time in environmental reservoirs. For example, coarse, heavy particles tend to have shorter residence times in the atmospheric reservoir than gases or extremely fine particles, which can be maintained aloft more easily.

Another characteristic that influences the residence time of a contaminant is its chemical and/or physical stability. If a substance breaks down, alters, decays, or reacts readily, it will tend to disappear more quickly than a more stable substance. Substances are broken down or transformed by many processes in the environment, including **biodegradation**, the biologic decay of materials; chemical and biochemical reactions; physical disintegration; and radioactive decay (discussed in chapter 11). Substances that take a long time to break down or degrade are **persistent**. A well-known persistent toxin is the insecticide DDT (dichloro-diphenyl-trichloroethane), an organochloride compound that was widely used for pest control in the 1960s. Although its use was banned in North America in the late 1970s, traces of DDT are still detectable in the environment, even in breast milk. (The publication in 1962 of the book *Silent Spring* by Rachel Carson is often cited as one of the landmark events in the modern environmental movement. The book detailed the harmful effects of DDT and other persistent chemicals on songbirds. It was a best seller.)

The nature of a reservoir and of the transporting medium and how it interacts with the toxic substance are important. For example, toxic substances may remain relatively stable for a long time within a pile of waste at a landfill site. But if rain falls on an unprotected surface of the landfill, the toxins may be dissolved and carried away by groundwater flow. Environmental conditions also influence the stability and persistence of substances. For example, organic garbage in a landfill in a cool, dry environment may remain intact for many years, but if the environment within the landfill is warm or oxygenated or wet, the organic materials will biodegrade more quickly.

The presence of other substances in a reservoir can affect the behavior of a toxin. If two substances of low toxicity interact in a **synergistic** manner, it means that their toxic properties are mutually reinforcing—the sum of their toxic properties is greater than would be expected by just adding them together. Conversely, two substances in a reservoir may interact such that they mutually reduce each other's toxicities; this is called an **antagonistic** interaction.

When a toxin resides in a biologic reservoir—that is, in an organism—its residence time is affected by processes that are specific to the reservoir. Some toxins accumulate in the tissues of animals (Figure 13.1). This happens when a substance has a chemical or biochemical affinity to the materials that compose the tissues of organisms such as fats, proteins, or bone. When an organism is continuously exposed to such a toxin or ingests it repeatedly, the toxin will build up over time because it is not excreted. This buildup process is called **bioaccumulation** (or **bioconcentration**). For example, if a child ingests lead on an ongoing basis by eating lead paint or playing in lead-contaminated soil, the lead will accumulate in the child's body, where it may eventually reach a harmful (even lethal) concentration. Special medications must be given to induce the child's body to excrete the lead.

There is another way that bioaccumulative toxins can build up to harmful levels in organisms. Consider the case of mercury contamination in lake-bottom

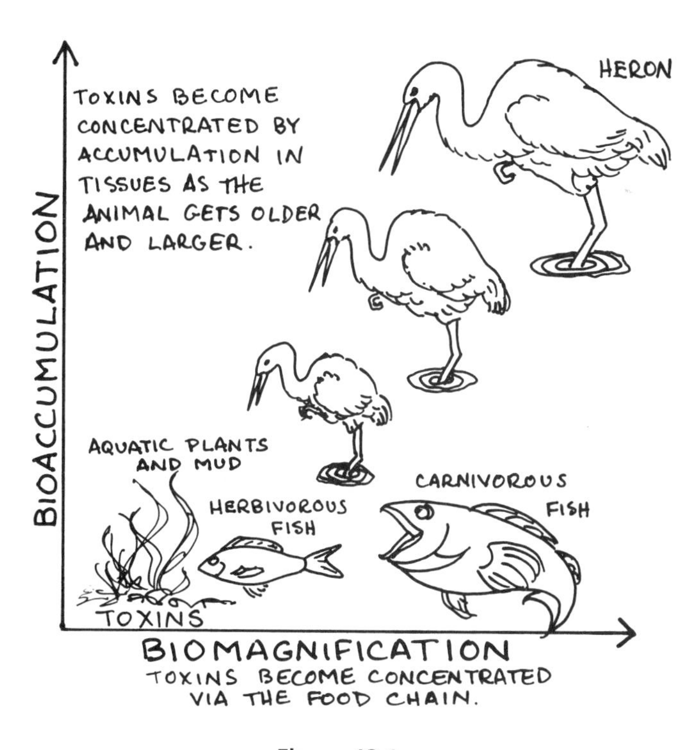

Figure 13.1.

sediments. Metallic mercury in this type of environment can be altered through bacterially mediated reactions, by which it is transformed into a much more toxic form called methyl-mercury. As methyl-mercury, it is ingested by bottom-feeding fish, where it bioaccumulates in their tissues over time. Imagine that a larger fish then eats a number of mercury-contaminated bottom-feeders. The older and larger the fish, the longer it has been eating smaller fish and accumulating mercury (Figure 13.1). This is called **biomagnification** (or **food chain concentration**) because the concentration of mercury is magnified with each step up the food chain. If the large fish is then eaten by a bird of prey, or by a person, the predator will consume all of the accumulated mercury from lower on the food chain. The most infamous example of methyl-mercury poisoning happened in Minamata, Japan, where inhabitants routinely consumed fish from a mercury-contaminated bay. Eventually, 730 people died and as many as 1,470 others were affected by neurologic problems and disfiguring birth defects from the high concentrations of mercury acquired through the contaminated fish.

What is the difference between bioaccumulation and biomagnification?

Answer: Bioaccumulation (bioconcentration): buildup, over time, of a persistent toxic substance in the tissues of an organism. Biomagnification (food chain concentration): increase in the concentration of a persistent toxic substance higher up the food chain.

Environmental Health and Toxins

Environmental health is the field of study that is generally concerned with the relationship between human health and the environment. **Toxicology** is the study of the chemistry, occurrence, and health effects of toxins. A substance that has a toxic effect immediately or very shortly after exposure is said to exhibit **acute toxicity**. Substances that are not immediately harmful can still be toxic; toxicity that is exhibited after long-term exposure or many years after the exposure is called **chronic toxicity**. In some cases a low level of exposure to a particular toxin may not cause obvious harm, but the **cumulative** (or additive) **effect** of repeated low-level exposure over time becomes damaging to the organism.

The specific health effects of toxic substances vary widely depending on the nature of the toxin and the organism's exposure to it. Some toxins cause **somatic damage**—that is, bodily damage to the living organism. For example, some pesticides, including DDT, are **carcinogens**, meaning they cause cancer in the living organism, although the cancer may not develop or become apparent until years after exposure. Many common toxins, including lead and methyl-mercury, cause damage to the nervous systems of organisms; these are called **neurotoxins**. Other toxins cause **genetic damage**—that is, damage to the organism's genes and therefore to its offspring. Toxic substances that cause genetic damage leading to mutations in offspring are called **mutagens**. An emerging environmental health concern associated with exposure to persistent toxins, particularly some pesticides, is the possibility that these chemicals may be endocrine disrupters. This refers to the ability of some chemicals to interfere with hormones, the chemicals that regulate reproduction, growth, sexual maturity, and other processes in animals as well as humans.

Dose, the amount of toxin that actually enters the body by ingestion, respiration, or absorption, is crucial in determining the health effects of the substance. Dose is described in terms of the amount of toxin relative to the body weight of the organism, in mg/kg (milligrams of toxin per kilogram of body weight). Note that dose is different from **exposure**, which refers to the amount or concentration of toxin in the surrounding environment to which the organism is exposed. This helps to illustrate why children are particularly susceptible to the health effects of environmental toxins. Children and adults can be exposed to the same amount or concentration of toxin in the environment, but the dose will be greater for a child because the body weight is lower than the adult's. It is also particularly damaging for children to be exposed to harmful toxins while their bodies and nervous systems are still developing.

A **dose-response curve** illustrates the effects of varying doses of a toxin on a particular organism. Figure 13.2 shows a substance that is harmful to the organism in high doses but beneficial or even necessary in moderate doses. The curves for many compounds and elements, such as selenium, copper, fluoride, and others, look very similar, although the specific doses differ. Let's say this dose-response curve illustrates the effects of human exposure to fluoride. A lack of fluoride (a dose that is *too* low) is potentially harmful to people and can lead to health effects such as susceptibility to cavities or increased risk of osteoporosis. With increasing doses, the health benefits of fluoride intake increase. After a certain dose is reached, however, no further health benefit accrues from higher doses. At still higher doses, the health benefits begin to decline, and at some point the dose begins to cause harm. In the case of fluoride, the main harmful effect is mottling or discoloration of the teeth. There are many other substances whose dose-response curves show that they are vital for health at low doses but harmful or even lethal at higher doses.

Scientists determine the harmful or lethal doses of toxins through experimentation. For example, toxins can be administered to a population of laboratory animals. If the population shows no response to the toxin, the dose is increased little by

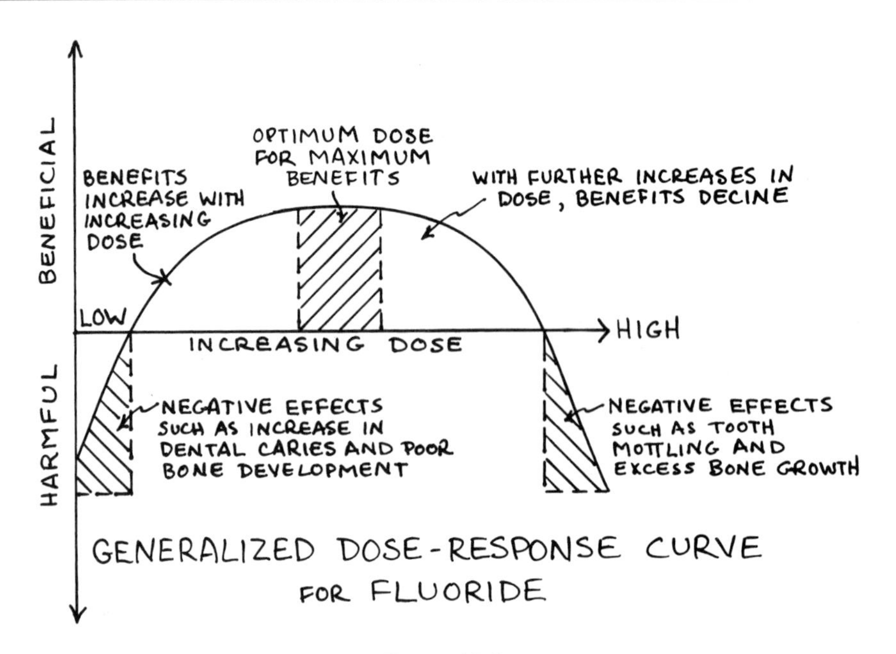

Figure 13.2.

little. A dose that is lethal to 50 percent of the population is called the **lethal dose-50**, usually abbreviated **LD-50**. A substance with a very low LD-50 would most likely be highly toxic for humans as well as for lab animals. The highest concentration of a toxin that shows no discernible harmful effects on laboratory animals is called the **NOAEC**, or **no observable adverse effects concentration**. Scientists and regulators commonly determine the legally permissible environmental concentration of a toxin by taking a fraction of the NOAEC for lab animals, usually one-millionth, and setting this as the permissible limit for humans. Although this method seems somewhat arbitrary, it is difficult to envision another way to determine the "safe" limits of toxins, short of laboratory testing on humans.

In addition to innumerable naturally occurring chemicals, some of which are highly toxic, there are at least 4 million manufactured synthetic compounds; tens of thousands of these are in common use. Some highly toxic substances, including cyanide and chlorine, have been used in industrial processes since the 1800s or even earlier. Scientists don't have complete information about the potential human impacts of all of these compounds at low and high doses, acute and chronic exposures. In particular, there is a lack of information about the potential synergistic or antagonistic interactions of compounds, whether natural or synthetic, in the environment.

Table 13.1 Common Soil and Water Pollutants

Pollutant Category	Examples	Common Sources	Potential Environmental Impacts	Potential Human Health Impacts
Sewage and animal wastes	*E. coli*, other infectious agents	Municipal sewage discharge; farm runoff; floodwaters	Eutrophication; high BOD	Bacterial infection, intestinal diseases (e.g., cholera)
Plant nutrients	Phosphorus, nitrogen	Excess fertilizers from farm runoff and suburban runoff; detergents	Algal blooms; high BOD; eutrophication	Blue-baby syndrome (nitrates)
Sediment	Clay, silt	Farm runoff; erosion from cleared land	Block light, inhibit photosynthesis in water bodies; clog organisms	Can carry toxic contaminants
Synthetic chlorinated organic compounds	DDT, mirex, dioxins, furans, PCBs; solvents, lubricants; TCE	Pesticides from farm runoff and suburban runoff; industrial effluents	DNAPL contamination of groundwater, soils	Cancer; neurologic damage
Hydrocarbon components	Benzene, toluene, xylene, ethylene	Oil from industrial effluents and spills; incomplete combustion	LNAPL contamination of groundwater, soils	Cancer; neurologic damage
Acids, caustics, salts	Sulfuric acid, nitric acid, hydrochloric acid; caustic effluents (lye)	Acid precipitation; acid mine drainage; soils; industrial effluents; irrigation	Acidification of lakes, forests, salinization	Main impacts on environment, organisms, habitat
Heavy metals	Lead, mercury, arsenic, selenium, cadmium	Mining; refining; paper mills; other industrial processes	Bioaccumulation, biomagnification in organisms	Neurologic damage
Radioactive substances	Radioisotopes; radwaste	Radon gas; nuclear reactor waste products; nuclear weapons	Cancer; genetic mutations in organisms	Cancer; genetic mutations
Thermal pollution	Heat, hot water	Hot industrial effluents; forest fires, volcanism	Impact on aqueous species and habitat; decreased oxygen levels in water	Main impacts on fish species

Examples of problematic toxins include **heavy metals** such as mercury, lead, cadmium, selenium, and arsenic. Many synthetic **chlorinated organic compounds**—manufactured compounds that contain chlorine and carbon-based molecules—are highly toxic, persistent, and readily dispersed in the environment. Other pollutants that are particularly toxic are polycyclic aromatic hydrocarbons (PAHs), which include the various components of oil, notably benzene, toluene, ethylene, and xylene, together known as BTEX. The most common types of water and soil pollutants, both naturally occurring and synthetic, are summarized in Table 13.1, along with examples, sources, potential environmental impacts, and potential health impacts. Most will be covered in greater detail as we look more closely at the details of surface water pollution, groundwater pollution, and soil pollution.

What is LD-50 and how is it used to determine the potential impacts of toxins on human health? ______________

Answer: LD-50 is the dose that is lethal to 50 percent of a test population of laboratory animals. A substance with a low LD-50 is likely to be highly toxic to humans as well as to lab animals.

Surface Water Pollution

Surface water bodies are important freshwater sources for human use. However, the accessibility that makes them so useful also renders them highly susceptible to contamination. In general, pollutants come from one of two types of sources. **Point sources** of contamination are discrete locations where emissions are occurring; they are points on a map. Examples are outflow pipes for industrial effluents, slaughterhouses, oil spills, and smokestacks. **Nonpoint sources** are broader regions of contamination; they are areas on a map rather than points. Examples of nonpoint sources include agricultural areas, urban areas, and suburban areas.

Point sources of pollution can become nonpoint sources when there are many of them or when they are mobile, such as automobiles, or when emissions become airborne masses that are eventually deposited over a broad area of land or water. Point sources tend to be more obvious than nonpoint sources; a pipe spewing waste is hard to ignore. But it is inherently easier to manage point sources because they can be identified, tested, and monitored. Nonpoint-source pollution may originate from countless individual sources, each of which contributes only a very small amount of the contaminating substance. The cumulative effects of minor point sources, when taken together over a large area, can add up to a very large pollution problem.

Contaminants in surface water come primarily from urban, suburban, and agricultural runoff. Airborne acid-forming substances contribute to acidic deposition, which can damage lakes and other surface water bodies and aquatic ecosystems. (Acid deposition will be discussed in greater detail in chapter 14.) Industrial **effluents** (contaminated liquid runoff), particularly those produced by the pulp and paper industries and chemical manufacturing, are significant point-source contributors. So are discharges related to resource extraction, among which mining, logging, and the petroleum industry are important. Sewage discharge and effluents from poorly engineered landfills are other important sources of surface water contamination. Effluents from landfills can contain numerous toxic substances; they will be discussed in chapter 15.

Many wastes that end up in surface water bodies contain organic material. Lots of cities, even in wealthy countries, still discharge sewage directly into nearby water bodies with little or no treatment. When organic matter biodegrades in an aerobic environment, oxygen is used up in the process. Oxygen, usually measured in terms of the amount of dissolved oxygen in the water, is critical for the maintenance of most aquatic life forms. Effluents that cause the depletion of dissolved oxygen in a water body are said to place a **biochemical oxygen demand**, or **BOD**, on the system (Figure 13.3). The greater the amount of organic matter discharged into the system, the higher the BOD. If sewage is discharged directly into a water body, the dissolved oxygen in the water downstream from the source of the sewage responds by becoming depleted.

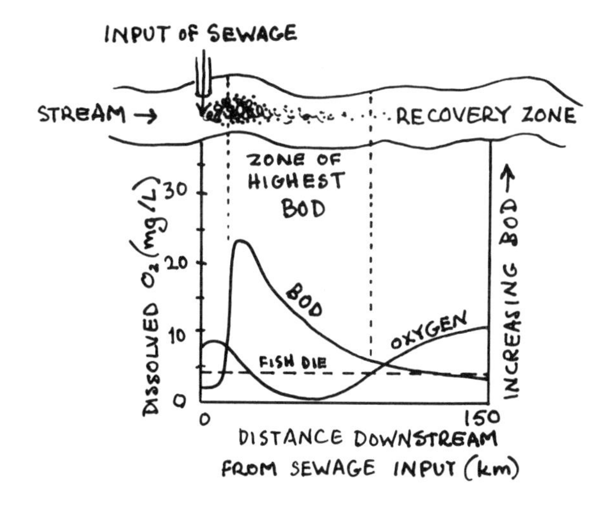

Figure 13.3.

One of the most common forms of surface water contamination results from an excess of plant nutrients, especially phosphorus and nitrogen from fertilizer runoff, phosphorus-containing detergents, and sewage. Nutrients are, of course, necessary to sustain plant growth. But when an aquatic system is overloaded with nutrients, the growth of algae, plankton, and aquatic weeds can get out of control; this is called an **algal bloom**. When the algae and other excess aquatic plants die, their biodegradation causes a depletion of oxygen called **anoxia**. Other organisms in the water begin to die, and the water turns mucky and sickly green. This process, called eutrophication, can occur naturally, as discussed in chapter 12. When eutrophication is accelerated by the addition of anthropogenic pollutants, it is called **cultural eutrophication**. If the water becomes completely oxygen depleted, it may turn crystal clear. This is *not* a sign of a healthy water body—it is a sign of a water body in which nothing is able to survive.

During the 1960s, Lake Erie began to turn an unhealthy-looking green color. This was caused by a bloom of blue-green algae resulting from an excess of phosphorus in the water. Beaches were covered with green, slimy masses of rotting algae. The lake became anoxic, and many aquatic species began to die. Although oxygen depletion in deep waters is still a concern, Lake Erie has now largely recovered from the effects of cultural eutrophication. The recovery was an outcome of the Great Lakes Water Quality Agreement (1972) between Canada and the United States that limited the discharge of phosphorus from point sources.

The presence of infectious agents is another form of surface water contamination associated with organic-rich effluents. Bacteria, viruses, and other disease-causing organisms may be found wherever human or animal wastes are contained in effluents. Measurements of the level of harmful microorganisms in water are commonly based on the abundance of a particular fecal coliform bacterium, ***Escherichia coli***, commonly known as ***E. coli***, which lives in the intestines of humans and some animals. In many parts of the world, the consumption of water contaminated with infectious agents is linked with outbreaks of diseases such as cholera. In North America, cholera and other waterborne diseases are less of a problem than they were a hundred years ago; still, high levels of *E. coli* routinely cause beach closures in many coastal areas. In Walkerton, Ontario, in 2000, the deaths of seven people and many serious illnesses were caused by a deadly strain of *E. coli* that entered the municipal groundwater supply through a cracked wellhead. The source of the bacteria is thought to have been fecal-contaminated runoff from a nearby pig farm.

Several major groups of substances occur as anthropogenic toxic contaminants in both surface and groundwater (see Table 13.1). The chemical names of these substances look very complicated, but many of them will be familiar to you from news reports about incidences of contamination. Some of the most significant toxic contaminants in surface water bodies are chlorinated organic pesticides such as DDT and Mirex. Other toxic organochlorides include pesticides, polychlorinated biphenyls (PCBs), dioxins, and furans. Some of these are highly toxic (carcinogenic), persistent, and easily dispersed in the environment. PCBs have been widely used as lubricants and heat exchangers in electrical transformers and capacitors, among other applications; their use is somewhat more limited since evidence emerged linking PCBs with the incidence of cancer in laboratory animals. Heavy metals, including cadmium, lead, tin, plutonium, and mercury, are released during mining and ore processing, metallurgical procedures (such as electroplating), paper manufacture, petroleum refining, and other industrial processes. Of these, lead and mercury have received the most attention. Long-term exposure to even very low levels of these substances can cause neurologic damage (especially in children), kidney damage, and a variety of other problems.

Another type of pollution that occurs mainly in the surface water environment is **thermal pollution**, the release of heat or (more commonly) heated water into the environment. Many industrial and energy-producing processes release effluents that are warm or hot compared to the receiving environment. Thermal pollution can harm or kill both plants and animals; it can also interfere with the reproduction of aquatic species. When the temperature of water rises, the amount of dissolved oxygen also decreases. Thus, another negative impact of thermal pollution is that it lowers the oxygen content of surface waters. It is usually possible to control the effects of heat by allowing effluents to cool before they are released into the environment, but this will not necessarily raise the oxygen content of the effluents.

What is anoxia? ______________

Answer: A condition in which a water body has become depleted in dissolved oxygen.

Sediment Pollution

Sediment pollution has two very different meanings. It can refer to the contamination of sediment by hazardous substances or to situations in which the sediment itself acts as a pollutant. The latter is one of the negative impacts of the erosion of soil from cleared land (discussed in chapter 9). The topsoil that washes off agricultural lands has to end up somewhere, and most often it ends up in adjacent waterways. Problems caused by eroded soil include reduced channel capacity because of silt accumulation; destruction of fish habitat; acceleration of algal growth; buildup of heavy metals, pesticides, and other toxic substances; loss of recreational value; and dredging and cleanup costs. It has been estimated that the economic impacts of topsoil loss from agricultural land may be as great *off* the farm as on it. Rates of erosion and resulting siltation also tend to be very high in areas with high rates of deforestation. In Malaysia, for example, most of the estuaries are heavily silted. As a result, the number of fish in coastal waters has decreased, and corals have died.

Where sedimentation and siltation are a problem, it is likely that the buildup of contaminants will also be a concern. Eroding sediments represent one of the primary mechanisms by which contaminants are transported by runoff as suspended solids into surface water bodies. For example, phosphorus, heavy metals, pesticides, and other organic compounds can bind themselves to soil particles. This is because clay particles in the soil have a negative electric charge (that is, they are anions), which gives them a strong affinity for contaminants that carry a positive electric charge, notably heavy metals (cations). The eroding soil carries the contaminants along to the water body.

Contaminated sediment at the bottom of a body of water is of particular concern because it can act as a reservoir, holding toxic substances and slowly releasing them into the water. Bottom-dwelling organisms may be most affected, as in the case of methyl-mercury contamination discussed previously. If the contaminated sediment at the bottom of a water body is disturbed, for example, by dredging, it may be stirred up, releasing a large quantity of the contaminant into the water. In some cases, the decision is taken to leave such a mass of contaminated sediment undisturbed; a stable cap may be placed over the contaminated mass to hold it in place and isolate it.

Soil degradation resulting from chemical changes is a growing problem. Soil contamination encompasses a variety of destructive processes, including salinization, contamination by pesticides and other agrochemicals, waterlogging, and acidification. In Europe alone, industrial and urban pollution, pesticides, and other chemicals have contaminated almost 14 million ha (more than 34 million acres) of land.

What are the two different meanings of sediment pollution? ____________

Answer: (1) Chemical contamination of sediment. (2) Sediment *as* a contaminant—that is, excess siltation that clogs waterways and channels.

Groundwater Contamination

Many of the pollutants that affect surface water and soils also cause groundwater contamination (Figure 13.4). Because of its hidden nature, however, groundwater contamination can be more difficult to detect, control, and clean up. Harmful

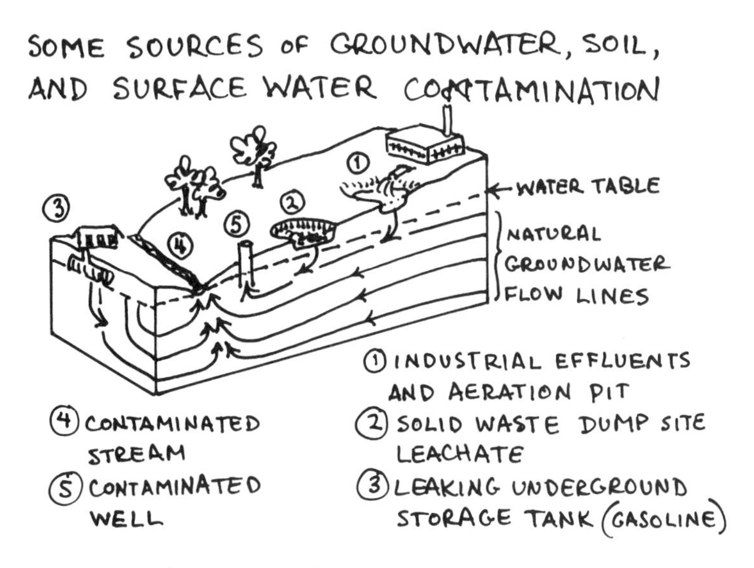

Figure 13.4.

chemicals from leaking underground gasoline storage tanks and pipelines, poorly designed or abandoned landfills, and toxic waste disposal facilities can contaminate groundwater reservoirs. Many industries that generate toxic wastes dispose of them (legally, in many cases) in open, on-site disposal receptacles called **aeration pits** or **aeration lagoons**; the theory is that the chemical wastes will oxidize and degrade over time. Aeration lagoons typically are underlain by an impermeable layer (an aquitard or an aquiclude) or liner, but if leaks develop, a dangerous mix of contaminants may escape.

Twenty of the top twenty-five toxic groundwater contaminants are volatile organic compounds, like benzene, toluene, ethylene, and xylene (BTEX), all of which are constituents of gasoline. Chlorinated organic solvents such as trichloroethylene (TCE), commonly known as dry-cleaning fluid, also pose a significant threat to groundwater quality in many sites. However, the most common pollutants in groundwater are untreated sewage and nitrate from agricultural chemicals; they are perhaps less toxic than other contaminants but no less problematic because they are so widespread. An additional problem is saline contamination of groundwater that is caused by the intrusion of seawater into coastal aquifers.

It has been estimated that as many as 25 percent of underground storage tanks at gas stations in North America are leaking. At an intersection with a gas station on each corner, there is likely to be one with a tank that is leaking gasoline into the soil and groundwater.

Contaminants can be present in the subsurface in many different forms. If a contaminant vaporizes easily, it is likely that much of it will be present in the form of a gas. Contaminants can also occur as dissolved constituents or as suspensions of **colloids**—submicroscopic particles—in groundwater. Some contaminants are not miscible with water (they do not mix easily, like oil and vinegar) and may occur as a separate, **nonaqueous-phase liquid** (**NAPL**). Physical and chemical interactions between the contaminants, the water, and the surrounding rock or sediment combine to determine the eventual fate of the contaminating substance.

Recall from chapter 12 that groundwater flow is very slow and is typically measured in centimeters per day or meters per year. This is because groundwater must move through small, constricted passages between the grains in soils and sediments or through tiny cracks in rocks. Therefore, the flow of groundwater depends to a great extent on the porosity (holes, or pore spaces) and permeability (interconnectedness of the pore spaces) of the aquifer through which the water is flowing.

The water table (the interface between the saturated zone and the zone of aeration) tends to mimic the overlying topography—high underneath hills, low in valleys. Responding to gravity, groundwater percolates from areas where the water table is high toward areas where it is lower; it typically flows toward surface bodies, which occur where the water table intersects the surface. Part of the groundwater flows directly down the slope of the water table. The rest flows along curving paths that go deeper through the ground under a hydraulic pressure gradient, eventually turning upward to enter the surface water body from underneath. When development, effluents, or waste discharge occurs in a recharge zone, there is potential for groundwater contamination. When the contaminated water reaches a discharge zone, the contamination may then enter the adjacent surface water body.

Pollutants typically travel from their source in the form of a spreading mass, called a **plume**, of contaminated groundwater (Figure 13.5). The direction of travel

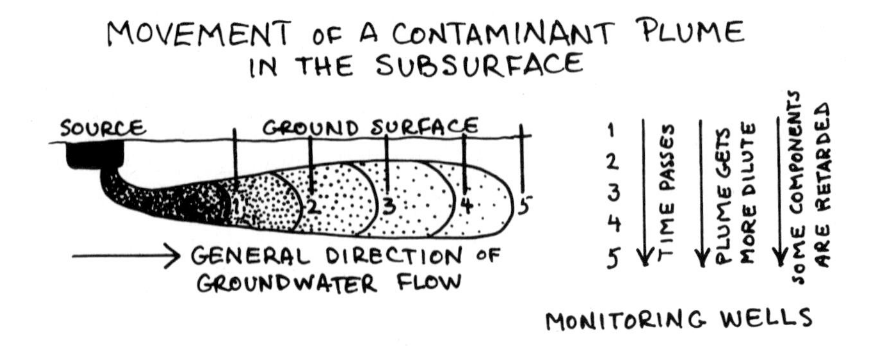

Figure 13.5.

depends on the regional groundwater flow patterns. The normal flow of groundwater along with any dissolved constituents is called **advection** (or **advective flow**). A contaminant that travels at the same rate as the groundwater with which it is being transported is said to be an *unretarded* contaminant: the contaminant has not been slowed down relative to the rate of movement of the groundwater. **Retardation**, therefore, is a measure of how much a contaminant has been slowed down relative to the groundwater flow.

There are three main mechanisms whereby contaminants can become retarded (Figure 13.6). In **sorption**, the contaminants adhere or stick to particles in the soil, sediment, or rock. **Dispersion** alters the simple advective flow of the contaminant by forcing it to take a less direct route. When a contaminant is subject to dispersion, its molecules follow a winding, indirect pathway around and between the mineral particles. Sometimes this occurs because contaminants flow into extremely tiny spaces between sediment particles as a result of **capillary attraction** (or **capillarity**), an adhesive force between liquids and solids that causes the liquid to be drawn into small openings in the solid. As time goes by, a third retardation mechanism may occur, which is breakdown, often through biodegradation. Sorption, dispersion, and chemical or biologic breakdown operate in concert to retard the movement of a contaminant, spread it out over a broader area, and eventually reduce its concentration in the groundwater.

Monitoring an unretarded contaminant is a good way to trace the movement of a contaminant plume and determine the maximum rate of groundwater flow within the plume. The most frequently used **tracer** for this purpose is chloride ion (Cl^-). Chloride is a common constituent of many contaminant plumes, and it is typically

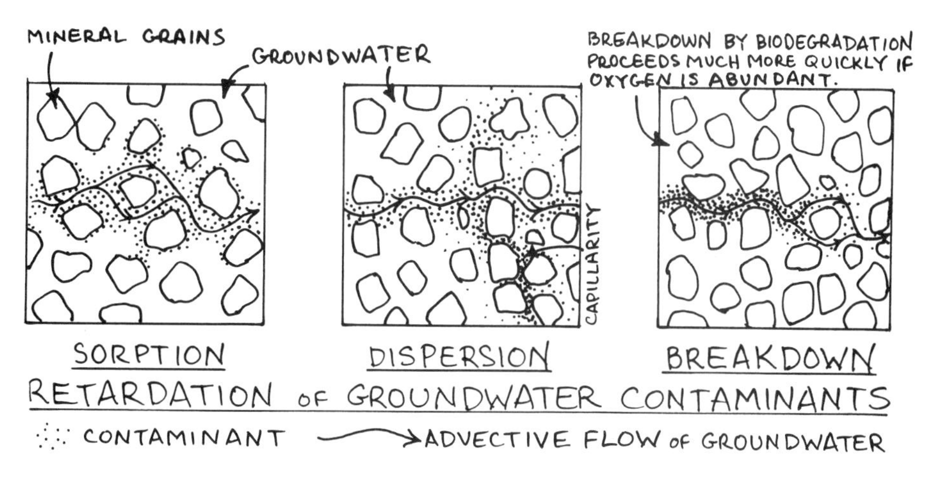

Figure 13.6.

unretarded. If the concentration of chloride ions in a "clean" water well suddenly increases, it may indicate that the rest of a contaminant plume (the retarded portion) is following some distance behind, moving more slowly than both the groundwater and the unretarded chloride tracer. Chloride ion is thus a harbinger of bad news; it represents the fastest-moving constituents—the front line—of the contaminant plume.

Among the most important physical characteristics used in determining the behavior and predicting the flow of a groundwater contaminant is the density of the contaminant relative to the groundwater. Most contaminants contain some components that are soluble in water and some that are not (that is, they are immiscible). If an immiscible liquid contaminant is less dense (lighter) than water, it is referred to as a **light nonaqueous-phase liquid**, or **LNAPL**. Petroleum is the most common LNAPL groundwater contaminant. In contrast, **dense nonaqueous-phase liquids**, or **DNAPLs**, are denser (heavier) than groundwater. Trichloroethylene (TCE) is an example of a DNAPL contaminant.

A contaminant's behavior in the subsurface depends on whether it is lighter or denser than the groundwater (Figure 13.7). For example, a contaminant plume composed of petroleum (LNAPL) will infiltrate the soil layers; upon reaching the saturated zone the petroleum will float on top of the groundwater, where it will tend to spread out in a lens-shaped body. In contrast, a plume of TCE (DNAPL) will sink through the zone of aeration, then continue to sink through the water in the saturated zone to the bedrock below, becoming mixed to a limited extent with the groundwater as it passes through. DNAPL plumes can be extraordinarily difficult to monitor and control. Because they are not always affected by lateral flow

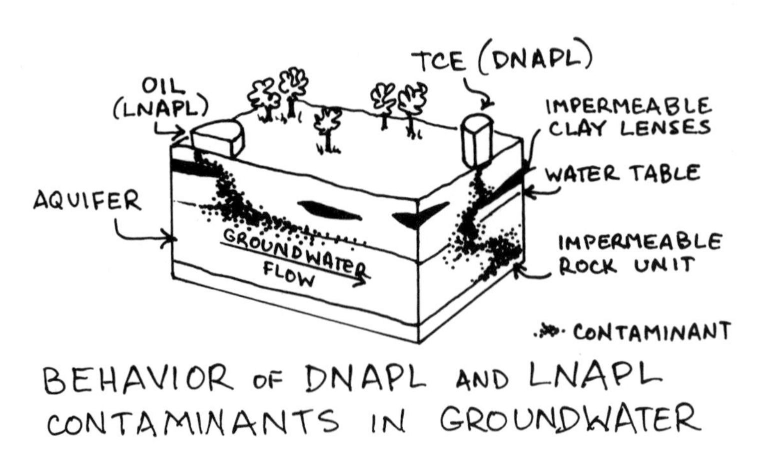

Figure 13.7.

in the groundwater, their movement is not very predictable. They pass easily through fractures and may contaminate deep-seated aquifers or seep between layers of bedrock. It is often difficult or impossible to locate the original source of the DNAPL plume.

What is a DNAPL? What is an LNAPL? ______________

Answer: DNAPL: a dense nonaqueous-phase liquid—a liquid that is not water-based and is denser than water. LNAPL: a light nonaqueous-phase liquid—a liquid that is not water-based and is less dense than water.

The Remediation of Contaminated Sites

Contaminated sites that have the potential to be cleaned up and used again for other purposes are called **brownfields**. The specific cleanup or **remediation** approaches and technologies chosen for a particular brownfield site will vary according to the type of contaminants present, how hazardous the contaminants are, the severity or degree of contamination, and the natural environment at the contaminated site. The response will also be influenced by the contamination event itself. For example, a spill of a known quantity of a known substance, such as an oil spill or a chemical spill in a laboratory, requires a different approach than a cleanup in which the extent and nature of the contamination are more complicated to determine, such as an unknown contaminant plume discovered at an old, abandoned industrial site. Flammable, highly toxic, or radioactive materials, in particular, obviously require a very cautious and highly skilled response.

Every site remediation begins with the obvious: identify the contaminants, shut off the source of the contamination, remove any residual wastes or contaminants that are still present at the source, and prevent the worst of the spilled material from dispersing any further into the environment. This process is called **containment**. In an oil spill, the vessel, storage tank, or well from which the oil is flowing must first be contained or removed. Then the main mass of spilled oil is enclosed to prevent it from dispersing. If the spill occurred on water, a variety of types of booms, like floating fences, are used for containment. Absorbent materials can be floated on top of the water to soak up the main body of oil. If the spill is on or under the ground, containment can be more difficult. Usually it is necessary to install **monitoring wells** to delineate the contaminant plume; these are similar to water wells, but they are used for the purpose of extracting samples for chemical testing and to detect the extent, the depth, and the direction and rate of movement of the contaminants.

Sometimes, in-ground barriers or fences are installed, using special impermeable materials to stop the contaminant plume from moving any farther.

Containment of the source is often difficult and sometimes impossible. For example, there might be mercury-contaminated soil at a building surrounded by several factories and a municipal incinerator. Any of these might be a source of vaporized mercury that was deposited on the soil; or the mercury might have come from farther away; or the source of the mercury might have ceased operations many years ago; or the site could even be underlain by a natural source of mercury in the bedrock. Sometimes it is possible to identify the source by detailed chemical testing and tracing of the pathways of the contaminant, but often it is not. Even if the source is located, there may be legal entanglements; for example, perhaps the company that owns the factory from which the contaminant originated went out of business forty years ago, and there is no one who is currently legally responsible for the site.

Assuming that the origin of the contaminant has been located and contained and any residual material at the source has been removed, the cleanup can proceed. There are two basic approaches to the remediation of polluted sites. The first is **passive remediation**, which involves relying on natural environmental processes to clean up the site. This may sound "lazy" or ineffective, but it is sometimes a viable approach. The natural environment can be very effective at filtering and removing contaminants from water and soil. If there is no immediate risk to people living or working nearby, there is no urgent need to prepare the land for new uses, and the environment itself is appropriate, it can save a lot of money and effort to simply wait and let nature take its course. Sometimes passive remediation can be helped by interventions such as turning over the soil to allow volatile (easily vaporized) contaminants to evaporate.

However, passive remediation of even a mildly contaminated site can take years or even decades. If the need for the land is urgent or the risk to people or adjacent environments is severe, it is advisable to undertake **active remediation**—that is, to actively intervene in the cleanup of the site. Soil and water remediation are obviously closely linked; if soil is contaminated, the water that sits on top of it and the groundwater that passes through it will likely be contaminated, too, and vice versa. Site remediation plans must take both soil and water contamination into account.

If the soil has good drainage and the contaminant is water soluble and nontoxic, it may be possible to pump large quantities of water through the site to dilute the contaminant. For example, this **dilution** approach is sometimes used with salinized (salty) soils. Dissolving a contaminant in water is only an option if the water used for dilution can then be extracted from the ground; otherwise, it will end up contaminating the groundwater. A related approach is to establish a **reconstructed** or **engineered wetland** at the site to promote natural filtration processes if the environment is appropriate; wetlands are particularly effective at filtering and removing contaminants through natural processes. If the contaminant is volatile, like oil or gasoline, **vapor extraction** can be used; this involves using something similar to a large vacuum cleaner to draw out the contaminated soil gases. If necessary, air or carbon dioxide can be pumped into the soil before extraction to mobilize the contaminants.

Many contaminants contain organic compounds, so biodegradation (organic decomposition of carbon-bearing compounds) is an extremely important process in site remediation. A cleanup technique that relies on biodegradation to break down contaminants is called **bioremediation**. If the technique is specifically designed to facilitate or speed up the biodegradation process, it is referred to as **biostimulation**. Biodegradation occurs more quickly and easily in an aerobic (oxygenated) environment; this means that the pumped-in oxygen used for vapor extraction can also be used to stimulate aerobic biodegradation. Sometimes specific microbes are introduced to stimulate either aerobic or anaerobic (nonoxygenated) biodegradation.

A related approach utilizes plants that selectively extract contaminants from the soil and soil moisture in much the same way that plants draw nutrient chemicals from the soil water; this is called **phytoremediation**. If soil is so highly contaminated that it cannot be cleaned in situ (in place), it may have to be excavated and brought to a hazardous waste facility for ex situ (out-of-place) treatment. This tends to be an expensive solution; sometimes the only option is to label the material as hazardous waste in a secure landfill.

If the groundwater is also contaminated and active remediation is necessary, there are several approaches that can be taken. For LNAPL contaminants such as oil, the low-density components will tend to float in a pool at the level of the water table—that is, on top of the saturated zone. If the main pool can be located, it may be possible to use vacuum extraction techniques to remove the LNAPL pool from the groundwater. This typically involves digging a trench or a pit or intercepting the pool in a monitoring well as it moves along with the groundwater flow. For DNAPL components that sink through the groundwater and for contaminants that are water soluble, it may be necessary to remove the groundwater itself for ex situ treatment; this is called **pump-and-treat**. The extracted groundwater is chemically treated in an aboveground facility, then may be reinjected into the ground.

The final step in site remediation is to monitor the effectiveness of the cleanup and track the movement of any remaining contaminants with periodic sampling and testing from monitoring wells. When soil and water tests show that the levels of contaminants are below legally required limits, the site may be approved for housing, parkland, or industrial use. The legal limits for contaminants and the land-use approvals process vary widely. In Canada, provincial governments are responsible for environmental approvals. Where an interprovincial or international situation is involved, as with the management of the Great Lakes, for example, **Environment Canada**, a federal agency, takes the lead. In the United States, the federal-level **Environmental Protection Agency** (**EPA**) oversees this process.

The EPA assists with the cleanup of some highly contaminated sites through the Superfund Program. Through the EPA, Superfund offers financial support for the remediation of sites that pose the greatest risk to public health and the environment, as designated on the National Priorities List of Contaminated Sites. Of the estimated more than 400,000 sites in the United States that are contaminated with hazardous materials, about 1,300 are on the National Priorities List. So far, only about 200 sites have been remediated to the extent that they can be removed from the list.

What is a brownfield? ______________

Answer: A contaminated site with the potential to be cleaned up and used for another purpose.

This chapter introduced lots of new terminology and concepts. But what more important topic could there be than the state of our water and soils and the impacts of contaminants on human health? In the next chapter we will consider contaminants in the air we breathe.

SELF-TEST

These questions are designed to help you assess how well you have learned the concepts presented in chapter 13. The answers are given at the end. If you get any of the questions wrong, be sure to troubleshoot by going back into that part of the chapter to find the correct answer.

1. The process used to remove salt from water is called ______________.
   a. salinization
   b. chlorination
   c. desalination
   d. filtering

2. Which one of the following would *not* be considered a point source of pollution?
   a. effluent from a mine site
   b. runoff from neighborhood lawn fertilizers and pesticides
   c. a municipal sewage discharge pipe
   d. a leaking oil tank

3. A technique for remediating contaminated soils that utilizes plants to draw contaminants from the soil is called ________.
   a. bioremediation
   b. biomagnification
   c. biodegradation
   d. phytoremediation

4. The two federal agencies that oversee transboundary environmental concerns in Canada and the United States are ________ and ________.

5. ________ provides a good tracer for groundwater contaminant flow because it is present in many common contaminants, and it is unretarded relative to groundwater flow.

6. Cleanup techniques that are designed to facilitate or speed up bioremediation are referred to as ________.

7. A spreading mass of contaminant in groundwater is called a plume. (True or False)

8. These days, all municipalities treat sewage extensively before discharging it into a surface water body. (True or False)

9. What is BTEX, and why does it present a remediation challenge?

10. What is NOAEC and how do scientists and regulators use this concept in setting legally permissible limits for toxins?

11. Why do clay particles in eroding soils tend to "grab" and carry along pollutants, notably metals?

12. What active and passive remediation techniques can be used to clean up soils that contain vapor-phase contaminants?

13. What is the difference between passive remediation and active remediation?

ANSWERS

1. c 2. b 3. d
4. Environment Canada; the Environmental Protection Agency
5. Chloride 6. biostimulation
7. True 8. False
9. Benzene, toluene, ethylene, and xylene, which are volatile hydrocarbon compounds that are constituents of gasoline; they are difficult groundwater contaminants because they are toxic and partially soluble in water.
10. NOAEC: no observable adverse effects concentration; the highest concentration of a toxin that has no evident harmful impact on laboratory animals. Scientists and regulators typically set the permissible limits of toxins to a somewhat arbitrary fraction, usually one-millionth, of this concentration.
11. Clay particles carry a negative electric charge, which makes them electrically "sticky" for particles that carry a positive electric charge such as metals.
12. Vapor extraction (active remediation); turning over the soil to allow volatile gases to evaporate (passive remediation).
13. In passive remediation, nature takes its course, and a contaminated site is left for a period of years or even decades so that natural processes will eventually clean up the contamination. In active remediation, interventions such as pump-and-treat or soil excavation are undertaken to clean up a contaminated site.

KEY WORDS

active remediation
acute toxicity
advection (advective flow)
aeration pit (aeration lagoon)
algal bloom
anoxia
antagonistic
bioaccumulation (bioconcentration)
biochemical oxygen demand (BOD)
biodegradation
biomagnification (food chain concentration)
bioremediation
biostimulation
brownfield
capillary attraction (capillarity)
carcinogen
chlorinated organic compound
chlorination
chronic toxicity

colloid
containment
cultural eutrophication
cumulative effect
dense nonaqueous-phase liquid (DNAPL)
desalination
dilution
dispersion
dose
dose-response curve
Escherichia coli (E. coli)
effluent
Environment Canada
environmental health
Environmental Protection Agency (EPA)
exposure
fluoridation
genetic damage
heavy metal
lethal dose-50 (LD-50)
light nonaqueous-phase liquid (LNAPL)
monitoring well
mutagen
nonaqueous-phase liquid (NAPL)
neurotoxin
no observable adverse effects concentration (NOAEC)
nonpoint source
passive remediation
persistent
phytoremediation
plume
point source
pump-and-treat
reconstructed (engineered) wetland
remediation
residence time
retardation
sediment pollution
somatic damage
sorption
synergistic
thermal pollution
toxicology
toxin
tracer
vapor extraction

14 Air Pollution

Breathing in Bombay is now equivalent to smoking ten packs of cigarettes a day.

—Christopher Flavin

Objectives

In this chapter you will learn about:

- the major sources of air pollution;
- smog, its causes, and its health effects;
- the causes and health effects of indoor air pollution; and
- acid rain and its impacts on materials and natural ecosystems.

Air Pollution: Sources and Processes

You are probably most aware of the air when a yellow-brown haze is hanging over the city, or when the weather report refers to a smog watch or a bad air day, or when a truck passes by, belching blue-black smoke. A wide variety of contaminants cause pollution at low levels in the atmosphere, especially in urban centers. Some of the impacts of air pollution are local, but many are regional; they result from the **long-range transport of atmospheric pollutants**, or **LRTAP**.

One of the best-known examples of LRTAP occurred when a nuclear reactor in Chernobyl, Ukraine, experienced a meltdown and explosion in 1986. Radioac-

Factories darken the sky with pollution in Tehran, Iran.

tive particles from this disaster encircled the globe within two weeks, much of the contamination eventually settling out as **dry deposition** (settling to the ground as dry, solid particles). However, scientists who investigated the spread of the radioactive contamination discovered that the most significant contamination was experienced by localities that were near Chernobyl *and* had experienced rain in the days following the explosion. By dissolving some components and adhering to others, the precipitation had facilitated the settling out, or **wet deposition**, of the contaminants.

Another problem associated with LRTAP is **arctic haze**. This red-brown haze, which is composed of fine particles of sulfates, soot, dust, hydrocarbons, metals, and pesticides, occurs in the lower 1 to 2 km of the atmosphere during the arctic winter. It is caused by long-range transport of pollutants, primarily from Europe and Asia, by the prevailing wind systems, and it can have negative impacts on sensitive arctic ecosystems. Scientists believe that long-range transport is the principal reason for the presence of industrial pollutants such as mercury in remote arctic lakes and wildlife.

The behavior of an airborne contaminant is thus controlled by a variety of factors, including its own chemical and physical characteristics, the local topography, and prevailing weather conditions and patterns. The characteristics of the pollutant determine how long it can remain in suspension in the atmosphere and therefore how far it will be transported. Wind systems and precipitation obviously play an important role in the transport of airborne contaminants. When the air is very still, pollutants are not transported as far and tend to become concentrated over the source area. The height of the emissions stack is also important. Pollutants emitted from very tall stacks can stay in the atmosphere a long time and may be transported 500 km (300 mi) or more before being deposited.

Suspended particulates are pollutants that are carried aloft as extremely fine, dispersed liquid or solid particles, like the radioactive particles from the Chernobyl explosion. Particulates are produced by natural processes such as volcanic eruptions, smoke from forest fires, and blowing sea salt, dust, and pollen. Most anthropogenic particulates originate from industrial activities, transportation, and the generation of electric power. Waste incinerators and the burning of wood and agricultural wastes also contribute particulates to the atmosphere. There is growing concern about the contribution of biomass burning, especially slash-and-burn agriculture, to both particulate and gaseous air pollution; the smoke from biomass burning is even observable from space.

One pollutant that commonly occurs in particulate form in the atmosphere is lead, a heavy metal. Airborne lead has traditionally come mainly from automobile exhaust, but now that gasoline is mostly lead-free, the main sources of lead are industrial processes such as smelting. Lead-based paint, which is no longer used but is found in many older houses, is of particular concern because ingesting or breathing lead particles can cause brain damage. Other airborne heavy metals, including mercury and zinc, are derived from emissions from smelting and, to a lesser extent, old waste incinerators (the newer technologies are almost emission-free), as well as human and animal crematoria. (Consider the fact that many crematoria are old and are thus not subject to newer, more rigorous air quality regulations; as well, everything that is in or associated with the body, including lacquered coffins, hardware, and any sentimental items placed in the coffin, are all vaporized by the cremation process.) Heavy metal particulates eventually settle out as nonpoint source pollution. They tend to have fairly short residence times in the atmosphere, so they may become concentrated in soils in the area surrounding the source.

Particulates are not the only type of air pollution that is hazardous to human health (Table 14.1). There are many gaseous pollutants that are even more problematic. Carbon monoxide (CO), which occurs naturally in the atmosphere in minute quantities, is toxic and sometimes fatal; even low doses, with chronic exposure, can be fatal. Carbon monoxide is generated by the combustion of carbon-bearing materials (fossil fuels) in a low-oxygen environment. Cars and small trucks are the main source of CO pollution, with some contribution from space heating, industry, and the burning of wood and wood wastes. Other carbon compounds, such as carbon dioxide (CO_2) from the burning of fossil fuels and methane (CH_4) from rice cultivation and cattle ranching, have contributed to significant changes in the chemistry

Table 14.1 Some Major Air Pollutants and Their Health Effects

Type of Pollutant	Examples	Common Sources	Health Effects
Particulate matter (PM)	Dust, soot, lead	Industrial emissions; motor vehicle emissions	Respiratory problems
Sulfur oxides (SO_x)	Sulfur dioxide, sulfur trioxide (reacts to form sulfuric acid)	Coal-fired power plants; other industrial sources	Acid deposition; respiratory irritant
Nitrogen oxides (NO_x)	Nitric oxide, nitrogen dioxide, nitrous oxide (react to form nitric acid)	Fuel combusion; high-temperature processes	Smog; respiratory irritant
Carbon monoxide (CO)		Vehicle emissions; industrial emissions	Headache, fatigue; mental impairment; death at high doses
Volatile organic hydrocarbons (VOCs)	Benzene, methane, toluene	Industrial emissions	Smog; some are carcinogenic
Ozone (O_3)		Secondary pollutant (produced by photochemical reaction)	Respiratory irritant; eye irritant; major component of smog

of the atmosphere, with the potential for long-term impacts on the climate system; these will be discussed in chapter 16.

Emissions of sulfur oxides (**SO_x**) and nitrogen oxides (**NO_x**) result from industrial activities that involve the burning of fossil fuels. (The subscript x refers to the fact that several different nitrogen compounds are involved.) Nitrous oxide (NO) emissions combine with oxygen in the atmosphere to form NO_2, a major component of smog (see next section), which causes respiratory problems. Nitrogen and sulfur oxides also combine with water vapor in the air to form nitric and sulfuric acids.

In addition to the nitrogen oxides and sulfur oxides, a number of **volatile organic compounds (VOCs)** come from incomplete combustion of fossil fuels, as well as from solvents and paints. **Dioxins** are a group of chemicals that derive from the combustion of compounds containing chlorine. A host of industrial processes, including milling and smelting, pulp and paper manufacture using chlorine for bleaching, and, in particular, medical and municipal waste incineration, result in the production of these chemicals. Dioxins are known to cause cancer in lab animals and may have additional impacts on human reproductive, immune, and nervous systems. Some volatile compounds are polycyclic aromatic hydrocarbons (PAHs), which are mentioned in chapter 13; these include a number of volatile compounds, such as benzene and isoprene, which can also be problematic water pollutants.

What is arctic haze and where does it come from? ___________

Answer: A red-brown haze of pollutants that are worst in the arctic winter; carried to the arctic by long-range transport of pollutants from industrial centers, primarily Europe and Asia, by prevailing wind systems.

Cities and Pollution

If you live near a large urban center, you are undoubtedly familiar with **smog** (derived from a combination of *smoke* and *fog*). Smog is the brownish or grayish haze that often blankets city skylines, especially on hot summer days. It results from **photochemical reactions**—chemical reactions that are facilitated, or catalyzed, by sunlight—among airborne pollutants. Smog is usually at its worst in urban areas near the source of the contaminants; on sunny days when smog-forming reactions are facilitated; and during atmospheric conditions that keep the smog trapped in the city and close to the ground. Various VOCs also play a role in the formation of photochemical smog. Together, NO_x, SO_x, and VOCs are the main chemical precursors of photochemical smog.

In the process of urbanization, large expanses of asphalt, concrete, and other artificial surfaces are created. These contribute to the warming of cities by absorbing and reradiating solar energy to the air above them. The warming is compounded by the release of heat from sources such as vehicles, factories, and heated or air-conditioned buildings. In addition, some airborne contaminants that are typically found in higher concentrations in urban environments directly absorb solar energy

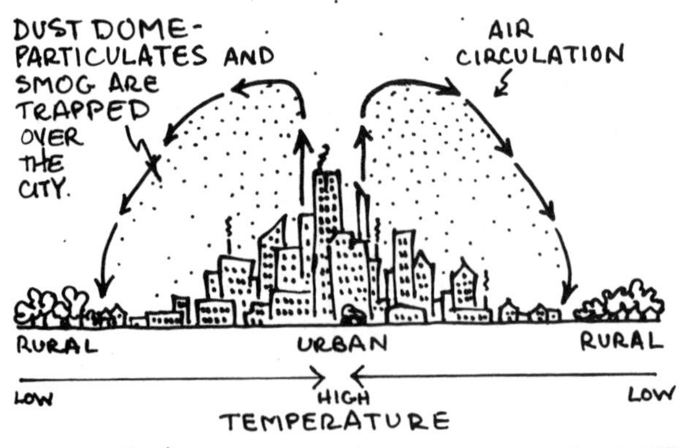

Figure 14.1.

and reradiate heat. The net result is that air temperatures in urban areas tend to be higher than those in the surrounding countryside. This is called the **urban heat island** effect (Figure 14.1). Tall buildings promote turbulence and the convection of heat above the city, and pollutants provide nucleation sites for the formation of water droplets in clouds. This can lead to increased cloudiness and greater frequency of precipitation, especially in summer downwind of the urban core. The pattern of air circulation that typically develops in an urban heat island also tends to trap and recirculate contaminants over the city, especially fine solid particles, contributing to a poor air quality situation called a **dust dome**.

Sometimes weather and topography join forces to exacerbate local air pollution. Under normal circumstances, temperatures in the lower part of the atmosphere decrease steadily with increasing height. However, if a warm air mass moves into the area, it may trap a pocket of cool air underneath it—a phenomenon known as a **thermal inversion** (Figure 14.2). When warm, contaminant-laden gases are released from factories or vehicles, the plume of polluted air rises because it is less dense (more buoyant) than the cooler air around it. In a thermal inversion, however, the plume of polluted air will rise until it hits the overlying warm air

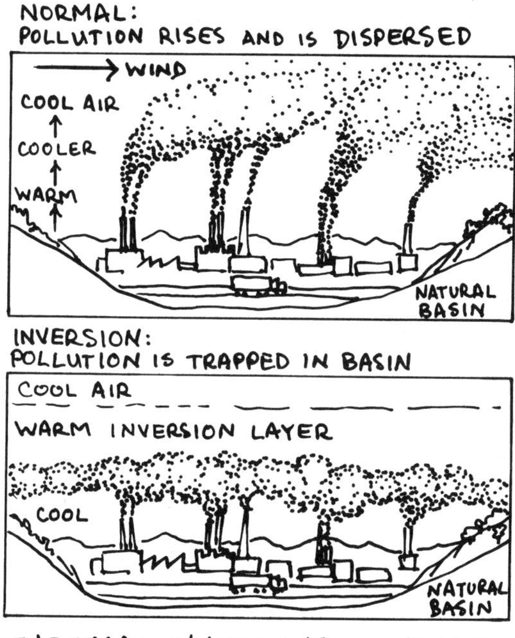

Figure 14.2.

mass, then spread laterally. Most episodes of acute air pollution are associated with thermal inversions. Local topography can intensify the effect of a thermal inversion. In Mexico City and the Los Angeles area, air pollution and thermal inversions are a particular problem because these cities are built on geographic basins that tend to trap cool air and concentrate pollutants.

The VOCs, SO_x, NO_x, and particulates that are emitted directly into the air are called **primary pollutants**; they are the chemicals that are transformed into the **secondary pollutants** that constitute smog. One of the main products of the photochemical reactions that create smog is ozone (O_3). Most of the ozone in the atmosphere is in the stratospheric ozone layer, where it blocks harmful incoming solar radiation. When it occurs in the lower part of the atmosphere, ozone is called **ground-level** or **tropospheric ozone**. Ozone is an eye irritant and can cause respiratory problems as well as damage crops and vegetation. Ground-level ozone is not generated directly by human activities but occurs as a product of photochemical reactions in air polluted with primary pollutants.

Another category of pollutant that can be an unpleasant and even hazardous aspect of city life is **noise pollution**. Sounds are called noise pollution when they become so loud that they are unpleasant, even becoming psychologically or physiologically damaging. Long-term exposure to loud noises can cause hearing loss, as well as a number of physiological changes that are not dissimilar to the physical effects of stress. Most of the noise that we are exposed to on a regular basis comes from human sources, primarily vehicles and machinery.

How do cities exacerbate atmospheric pollution? ______________

Answer: Large expanses of asphalt and concrete, with heat from vehicles, air conditioners, and factories, lead to the urban heat island effect. The air circulation tends to trap and recirculate contaminants over the city, especially fine particles, in a dust dome. Urban activities also contribute to noise pollution.

Indoor Air Pollution

We spend a lot of time thinking about air pollution outdoors, but we don't often give much thought to the pollutants that surround us in our homes, schools, and offices. With the advent of more thermally efficient buildings as a means of energy conservation, air leakage has been minimized. Unfortunately, a consequence is that the air inside buildings is not refreshed very often. Pollutants introduced into indoor air can become trapped and concentrated within the building. This may exacerbate existing allergies or asthma. Cases of extreme allergic sensitivity to indoor air pollutants are sometimes referred to as **sick building syndrome.**

There are three main sources of indoor air pollutants: pollutants that are introduced from outside; emissions from materials inside the building; and processes occurring inside the building such as smoking or cooking. One of the most hazardous of all indoor air pollutants is one that some people willingly and knowingly carry around with them, subjecting themselves and others to a lethal mix of hundreds of toxic air pollutants; it is, of course, cigarette smoke. Smoking and exposure to secondhand smoke are known to cause lung cancer, emphysema, and heart disease, and some evidence is emerging that they may contribute to the development of other cancers as well. This may be partly a synergistic effect; for example, women who smoke *and* take birth-control pills may be more susceptible to disease than women who do one or the other. Some of the toxins contained in cigarette smoke are cyanide, carbon monoxide, carbon dioxide, volatile hydrocarbons such as benzene, particulate matter, and even a small amount of radioactive material.

One of the most interesting pollutants introduced into buildings from outside is **radon**, a colorless, odorless, radioactive gas produced by the radioactive decay of uranium in soil derived from uranium-bearing rocks. As a soil gas, radon diffuses naturally through cracks and openings in permeable soils (Figure 14.3). Once it

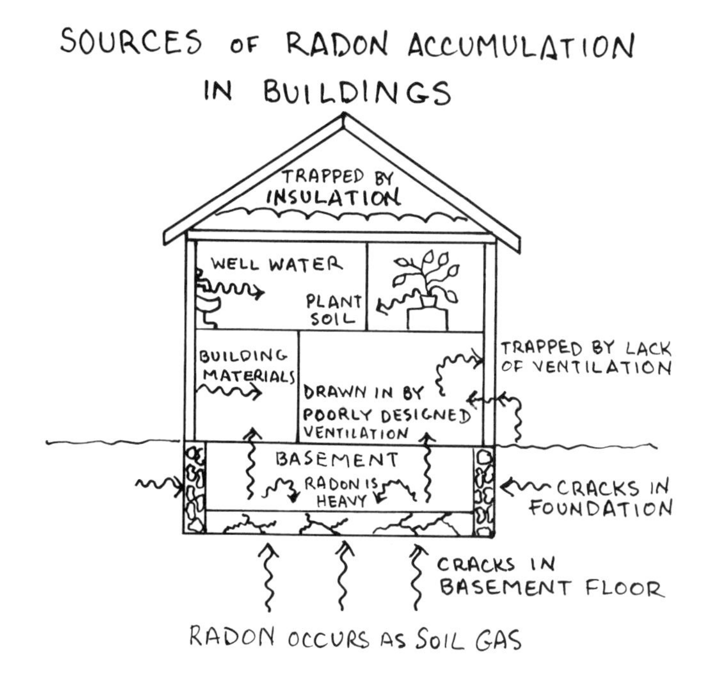

Figure 14.3.

enters the atmosphere, it is normally diluted and dispersed. The concentration of outdoor radon is therefore quite low. However, problems occur when radon enters the lower levels of buildings. Soil gas can enter a building via a number of pathways, including cracked foundations, pipes or windows with loose fittings, sump pumps, taps, and well water. Radon can also be emitted directly into the building from construction materials, such as concrete that contains uranium-bearing gravels, and even from the soil in potted plants.

Because radon is very heavy (it is the densest known gas), it tends to become concentrated in basements. Once inside a building, the radon gas continues to decay radioactively. Some of its decay products, or *daughters*, are also radioactive and emit harmful radiation as they decay. Some of the daughter products attach themselves to dust particles and thus become heavier and even more likely to be concentrated in the lower levels of the building. The real problem occurs when air contaminated by radon or one of its daughter products is inhaled. The radioactive decay emits alpha particles that are particularly damaging to biologic tissue. Direct doses of alpha radiation to the lining of the lung are thought to cause more deaths from lung cancer than any other cause except smoking.

It is exceedingly difficult to predict the level of radon in a building. If the building is constructed on a known uranium-bearing rock formation, chances are that it will have an elevated level of radon. But many other factors come into play, including the construction materials, weather (partly because the problem tends to be worse in the winter when buildings are shut tight), chemical and physical characteristics of the underlying soil and rock, and a host of other factors. Radon home testing kits are available; they can be used to identify areas of potential concern where more detailed measurements and remedial steps should be undertaken.

Other materials and activities within buildings can emit harmful pollutants. Smoke from wood fires have been a problem ever since people began cooking indoors, and it is still a major source of respiratory problems in developing countries, especially for women. Among the most common indoor air pollutants are cigarette smoke, carbon monoxide (mainly from furnaces), and formaldehyde, a VOC used in blown foam insulation. Formaldehyde and other VOCs can also be emitted directly from materials within the home or office, especially new carpets and new furniture (that "new car" smell). Fumes from cleaning solvents and paints are other common sources of indoor air pollutants.

Another particularly challenging indoor air pollutant is **asbestos**, the industrial name given to a group of minerals with fire-retardant characteristics. There are several asbestos minerals, including crysotile, amosite, and crocidolite. They have been widely used as insulating materials in a variety of settings; ceiling and floor tiles may contain asbestos. The asbestos minerals grow naturally with a fibrous or "asbestiform" crystal form, which makes them look more like very fine threads than rocks. These extremely fine, tough fibers that give the asbestos its special properties have led to concerns about the health impacts of asbestos use. If they are disturbed and mobilized, the fibers can become airborne; if they are inhaled, they may become embedded in the lining of the lung. Scientific evidence has shown that some asbestos fibers are carcinogenic; this has led to the widespread removal of all asbestos materials from buildings at great expense. However, the evidence suggests that the most widely used asbestos mineral, crysotile, may not be highly carcinogenic and that some of the less widely used forms of asbestos are much more carcinogenic.

Another ongoing controversy concerns the risk to humans of living in close proximity to an **electromagnetic field** (**EMF**)—an invisible field of long-wavelength radiation. Exposure is typically associated with high-voltage power lines or with small appliances such as microwave ovens, electric blankets, clock-radios, and cell phones. Some studies have suggested that long-term exposure to EMFs may lead to increased incidence of cancer. So far, both lab studies and epidemiological studies (studies of the geographic distribution of diseases) have proven inconclusive on this issue.

What is radon and why does it pose a health risk? ______________

Answer: The heaviest known gas. It is radioactive, as well as colorless and odorless, and occurs naturally as a soil gas. If it becomes concentrated inside a building, it can be inhaled, where it will continue to decay, emitting alpha particles directly to the lung. This is a significant cause of lung cancer.

Acid Deposition

Acid precipitation, or **acid rain**, is rain that has a pH less than 5. As discussed in chapter 9, pH is a measure of the concentration of hydrogen ions in a solution. The pH scale ranges from 0 (highly acidic) to 14 (highly alkaline), with 7 indicating a neutral solution (Figure 14.4). It is a logarithmic scale, which means that a solution with a pH of 4 is ten times more acidic than a solution with a pH of 5. For comparison, the pH of battery acid is 1; lemon juice, 2; vinegar, 3; distilled water, 7; baking soda, 8; and ammonia, 12.

All rain is somewhat acidic; the pH of "natural" rain is around 5.6. Rain is naturally acidic because there are trace gases naturally present in the atmosphere, especially CO_2, which combine with water vapor to make acids. Carbon dioxide

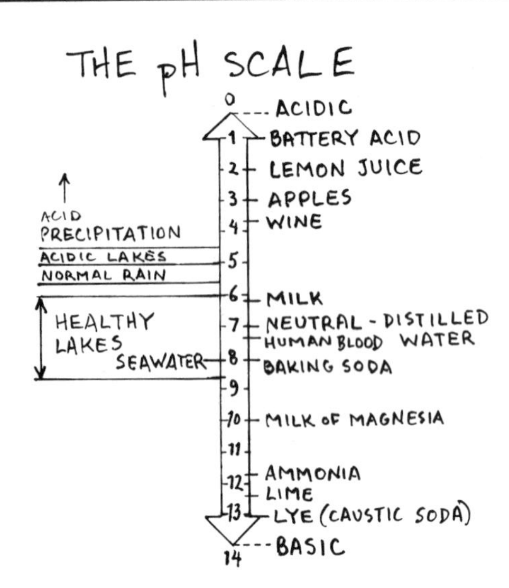

Figure 14.4.

gas reacts with water vapor in the atmosphere to make carbonic acid (HCO_3). Groundwater is naturally slightly acidic for the same reason. Rain can also become acidic if it interacts with smoke from a forest fire or ash from a volcanic eruption. Rain that is more acidic than natural rain commonly has a pH in the range of 3 to 5, although rains with a pH as low as 1.5 have been recorded. Acid rain is a form of **acid deposition**, which may be wet or dry, solid or liquid. There have been incidents of acidic snow and even acidic fog. Acidic particulate matter can also fall as dry deposition.

Acids form in the atmosphere when precursor pollutants interact with water (Figure 14.5). The most common precursors are CO_2, SO_x (sulfur oxides, especially SO_2 and SO_3), and NO_x (nitrogen oxides). All of these chemicals are present in the atmosphere naturally, but their concentrations have increased as a result of anthropogenic emissions. For example, the amount of sulfur mobilized by anthropogenic processes is now thought to exceed the sulfur that comes from natural sources such as forest fires and volcanic eruptions. Anthropogenic carbon dioxide and sulfur dioxide come from the burning of fossil fuel, mainly coal. Coal-fired power plants are often installed with **scrubbers**, which remove some of the harmful compounds from stack emissions before they are released to the environment. The smelting of sulfide-bearing ores is another important source of sulfur emissions. Nitrogen oxides also come from power plants and from motor vehicle emissions; natural sources include forest fires, soil gases, and lightning.

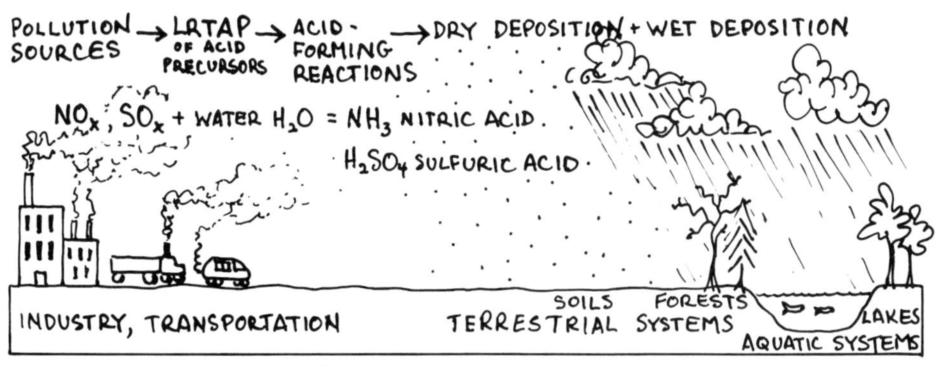

ACID DEPOSITION
FORMATION, DEPOSITION, AND IMPACTS

Figure 14.5.

Once in the atmosphere, the acid precursors are transported by the wind (see Figure 14.5). They become oxidized, and some of the resulting sulfate and nitrate particles may fall as dry deposition close to the emission source, becoming acidic when they interact with moisture on the ground. The remainder of the airborne material interacts with water vapor in the atmosphere, reacting to become carbonic acid (HCO_3), sulfuric acid (H_2SO_4), or nitric acid (HNO_3). Depending on local and regional weather conditions and circulation patterns, gaseous pollutants and acids can stay in the atmosphere up to two weeks.

To summarize, acid precursors are airborne pollutants that come primarily from the burning of fossil fuels either in industrial processes or in transportation. They form acids through oxidation and chemical interaction with water vapor in the atmosphere. The precursor pollutants and resulting acids are transported and distributed by atmospheric circulation and fall to the ground as either dry or wet deposition. Therefore, it is not difficult to understand why the areas that are hardest hit by acid precipitation are in the immediate vicinity and downwind of heavily industrialized areas. Areas most affected by acid precipitation include the northeastern United States and southeastern Canada, eastern and western Europe, and newly industrialized countries in Asia.

Acid deposition has negative impacts on both natural and human systems. For example, acids can speed the degradation of rubber, paint, and plastics, as well as building materials such as steel, cement, masonry, and building stone. Acid precipitation has a negative effect on trees, rendering them more susceptible to the effects of cold and disease. In areas with highly acidic precipitation, soils and surface waters can become acidified. In aquatic ecosystems, particularly lakes, the acidification of water can cause reproductive failure and death in fish.

The acidification of lakes and streams is influenced by underlying rocks, sediments, and soils. Some surface water bodies are particularly susceptible to acidification; others are more resistant. Rocks, soils, and other materials that have the chemical ability to neutralize acids are called **buffers**. In particular, carbonate-rich materials, such as limestone, chalk, and calcium carbonate-rich soils, are efficient buffers that increase the resistance of a water body to acidification. In contrast, water bodies that are underlain by silica-rich rocks such as granites, which do not function as buffers, are highly susceptible to acidification. These lakes are said to be sensitive to acidification.

The acidification of water has an impact on the solubility of other components in the water. For example, some toxic metals—notably aluminum, lead, and mercury—are more soluble in acidified water, which means that the aqueous concentration of these toxic metals typically increases as a lake becomes acidified. This may be the cause of the death and reproductive failure of fish in acidified lakes.

In some cases, it is possible to remediate an acidified water body. This is usually done by the addition of buffering materials such as lime. It is an expensive and temporary solution that must be repeated periodically. Ultimately, the best solution for acid precipitation is to eliminate or drastically reduce the emission of acid precursors from industrial and other sources.

What is a buffer? ______________

Answer: A rock, soil, or other material that has the chemical ability to neutralize acids.

SELF-TEST

These questions are designed to help you assess how well you have learned the concepts presented in chapter 14. The answers are given at the end. If you get any of the questions wrong, be sure to troubleshoot by going back into that part of the chapter to find the correct answer.

1. The concentration of particulate pollution over cities is called a(n) ______________.

   a. thermal inversion
   b. suspended particulate
   c. heat island
   d. dust dome

2. Which one of the following is *not* a health or environmental impact of ground-level ozone?
   a. respiratory problems
   b. eye irritation
   c. damage to crops and vegetation
   d. All of the above are impacts related to ground-level ozone.

3. Normal rain typically has a pH of approximately ______________, whereas acid rain typically has a pH of ______________ or less.
   a. 5.6; 5 b. 7; 5 c. 8.5; 3 d. 7.5; 2

4. A technology that removes harmful substances in stack emissions from coal-fired power plants before they are released into the environment is called a(n) ______________.

5. ______________ is the heaviest known gas; it is radioactive, and it occurs as a soil gas.

6. Indoor air pollution has been more of a problem since buildings have become more energy efficient and leakproof. (True or False)

7. Sulfuric acid and nitric acid are primary air pollutants. (True or False)

8. Suspended particulates come from natural as well as anthropogenic processes. (True or False)

9. What are dioxins, where do they come from, and what are their potential health impacts?

10. Why is a thermal inversion inherently unstable?

11. What is the impact of lake acidification on the solubility of toxic metals?

12. Describe how radioactive contaminants from the Chernobyl nuclear disaster were transported and deposited in the days and weeks following the event.

ANSWERS

1. d 2. d 3. a 4. scrubber

5. Radon 6. True 7. False 8. True

9. A group of chlorinated hydrocarbon compounds that result from a number of industrial processes that use chlorine, including milling, smelting, pulp and paper manufacture, and medical waste incineration. They are carcinogenic and may also have impacts on human reproductive, immune, and nervous systems.

10. In a thermal inversion, warm air ends up trapped beneath cooler air; this is backward with regard to density stability—the warm air will tend to rise and the cooler air to sink, which leads to an unstable situation.

11. Some toxic metals, notably aluminum, lead, and mercury, are more water soluble in acidified water than in regular lake water, which means the amount of metals dissolved in the lake water will likely increase if the lake becomes acidified. This can have a negative impact on organisms living in or near the lake.

12. Radioactive particles were transported as suspended particulates by long-range atmospheric transport; they encircled the globe within about two weeks. Areas that were close to the accident site and had experienced rainfall within a few days of the event had the most intense deposition of contaminants.

KEY WORDS

acid deposition
acid precipitation
acid rain
arctic haze
asbestos
buffer
dioxin
dry deposition
dust dome
electromagnetic field (EMF)
ground-level (tropospheric) ozone
long-range transport of atmospheric pollutants (LRTAP)
noise pollution
NO_x (nitrous oxides)
photochemical reaction
primary pollutant
radon
scrubber
secondary pollutant
sick building syndrome
smog
SO_x (sulfur oxides)
suspended particulates
thermal inversion
urban heat island
volatile organic compound (VOC)
wet deposition

15 Cities and Waste Management

This was among my prayers: a piece of land not so very large, where a garden should be and a spring of ever-flowing water near the house, and a bit of woodland as well as these.

—Horace

Historians may refer to this as the Throwaway Age.

—Vance Packard

Objectives

In this chapter you will learn about:

- urbanization and the physical and socioeconomic impacts of cities on the environment;
- environmental risk;
- cities as ecosystems; and
- waste types, sources, and management.

Urbanization

Our world and the people who live in it are changing. In 1900, only one in ten people lived in cities. By 2000, more than half of the world's people—approximately 3.3 billion—were city dwellers. The trend is continuing; by the year 2025, about two-thirds of the world's population will live in an urban center. The world's people are becoming increasingly urbanized, and cities are growing in size as well as in economic importance. Much of the future growth of urban centers—perhaps as much as 90 percent of it—will occur in cities in the less economically developed

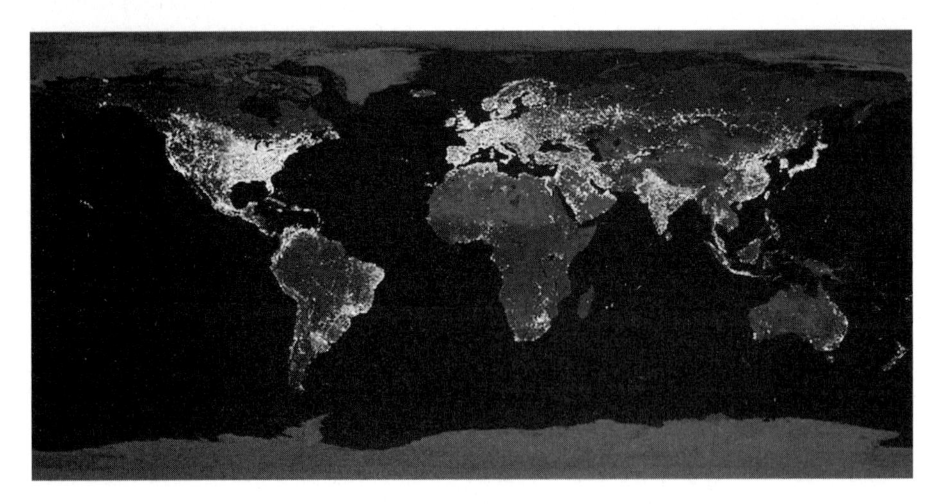

In this composite satellite image made by the U.S. Defense Meteorological Satellites Program, the U.S. east coast is ablaze with nighttime light, with its cities growing closer and closer together.

countries of the world, where the average population growth rate is 5 percent per year. In countries of the industrialized world, in contrast, average growth rates are 0.8 percent; some countries are even at the point of experiencing negative population growth—their populations are contracting rather than expanding.

Do you recall how to calculate the doubling time of a population with a growth rate of 5 percent per year? (Check chapter 7.)

As a result of rapid urbanization, cities in less developed countries are characterized by three features: (1) Very high rates of growth and the "unintended city" phenomenon, in which squatter settlements spring up, and many urban areas lack basic infrastructure and services such as roads, piped water, and electricity. (2) Absolute size and the "mega-city"; by the year 2015, there will be more than thirty cities with populations greater than 8 million, and twenty-seven of them will be in developing nations (Table 15.1). (3) Urban poverty; half of the world's "absolute poor"—those who live on less than $1 per day—now live in cities. Altogether, this is a very different world compared to a few decades ago. These changes have important implications for people's lifestyles as well as for the biophysical environment.

There are two basic reasons why the population of a city increases. The first is the natural population increase from the fertility of those who already live in the city; the second is rural–urban migration. In the 1960s in North America, for

Table 15.1 Mega-Cities of the World: 2005 and 2015 Population (in millions)

	2005	2015
Tokyo, Japan	26.8	27.2
São Paolo, Brazil	19.6	21.2
Mexico City, Mexico	18.9	20.4
Mumbai (Bombay), India	18.3	22.6
New York, U.S.A.	17.1	17.9
Dhaka, Bangladesh	15.9	22.8
Delhi, India	15.3	20.9
Calcutta, India	14.3	16.7
Los Angeles, U.S.A.	13.8	14.5
Jakarta, Indonesia	13.2	17.3
Shanghai, China	12.7	13.6
Buenos Aires, Argentina	12.4	13.2
Karachi, Pakistan	11.8	16.2
Rio de Janeiro, Brazil	11.2	11.5
Lagos, Nigeria	11.1	16.0
Osaka, Japan	11.0	11.0
Beijing, China	10.8	11.7
Metro Manila, Philippines	10.7	12.6
Cairo, Egypt	10.1	11.5
Istanbul, Turkey	9.9	11.4
Seoul, Republic of Korea	9.9	9.9
Paris, France	9.8	9.9
Tianjin, China	9.4	10.3
Lima, Peru	8.2	9.4
Moscow, Russian Federation	8.1	< 8.0
Bangkok, Thailand	8.0	9.8
Kinshasa, Dem. Rep. Congo	< 8.0	9.9
Bogotá, Colombia	< 8.0	9.0
Lahore, Pakistan	< 8.0	8.7
Bangalore, India	< 8.0	8.4
Teheran, Iran	< 8.0	8.2

Population projection data from United Nations Population Division

example, many large cities underwent a major migration from urban centers to outlying suburban and rural areas. This led to a widespread decline of city centers. Recently, however, cities have made a comeback, as an aging population moves to urban areas for greater access to cultural activities, health care, jobs, and other services. It is worth noting that even cities in industrialized countries that are gaining people very slowly can continue to spread geographically. For example, the metropolitan New York area has only grown in population by about 5 percent in the past twenty-five years, but the developed area of the city has increased by more than 60 percent.

In developing countries, people are drawn to the city for the same reasons; statistically, people who live in urban centers earn more income than people in rural villages. Even so, the life they find when they arrive in the city may be one of grinding, unforgiving poverty, with few opportunities or services available to them. People in poorer countries may also be driven from rural areas by war, persecution, or environmental degradation. Desertification and famine in the Sahel region of sub-Saharan Africa over the past couple of decades has led to massive migration and **environmental refugeeism**, with the concomitant rapid growth of cities like Dakar (Senegal), Abidjan (Ivory Coast), and Lagos (Nigeria).

What are the three features that characterize urbanization in developing nations (as compared to urban growth in more developed nations)? ______________

Answer: (1) Very high rates of growth and a corresponding lack of infrastructure and services; (2) absolute size, the "mega-city"; and (3) urban poverty.

The Physical Impacts of Urbanization

In both industrialized and developing countries, uncontrolled urban growth—in numbers of people or space consumed—requires increasing land, water, food, and energy resources from surrounding regions to meet the urban residents' needs. Although we tend to think of cities as concrete jungles and pollution generators, the concentration of activities in urban centers can have both positive and negative environmental impacts. Let's look at some of the common physical impacts of urbanization.

From the earliest days of human society when urban centers were first established some 6,000 years ago, human settlements have had impacts on the biophysical environment. The initial environmental impacts of urban development result from the clearing of land. Deforestation may have begun as far back as 23,000 years ago in parts of Africa and Asia that were home to early humans. Indeed, it is possible that virtually no forested area remaining on Earth is completely pristine and free from the imprint of humans. Today deforestation is occurring more rapidly in the developing world than in North America and Europe; the older industrialized world has already been widely deforested over the past 200 years.

Another impact of urbanization is the draining of resources from surrounding areas. This type of impact has been observed in some of the earliest known human settlements. Cities grow at the expense of the surrounding countryside, which geographers refer to as the **hinterland**. The loss of farmland and forested area and the drain of water, energy, and food resources can lead to a loss of ecologic integrity and reduced carrying capacity in the rural hinterland as well as surrounding smaller settlements.

As human settlements grow to become large industrial centers, further modification of the physical environment occurs. Some of these modifications can lead to increased risk from **natural hazards**, which include floods, landslides, earthquakes, cyclones, and volcanic eruptions. The increased potential for exposure to natural hazards associated with the urban setting is part of **urban environmental risk**. Urbanization can increase the likelihood that a particular type of hazard, such as floods or landslides, will occur. It can also increase the likelihood that people will be exposed to a hazard like an earthquake or a cyclone not because the frequency or intensity of the hazard has increased but simply because of the high population density that is typical of cities.

One of the obvious changes associated with urbanization of a natural landscape is the increase in impermeable surfaces. Roads, parking lots, and buildings all modify natural surfaces, which has a significant impact on the hydrologic cycle. Areas that acted as groundwater recharge zones prior to urbanization may become completely impermeable; water that would have infiltrated must now run off over the land surface. The installation of storm sewer drainage and channelization of waterways are typical engineering interventions designed to control surface runoff in cities. With these modifications, however, precipitation and surface runoff drain into the nearest river much more rapidly than they would from a natural, vegetated, permeable surface. This can contribute to an increased risk of urban flooding.

A further impact of urbanization on the hydrologic cycle comes with the depletion of groundwater resources. If aquifers become depleted, they may undergo compaction. This is particularly true of aquifers in which the rock or sediment is silty, meaning it consists of very fine clay particles. Silty sediments are typical of riverbanks and estuaries—perfect sites for the establishment of a settlement because of the access to water. Clay minerals typically occur as tiny platelets, which can collapse on one another if the interstitial water that supports the platelets is withdrawn. When aquifers undergo compaction, the entire ground surface may respond by subsiding; this process can be greatly exacerbated by the overlying weight of a large urban center. Subsidence combines with rapid runoff from impermeable surfaces, storm drainage, and channelization to contribute further to urban flooding.

Channel modification, such as the straightening, widening, or concrete lining of stream channels as they pass through the city, is one type of modification to urban landforms, or **geomorphology** (from the Greek words *geo*, meaning "land," and *morph*, meaning "form"). Other geomorphologic changes include slope modifications for construction or for road-building purposes. Changing a natural slope may stabilize it or destabilize it; geomorphologic changes can increase the risk of urban landslides.

Large urban centers can contribute to atmospheric and even climatologic changes in the local area. For example, the urban heat island effect (described in chapter 14) leads to higher temperatures, altered wind patterns, and a concentrated "dome" of pollution over the city center. Pollutant output leads to increased health risks from exposure to air, water, and soil pollution. The export of pollution and waste, or **residuals**—the "leftovers" of urban and industrial processes—represents another potentially negative impact on the environment in the hinterland. We will consider the issue of waste and waste management in greater detail in the latter part of this chapter.

What types of geomorphologic changes are typical in urban centers and what are their potential environmental impacts? ______________

Answer: Channel modifications (the straightening, widening, and concrete lining of streams) and slope modifications for construction and road building. They can contribute to increased risk of urban flooding and landslides.

The Socioeconomic Impacts of Urbanization

We mentioned that urbanization can have positive impacts on the environment, but so far we have not presented any real evidence of this. The positive environmental impacts of cities may be seen when we consider the socioeconomic changes that accompany urbanization. When people become city dwellers, they change not only their place of residence but their lifestyle and consumption habits. Statistically, people who live in cities have better access to jobs, higher incomes, better health care, higher literacy rates, and better access to information and cultural services. These factors contribute to some very significant differences between rural and urban people. On average, city dwellers have lower **fertility rates** (the average number of children born to each woman during her lifetime), as well as lower infant mortality rates. Thus, even though cities themselves are crowded, the urban setting contributes to lower fertility rates and therefore to a more rapid progression through the demographic transition into a phase of stable or even declining population.

Look back at chapter 7 to review the demographic transition and other concepts related to human population.

Cities can have other positive impacts on the environment. It is cheaper and more efficient to provide goods and services in bulk and to people who are living close together. This is why people who live in the country or in remote areas often pay more for goods and may not have access to services such as piped-in natural gas and cable TV; it is simply too expensive to supply these goods and services to individuals in remote areas. This is what economists call **economy of scale** and **economy of proximity**. The savings experienced by grouping people together in the urban setting are both economic and environmental; it costs less to deliver to people who are closely grouped, and it places less of a burden on the environment.

Consider transportation. It would cost a lot more—both financially and in terms of greenhouse gas emissions and other airborne pollutants from cars and trucks—to deliver goods to individual families in the country than to make one big delivery to a central location in a city. In fact, people who live in cities have lower per capita greenhouse gas emissions than people who live in rural areas because of the differences in their transportation needs and choices.

There is, of course, a downside to the socioeconomic impacts of cities. On average, people who live in cities generate considerably more waste and have higher energy consumption per capita than rural people. They also earn more money. They tend to make dietary choices that are more processed (review the diet transition, chapter 9), and they tend to consume more durable goods (like refrigerators, stereos, televisions, and so on). All of this is part of what is sometimes called the **consumption transition**, a change in lifestyle associated with increased income and a more urban setting. The nonbiodegradable wastes, pollution, energy use, and other environmental costs associated with urban consumption are part of the environmental impacts of the urban lifestyle.

An unavoidable aspect of city life is exposure to urban risk, as discussed earlier. This includes increased potential for exposure to natural disasters, whether the exposure is a result of high population density or the hazard is directly caused by modifications to the physical environment. There are enhanced risks from **technologic hazards**—hazards generated by human industrial activity. A particularly tragic example of a technologic disaster is the 1984 release of 36 metric tons (40 tons) of the toxic gas methyl isocyanate (MIC) from a storage tank at a pesticide-producing facility in the city of Bhopal, India. Approximately 600,000 people were exposed to the deadly gas; about 2,500 were killed directly, and another 60,000 have lingering health problems related to the exposure. Urban environmental risk contributes to health impacts such as the spread of communicable disease. For example, the SARS (sudden acute respiratory syndrome) outbreak of 2003 occurred because the virus was easily spread among people living and working in close proximity; in fact, the virus was probably first transferred to the human population as a result of people and animals living in extremely close quarters.

Increased environmental risk in the urban setting speaks to the concept of **vulnerability**, a combination of the individual or community potential for exposure to hazards, and the capacity to respond to that risk. For example, the people of Bangladesh are particularly vulnerable to the effects of cyclones not only because the Bay of Bengal is a cyclone-prone area but also because the country and the people are poor. They have no early warning system. Construction is of poor quality, population density is high, and buildings often collapse during frequent flood events. Emergency-response organizations are few, their funding is low, equipment is poor, and training is inadequate. All of these social, economic, and environmental factors contribute to the high vulnerability of the Bangladeshi people to this particular natural hazard.

What is the difference between a natural hazard and a technologic hazard? Give at least two examples of each. ______________

Answer: Natural hazard: a natural process that is hazardous to people; examples are landslides, earthquakes, tsunamis, volcanic eruptions, cyclones, and floods, as well as insect infestation. Technologic hazard: generated by human industrial activity; examples are release of toxins and pollutants, sick building syndrome, radon buildup, asbestos exposure in the built environment, and building collapse.

Sustainable Cities

In chapter 2 we defined a system to be any portion of the universe that can be separated from the rest of the universe for the purpose of observing changes within it. In environmental science, perhaps the most widely studied type of system is the ecosystem, which comprises a biologic community and its interactions with the abiotic components of the environment such as water, sunlight, and soil (chapter 4). These interactions include exchanges of both matter and energy within the ecosystem and with the surrounding environment. People who study the urban environment have also applied the systems concept to the study and management of cities.

If you think about it, a city is much like an ecosystem (Figure 15.1). There are individuals who live within the boundaries of the system. Energy enters the system

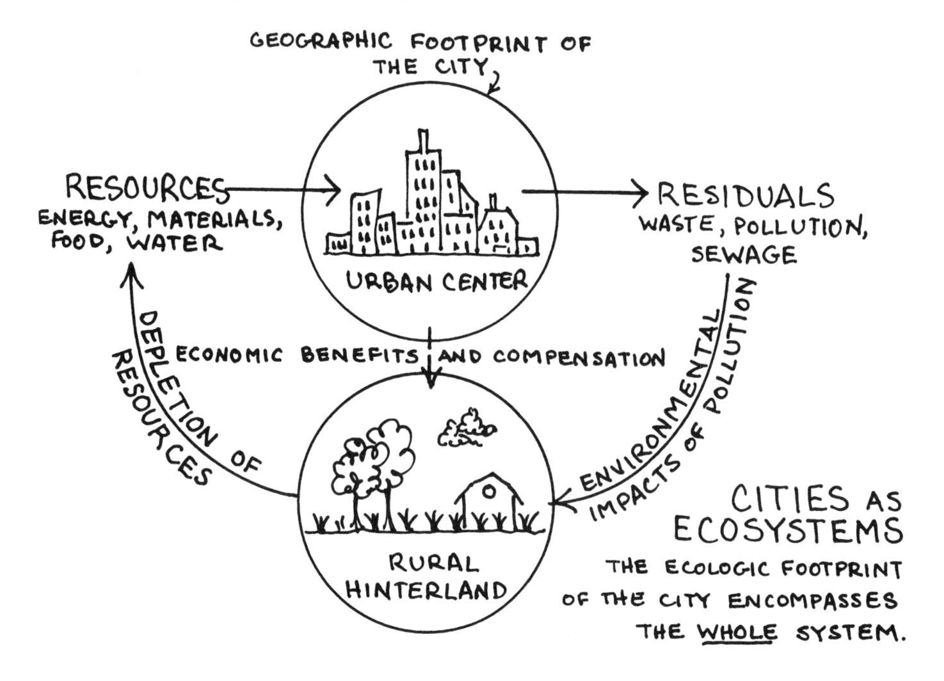

Figure 15.1.

(natural energy in the form of sunlight and technologic energy in the form of fossil fuels and other chemicals) and leaves the system (mainly as heat). Matter enters and leaves the system in the form of the importation of resources (water, food, fuels, products) and exportation of wastes. Within the urban system, various interactions occur by which basic resources are modified and wastes are produced. These interactions are similar to the metabolic processes that occur in natural ecosystems and, in fact, in individual organisms.

What kind of system do you think best approximates a city—open or closed?

Answer: Open, like most environmental systems, because both matter and energy cross the boundaries.

The ecologic footprint concept (chapter 7) can help us think of cities as open systems. Urban centers typically have ecologic footprints—land-related environmental impacts—that are much, much larger than the actual boundaries of the city. The large ecologic footprint is unavoidable; cities must draw resources from a wide area to support the needs of the many inhabitants within their borders. This fact requires us to recognize that the resources that support a city must come from somewhere—they are drawn from potentially finite sources, and there are environmental impacts associated with their extraction, transportation, and use. The wastes and pollution generated and exported by cities do not go to a reservoir with infinite waste-absorbing capacity; they go to the hinterland, where they have environmental impacts.

If these interactions are inevitable, can there be any such thing as a sustainable city? Perhaps not; a system that grows at the expense of surrounding areas, has a seemingly limitless appetite for the input of resources, and produces and exports an ever-increasing quantity of waste does not seem to fit anyone's definition of sustainability.

Recall the definition in chapter 7 of sustainable development as the cautious, planned utilization of resources to meet current needs without degrading ecosystems or jeopardizing the future availability of those resources. How do you think this concept can be most effectively applied to the management of urban systems?

Still, urban managers have begun to take the concept of the sustainable city seriously. The basic idea is to manage the city in partnership with its supporting rural hinterland, not in isolation from it. This allows for the planning of more efficient and effective transportation systems in metropolitan and associated suburban regions. Minimizing urban environmental risks within the city and enhancing the potential benefits associated with grouping people together in close proximity are important in aiming for sustainability of urban systems. Surrounding communities may suffer a loss of carrying capacity—that is, they receive the environmental impacts of resource depletion and the export of pollution and waste from the city. The economic growth and development offered by cities is partial compensation for this loss.

The city of Curitiba, Brazil, was designed to be a model of a sustainable city. High-density development in Curitiba is focused along high-speed bus lines to maximize the benefits of mass transportation and minimize traffic and pollution from cars. Do some investigating. What are some of the other urban design features of Curitiba? How does its population (over 2.5 million) compare to the population of the city you live in or near? What are some of the features that your city has in common with Curitiba? What are some of the differences? In your opinion, is Curitiba truly a sustainable city?

Solid Waste: Sources and Types

Virtually every human activity generates waste. Even the simple act of eating results in one of the most challenging waste problems of modern society: sewage. And this doesn't even take into account the wastes generated in the production and transportation of the food that was eaten. Waste disposal is a problem that increasingly demands the attention of scientists, engineers, policy makers, and the general public. This is partly because the volume of waste is increasing at an alarming rate and partly because our understanding and appreciation of the hazards associated with improperly handled wastes are growing.

The range of wastes produced by human activity is so vast in terms of sources, chemical characteristics, and physical properties that it is virtually impossible to come up with a simple, all-encompassing categorization scheme. In this chapter we focus mainly on two types of waste that present significant disposal problems, particularly in the urban context. These are municipal solid waste and sewage. Many of the waste-disposal concepts discussed in this chapter are also applicable to hazardous and radioactive wastes, but the particular challenges of dealing with those materials are discussed in somewhat greater detail in chapters 13 and 14 (water, soil, and air pollution, and the characteristics of toxins) and chapter 11 (nuclear energy and radwaste). In this chapter we focus primarily on solid waste. For more information about hazardous wastes, please review those earlier chapters.

Solid waste, which can be defined very broadly as waste material that cannot be easily passed through a pipe, comprises a very wide range of materials that come from a variety of sources. Although the difference between a solid and a liquid seems clear enough, the distinction between solid and liquid waste is not always quite so obvious. When solid wastes accumulate, water may pass through and pick up soluble components; as a result, the distinction between solid and liquid waste may become blurred. In contrast, a waste that is principally a **liquid waste** that can be passed through a pipe, such as sewage or watery mud, may become separated or concentrated into a more solid form as a result of treatment procedures or natural settling processes.

The principal sources of solid waste are agriculture and mining. More than half of all solid waste is generated by the agricultural sector, which includes farms, orchards, ranches, and animal feedlots. Sediment eroded from fields is an important type of agricultural waste. The eroded sediment can settle and clog adjacent waterways, and soils lost from agricultural lands often carry with them a wide range of chemicals such as pesticides, fungicides, herbicides, heavy metals, and chemical fertilizers. Aside from sediment and the associated chemicals, however, agricultural solid wastes are almost entirely organic, consisting of such materials as animal excrement, dead animals, and offal; stubble from fields; prunings from orchards; and other types of plant and animal debris. Because of its organic nature, most agricultural waste can biodegrade if it is collected in the right type of environment. However, it can also cause an excess of nutrients, oxygen depletion, and eutrophication in water bodies if it accumulates too quickly. Animal wastes can carry infectious agents, such as *E. coli* bacteria and parasites, which can present significant human health hazards.

The second largest generator of solid waste is mining. The largest volume of mine waste occurs as discarded piles of waste rock; most of this material is disposed of at the mine site. Another major component of solid mine waste is tailings, the slags and sludges that are left over after processing or smelting of ores. Piles of waste rock and tailings are an eyesore, and the minerals and chemicals in them may combine with rainwater to produce acid mine drainage or toxic effluents.

Industries other than mining and agriculture also generate solid waste. Most of this (aside from hazardous wastes, which are discussed in chapters 13 and 14) is in the form of paper, cardboard, scrap metal, wood, plastics, glass, tires, and rags. Many industrial waste materials are potentially reusable or recyclable. **Recycling** refers to the extraction of usable raw materials, such as metal, glass, and pulp, from waste products such as manufactured objects, paper and cardboard, cans, jars, and bottles. It is worthwhile for industries to consider recycling as a means of avoiding large **tipping fees** for the disposal of solid wastes at dumpsites. Some studies have shown that industrial recycling rates improve noticeably when tipping fees increase because high tipping fees make it economically beneficial for companies to recycle their wastes. **Source reduction**—cutting back on wastes through production efficiencies, increasing the useful lifetimes of products, and reducing packaging—is another important aspect of industrial waste management. Some companies are able to find industrial uses for their waste through **waste exchange** programs, in which a central agency endeavors to find a market for waste products; the leftovers from one industrial process can often be used as the raw materials for another.

Some waste from stores, offices, and small industries is set by the side of the road and collected along with residential (household) waste. Together, they comprise **municipal solid waste**, the smallest but fastest-growing source of solid waste. Essentially anything that is routinely collected by city trucks on garbage day is municipal waste. The main components of residential waste are paper and cardboards, organic household and yard wastes, nonbiodegradable materials, including glass, metals, and plastics, and many hazardous substances (Figure 15.2).

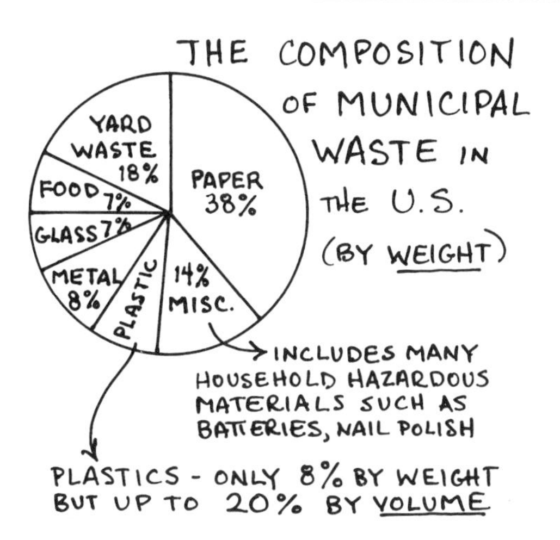

Figure 15.2.

Both the volume—more than 400,000 tons *per day* in the United States alone—and the composition of municipal solid waste have changed over the past few decades. Paper, cardboard, and other packaging now account for the largest amount of space in most landfills. Some materials take up a lot of room but are quite light, such as plastics. Other materials, such as glass and metal, are denser, so they weigh more but take up less space. If space in the landfill is an issue, it may be wise to cut down on the proportion of high-volume wastes. But if the waste of potentially useful resources is of primary concern, the percentage of waste types by weight is a better indicator. Wood waste, tires, ash from incineration, and household hazardous substances are growing disposal problems, too.

Hazardous wastes are waste materials that are toxic, caustic, acidic, explosive, infectious, or radioactive. In the household context, such substances commonly come from cleansers and solvents, nail polish, paint, batteries, pesticides, and many other common household substances. Even unused pharmaceuticals can become hazardous wastes if they are allowed to enter the natural environment.

Take a hazardous substance inventory of your home. What are the potential sources of hazardous wastes in your house and yard? Do you know where to find the local disposal facility that will accept household hazardous wastes? What are its hours and days of operation? Go on a field trip to the facility and drop off your leftover paints, batteries, and other hazardous substances; the trip will probably be an eye-opener.

As in the industrial context, homeowners have become increasingly aware of the need to manage their solid waste generation by making use of the "**4 R's**": **refuse** (that is, simply consume less and buy less); **reduce** (for example, waste generation can be reduced by avoiding products with excessive packaging); **reuse** (don't throw away anything that might still be usable); and **recycle** (most communities now have "blue box" programs for municipal recycling of glass, metal, paper, and plastics).

Another important component of household solid waste is the organic material commonly left over after family meals (fruit and vegetable peels, meat trimmings, egg shells, and so on) and yard work (tree and shrub prunings, grass clippings, autumn leaves, and so on). These wastes can be reduced through **composting**—that is, providing a contained environment in which organics can biodegrade quickly. Some municipalities now provide "green box" curbside pickup for yard wastes and household organics.

What is a waste exchange program? ____________

Answer: A program in which a central agency finds markets for industrial waste products, using the leftovers from one industrial process as raw materials for another.

Waste Disposal

Solid waste produced in homes, schools, offices, and small commercial establishments accounts for a relatively small proportion of total solid waste generation, but it presents a major disposal problem. In addition to being the fastest-growing category of solid waste, the majority of municipal solid waste is generated in areas of high population density, where land may not be available on which to dispose of the waste. Environmental and human health problems can result if municipal wastes are disposed of improperly, whether in open dumps, inactive landfills, or active landfills that are improperly engineered.

In the past, most towns discarded their waste in **open dumps**—usually little more than a hole in the ground where garbage was allowed to accumulate. Open dumps still exist in many rural areas and in most cities in the developing world. They tend to be eyesores, attracting birds and vermin, causing unpleasant smells, and creating potential health hazards. Sometimes people dispose of solid waste by *not* disposing of it properly; dumping by the side of the road or off the edge of a cliff is usually done to avoid the costs of proper disposal.

Incineration—the burning of refuse in a specially designed facility—is another widely used method of solid waste disposal. Incineration methods vary from backyard burning to modern, highly engineered incinerators. Modern incinerators are designed to burn garbage at very high temperatures, and they typically have very effective technologies to remove hazardous emissions before they are allowed into the atmosphere. All incinerators, even modern ones, generate residual ash, which may contain toxic metals and by-products of combustion such as dioxins and furans. The ash must be disposed of in a landfill and sometimes requires special handling. Some incinerators take advantage of the heat they generate to produce electricity through cogeneration.

Today **landfills**—waste disposal sites that are engineered and monitored to contain wastes within the site—are the most common approach to waste management and are used by municipalities around the world. Landfill types range from the old-fashioned dump to highly engineered, carefully located disposal facilities (Figure 15.3). Modern landfills are designed to confine the waste and prevent it from causing environmental and health problems in nearby areas. The typical procedure is to compact the waste as much as possible and periodically (usually daily) cover it with a compacted layer of soil and/or clay. The soil layer isolates the waste from birds and rodents and prevents some infiltration by precipitation. This type of disposal site is called a **sanitary landfill**.

When rainwater percolates through a pile of solid waste, it picks up soluble (easily dissolved) materials. The resulting contaminated solution, called **leachate**, represents one of the most significant environmental problems in the design and maintenance of landfills. Depending on the composition of the waste, leachate can be highly toxic, or it may contain infectious agents. One study of heavy metals in

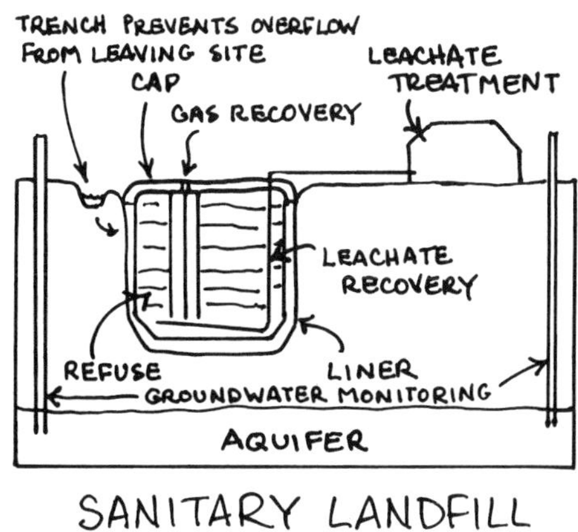

Figure 15.3.

leachate from a municipal landfill in North America found arsenic, cadmium, copper, lead, manganese, zinc, and other toxic metals, in some cases at concentrations a hundred times the maximum levels permitted in drinking water. Since leachate is itself a hazardous liquid waste derived from the interaction of rainwater with solid waste, if it manages to migrate out of the landfill and come into contact with surface water, groundwater, or any part of the biosphere, it can cause significant problems.

Perhaps the most important environmental constraint on the design of a landfill is the selection of an appropriate site that will prevent leachate from forming, or, once formed, prevent it from migrating very far from the site (Figure 15.4). The level of the water table, characteristics of the rock or sediment surrounding the landfill, amount of precipitation, and characteristics of the groundwater flow system are crucial factors in site selection. In a dry environment where the water table is low and precipitation is minimal, the rate of leachate production will be low. Groundwater flow rates will likely be low, too, so the leachate will have little opportunity to

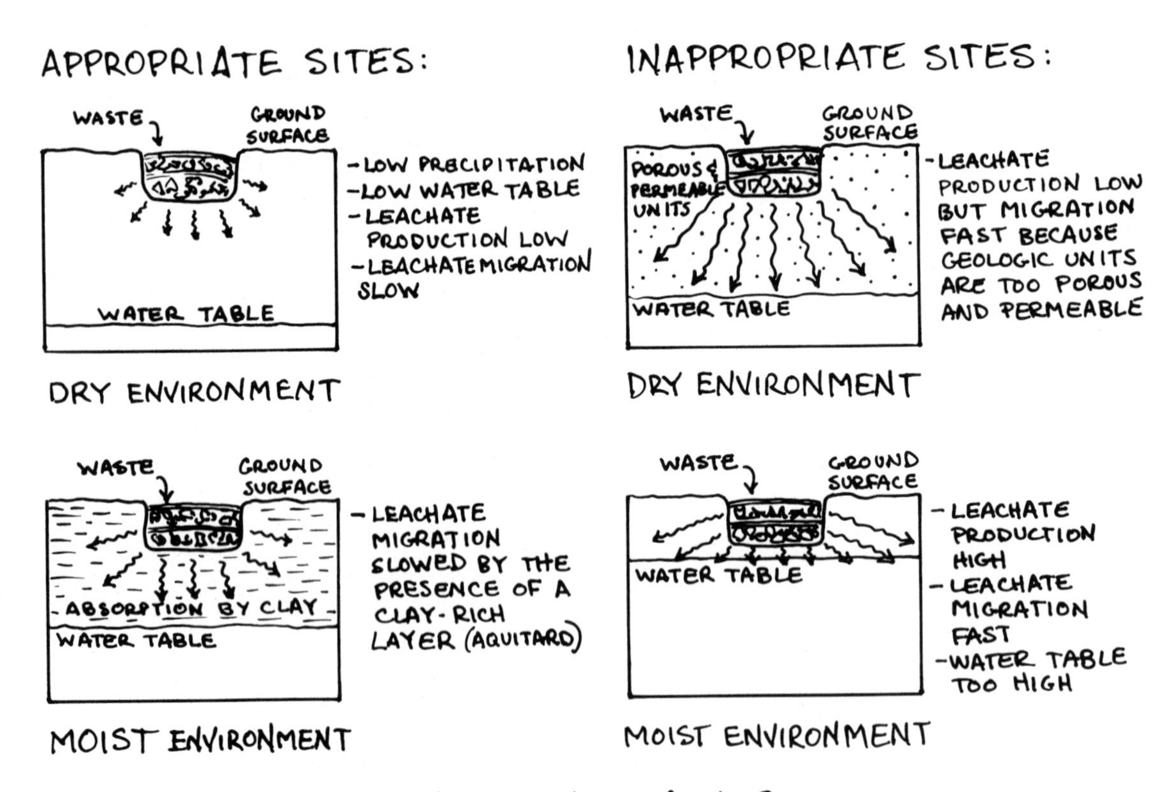

Figure 15.4.

migrate into the groundwater system before becoming dispersed by natural filtration, adsorption, or biodegradation. Thus, a dry environment is preferable to a moist one for landfill siting.

In a wetter environment, leachate production and migration will be faster. Therefore, it is important to site the landfill on an aquitard or an aquiclude unit (a rock or sediment layer that slows down or halts the migration of fluids). Clay layers are particularly effective for this purpose because they tend to be porous (lots of pore spaces), so they can hold and store fluids; they are also impermeable (the pore spaces are tiny and not interconnected), so they do not transmit fluids readily. Clays can bond readily with metals and some other contaminants, further slowing the migration of toxic materials from the site. An unfavorable site for a landfill would be in a wet environment, in which the water table and/or precipitation is high or the underlying rock or sediment units are too porous and permeable. Landfill siting can be particularly challenging in coastal areas (too wet) and arctic ecosystems (too cold for natural biodegradation).

Where is the municipal landfill nearest to your home? When is it expected to be full? Has the municipality started to search for a new landfill site yet? Is there any controversy surrounding the search?

A further constraint on landfill siting is the so-called **NIMBY syndrome**, which stands for "not in my backyard." Some municipalities have successfully overcome NIMBY by offering deep tax cuts and employment opportunities in exchange for the town's agreement to host a landfill; in other words, economic incentives are used to overcome social objections. One drawback of this approach is that the town that accepts the deal will not necessarily be able to offer the best site from an environmental or geologic perspective.

In addition to careful siting, modern sanitary landfills typically include a variety of technologies for leachate control. Impermeable caps and liners (either natural clay or synthetic fabrics) can help prevent both the production and migration of leachate. Drainage pipes and leachate collection systems are usually installed to contain the leachate within the site. Monitoring wells are located around the site so that any leachate migration can be detected and contained. Another common feature is a system for collecting the gas generated by the decay of materials within the refuse pile. Landfill gas is typically dominated by methane (CH_4) but may also contain carbon monoxide (CO) and other gases. Methane has a disagreeable odor, and it can be explosive when mixed with air in certain proportions. Some modern landfills collect and sell the methane gas as a fuel, thereby reducing pollution and generating extra income for the facility.

As discussed, many hazardous materials find their way into municipal landfills. But some landfills are specifically designed and engineered to contain hazardous materials. They are called **secure landfills**. The engineering features of secure landfills are similar to those of sanitary landfills, but the leachate drainage and collection systems are more sophisticated, caps and liners are typically more substantial, and the concentration of monitoring wells around the perimeter of the facility may be much greater. In addition, most secure landfills host an on-site facility for the treatment of hazardous wastes. These facilities permit the chemical stabilization, neutralization, or incineration of the wastes so that as little hazardous material as possible will actually be landfilled. Some liquid hazardous wastes are disposed of by a technique called **deep-well injection**, in which toxic liquids are injected deep underground into an appropriate reservoir rock. This approach makes use of the natural geologic environment to contain and isolate the wastes.

What is the difference between a sanitary landfill and a secure landfill?

Answer: A sanitary landfill is intended to accept mainly municipal waste. A secure landfill is specifically engineered to accept hazardous wastes. Many of the siting constraints and engineering features are similar, including dry environment, impermeable caps and liners, drainage and leachate collection systems, and monitoring wells; however, secure landfills are engineered and monitored more carefully.

Dealing with Sewage

Sewage is what we flush down the toilet and wash down the drain every day. It is a major municipal waste management challenge if only because of the volume generated. On average, each person in North America generates about 550 liters (145 gal) of sewage each day. Sewage is primarily a liquid waste composed of water, urine, detergent, grease, and other liquids. It also contains a significant amount of solid material—paper, feces, dirt, food waste, and other garbage.

In rural areas, domestic sewage is often managed on an individual household basis by **septic tank systems**. A septic tank is a holding tank designed to receive domestic sewage from a single household. In principle, a septic tank contains the waste long enough to allow the organic material in it to partially biodegrade and the solids to settle. The clear liquid component of the waste is then slowly released into the surrounding sediment either through a perforated tank or through a system of porous pipes called a leach field. In passing through the sediment, the effluent is purified by natural filtration and biochemical processes and by the adsorption of contaminants to sediment particles. Modern septic systems are now being engineered with materials that promote the rapid biodegradation of organic wastes.

In areas with higher population densities, sewage is commonly disposed of through municipal sewers combined with various technologies for treating the effluents and solid residues. In some communities, even in North America, sewage is released directly into rivers, lakes, bays, and the open ocean without undergoing any prior treatment to remove impurities. This can cause unpleasant smells, algal blooms, eutrophication, and ultimately significant human health hazards.

In most municipalities in North America, sewage is subjected to at least some treatment before being discharged into a surface water body. **Primary treatment** is a physical process involving the mechanical removal of solid materials such as trash and fine sediment from the effluent. Chemicals may be added to the sewage to cause the solids to precipitate and settle or gather on a filter. **Secondary treatment** is a biologic process, in which bacteria are used to assist in the biodegradation of dissolved organic material in the effluent. Sometimes the sewage is chlorinated at this stage to disinfect it. Finally, **tertiary treatment** is a chemical process designed to remove any remaining contaminants, such as heavy metals, and inorganic dissolved solids.

The quality of the water returned to the system depends on the extent and type of treatment used. Few municipalities can afford to offer tertiary treatment. In communities where water resources are under stress, treated sewage, called **gray water**, may be used for watering boulevards or city landscaping. The solid material that is removed from sewage during treatment is called **sludge**. Sludge is usually landfilled, incinerated, or placed in a large body of water. In some cases, because of its high organic content, sludge may be usable as a fertilizer for nonagricultural purposes (such as for parks or golf courses). However, sludge sometimes contains high levels of toxic chemicals or heavy metals and may itself constitute a hazardous waste.

How does a septic system work? ______________

Answer: A septic tank holds sewage from a single household long enough for solids to settle and organics to partially biodegrade. The clear liquid effluent is slowly released into the surrounding leach field, where it undergoes a natural filtration and purification process.

SELF-TEST

These questions are designed to help you assess how well you have learned the concepts presented in chapter 15. The answers are given at the end. If you get any of the questions wrong, be sure to troubleshoot by going back into that part of the chapter to find the correct answer.

1. The largest producer of solid waste by volume is the ______________ sector.
   a. agricultural c. mining
   b. municipal d. industrial

2. The "4 R's" are
   a. recycle, replenish, restore, and reuse.
   b. restore, reduce, recycle, and remove.
   c. reform, retain, remake, and recycle.
   d. refuse, reduce, reuse, and recycle.

3. Which one of the following would likely be an appropriate site for a sanitary landfill?
   a. coastal marsh
   b. permafrost in Alaskan tundra
   c. dry, clay-rich soil with very deep water table
   d. floodplain adjacent to a large river

4. When people are driven from their rural homelands by environmental degradation, they are called ______________.

5. The very first environmental impact associated with the establishment of a human settlement is typically ______________.

6. ______________ are leftover materials that are toxic, caustic, acidic, explosive, infectious, or radioactive.

7. People who live in cities typically have higher per capita greenhouse gas emissions than people who live in rural areas. (True or False)

8. Leachate often contains toxic materials dissolved from garbage. (True or False)

9. What causes rural-urban migration?

10. What is a hinterland?

11. What is vulnerability in a risk management context?

12. In what ways is a city like an ecosystem?

13. Why is there sometimes a confusing overlap between what constitutes solid waste and what constitutes liquid waste?

ANSWERS

1. a   2. d   3. c

4. environmental refugees

5. deforestation (or) changes in the natural vegetation (or) land clearing

6. Hazardous wastes

7. False (because rural inhabitants make much greater use of private cars to travel long distances)

8. True

9. People are drawn to cities for greater access to job opportunities, higher income, cultural activities, information resources, better health care, and other services. People may also be driven from the countryside (especially in developing countries) by environmental degradation, poverty, war, persecution, and other causes of displacement, refugeeism, and migrantism.

10. The rural areas and smaller communities that surround a large urban center, supporting it with resources and absorbing wastes and residual outputs from it.

11. A combination of the probability of exposure to a natural or technologic hazard and the capacity of the person or community to deal with or respond to the hazard.

12. Individuals live in the system. They interact with the other organisms and with the abiotic components of the system. Matter (water, food, fuel, and so on) is brought into the system and released from the system (heat, waste materials, pollution). Energy is used to run industrial (metabolic) processes within the system from which wastes are produced.

13. Solid wastes may be transformed into liquids, for example, in the production of leachate from a pile of garbage. Liquid wastes often contain a significant solid component, for example, the sediments, paper, feces, and other solids that can be removed from sewage.

KEY WORDS

composting
consumption transition
deep-well injection
economy of proximity
economy of scale
environmental refugeeism
fertility rate
"4 R's" (refuse, reduce, reuse, recycle)
geomorphology
gray water
hazardous waste
hinterland
incineration
landfill
leachate
liquid waste
municipal solid waste
natural hazard
NIMBY syndrome
open dump
primary (sewage) treatment
recycling
residuals
sanitary landfill
secondary (sewage) treatment
secure landfill
septic tank system
sewage
sludge
solid waste
source reduction
technologic hazard
tertiary (sewage) treatment
tipping fee
urban environmental risk
vulnerability
waste exchange

16 Global Change

Some say the world will end in fire,
Some say in ice.
From what I've tasted of desire
I hold with those who favor fire.
But if I had to perish twice,
I think I know enough of hate
To say that for destruction ice
Is also great
And will suffice.

—Robert Frost

Objectives

In this chapter you will learn about:

- environmental change in the geologic past;
- natural and anthropogenic factors that influence global climate;
- depletion of the stratospheric ozone layer; and
- the international policy dimensions of global change.

Global Change

The term **global change** refers to environmental change that affects the entire planet, causing long-term or potentially permanent alterations to the natural environment. Global environmental change can occur naturally—it has done so throughout Earth history. Human actions also can cause global change, often by accelerating or otherwise modifying natural processes. Some impacts of human activities are local or regional, but some have become so common and widespread that they are effectively global in extent. Examples include the impacts of land conversion, such as deforestation of tropical forests for ranching or farming; urbanization and solid waste generation; acid precipitation; depletion of non-renewable resources; destruction of habitat and loss of biodiversity; and eutrophication and toxic contamination of water bodies.

Let's take a closer look at the magnitude of human impacts on the global environment using the example of mineral consumption. The average per capita mineral consumption is about 10 metric tons per year. Multiplied times 6.2 billion people on Earth, we have a total annual consumption of mineral resources of about 62 billion metric tons per year. This doesn't include materials moved for construction purposes, and it doesn't include erosion caused by the building of dams, channelization of rivers, improper agricultural methods, or deforestation. It *only* includes mineral resources that are mined and used for industrial and manufacturing purposes. Compare 62 billion metric tons to the total amount of sediment moved by all of the great rivers of the world—16.5 billion metric tons per year—and you will find that rivers move only slightly more than one-fourth the amount of mineral resources mined for human use each year!

People have become an environmental force to be reckoned with on a global scale. Yet mineral consumption, deforestation, and habitat destruction are not usually what is implied by the term *global change*. More commonly, the term is used to refer to global *atmospheric* change, primarily changes in atmospheric chemistry. The two greatest concerns about global atmospheric change today are climatic change (global warming) and ozone depletion; they are the main focus of this chapter. We will begin by looking at the tools and approaches used by scientists to study ancient environments.

What is global change? ________________

Answer: Long-term or potentially permanent alterations to the natural environment that are global in extent; impacts associated with changes in atmospheric chemistry, especially global climatic change and ozone depletion.

Studying Past Environmental Change

Paradoxically, no science is more relevant to our understanding of present-day climatic change than the study of ancient climates, **paleoclimatology** (*paleo-* means "ancient"). Why is it important to know what Earth's climate was like thousands or even millions of years ago? Scientists need to establish a baseline against which to measure changes that are currently being observed in global climate, some of which may be attributable to human impacts. If we don't understand the natural influences on global climate and the natural time scales of global climatic change, we will have no way to assess the magnitude of human impacts upon them.

How do paleoclimatologists determine what Earth's climate and environment were like 1,000 or 100,000 or even 100 million years ago? Direct instrumental measurements of temperature, precipitation, and other aspects of weather have only been recorded on a regular basis since the mid-1800s. This record is not long enough to provide an understanding of climatic phenomena that may vary on time scales of thousands or tens of thousands of years. Scientists employ a variety of techniques to extend the instrumental records of temperature and precipitation.

If we wish to determine the temperature, precipitation, or other indicators of ancient climate for which no direct instrumental measurements exist, we must use a climate proxy. A proxy is a stand-in, or surrogate. Thus, a **climate proxy** is a surrogate for past climatic indicators—a measurement or record of something that is controlled by, is influenced by, or mimics an aspect of climate. For example, in the past, fishers in Iceland kept records of the number of weeks that sea ice reached the coast of Iceland. The presence of sea ice is clearly a proxy for temperature; the colder the temperature, the longer the ice remains. Icelandic fishers began keeping this record nearly 1,000 years ago—much longer than the existing record of direct measurements of temperature. Other examples of historic records that serve as climate proxies for the past 1,000 years or so include the date of the wine harvest in Germany (recorded since the ninth century); the freezing date of the lake at the emperor's palace in Japan (since 1450); the height of the Nile River at Cairo (since 622); and the severity of winters in England (since 1100).

These are examples of events influenced by climate and recorded by people. The natural world also keeps records of climatic change. Everything that grows or changes in response to seasonal variations is a potential indicator of past climates. For example, trees add new growth rings every year (Figure 16.1). The width and density of the rings are a reflection of temperature, precipitation, and other environmental factors, and the number of rings indicates how many years have passed. (It is very important to have chronologic indicators associated with climate proxies; if we know what a past climate was like but we don't know *when* it occurred, it has no real impact on our understanding of climatic change.) Corals also add periodic growth rings, reflecting variations in the temperature and composition of the water. Finely layered lake bottom sediments, called **varves**, also reveal detailed information about

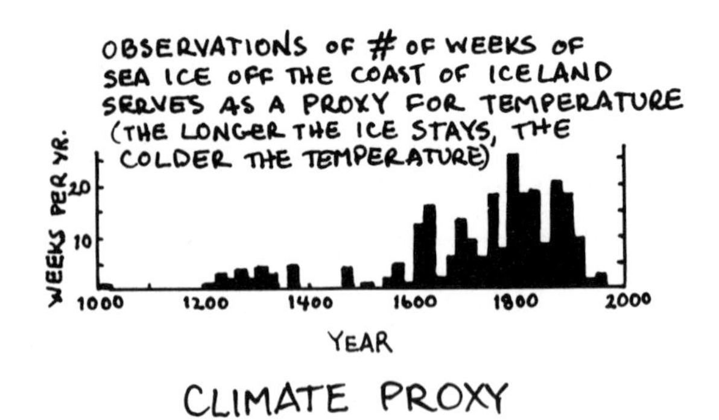

Figure 16.1.

the timing and duration of seasons. These natural proxies extend the climate record thousands of years into the past.

To extend the climate record to millions or billions of years ago, scientists study fossilized plants and animals. If a fossil assemblage from a particular time is dominated by species that can survive only in tropical climates, for example, then we can conclude that a tropical environment existed at that time and place. Microscopic fossilized pollen spores have been particularly useful in reconstructing past climatic changes. They are more resistant to weathering and decomposition than larger plant fragments and more likely to be preserved. From the pollen spores we can find out which plant assemblages flourished in the past at a particular location. In turn, this information can be used to infer temperature and precipitation.

Sediments and sedimentary rocks preserve a record of changes in the chemical compositions of water and air. For example, the amount of dust in past atmospheres can be deduced from the distribution of wind-deposited sediments. Dust is an important indicator of climate because a dusty atmosphere tends to block incoming solar radiation; periods in which the atmosphere is dusty generally correlate with cool surface temperatures. Ancient soil horizons, called **paleosols**, also provide information about climate and weather in ancient environments. Scientists can learn about the weathering environment by examining the mineral assemblages in paleosols, which indicate the chemical compositions of air and water at the time of soil formation.

Some climate proxies yield fairly specific estimates of past temperatures. These are called **paleothermometers** because they allow us to "take the temperature" of ancient environments. More commonly, however, climate proxies provide scientists with a range of possible temperatures for a given location and time. By comparing several proxy data sets, scientists can begin to narrow down the actual paleoclimatic conditions.

Scientists can determine past temperatures from studies of ice cores derived from polar ice sheets and mountain glaciers. The chemical characteristics of the ice can indicate whether temperatures, both locally and globally, were relatively warm or cool at the time the ice was deposited (Figure 16.2). If the ice is composed mainly of the light isotopes of hydrogen and oxygen (in H_2O), it is indicative of formation during a period of global coolness. Ice that is rich in the heavier isotopes of oxygen and hydrogen is indicative of formation during a period of global warmth. (*Isotopes* are atomic variants, with the same atomic number and chemical characteristics but different masses. Oxygen, for example, has three naturally occurring isotopes; the heaviest is ^{18}O and the lightest is ^{16}O. These isotopes have different mass numbers, but they are chemically the same. Hydrogen has two naturally occurring isotopic variants; deuterium is the heavier isotope, and hydrogen is the lighter one. Both water and ice can be isotopically heavy or isotopically light, depending on which

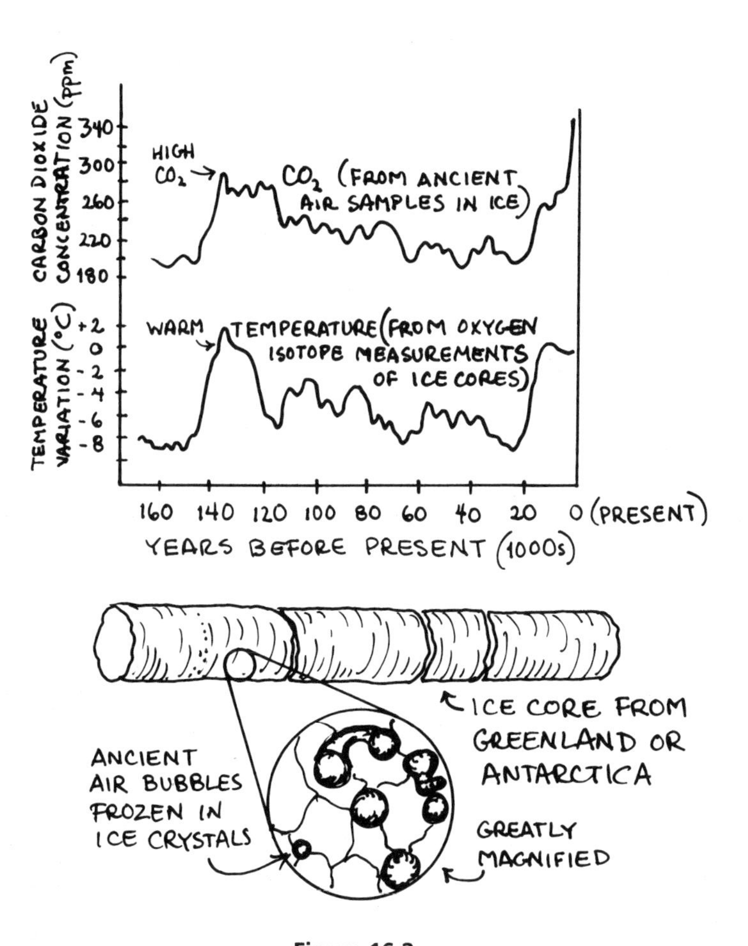

Figure 16.2.

forms of hydrogen and oxygen are present. Chemically, they are still H_2O—whether the isotopic composition is heavy or light.) The isotopic compositions of water and ice are extremely sensitive indicators of temperature because processes such as evaporation and precipitation are strongly influenced by the mass (or weight) of the water molecules. For example, evaporation favors isotopically lighter water, while precipitation favors isotopically heavier water.

It is even possible to obtain samples of ancient air from bubbles trapped in glacial ice. Chemical testing of this air can reveal its composition, including the presence of greenhouse gases (GHGs) such as carbon dioxide or methane. Studies of this type have been crucial in extending the detailed records of temperature and atmospheric composition to as far back as 400,000 years ago (see Figure 16.2). These records show that during periods of globally cold temperatures, atmospheric concentrations of GHGs typically were low. In contrast, warm periods correspond to high concentrations of GHGs in the atmosphere. This is a crucial piece of information in the global climate puzzle.

What is a climate proxy? Explain how tree rings can be used as a climate proxy.

Answer: Events or processes that are influenced by, controlled by, or mimic climate and for which a record (either human or natural) exists. The width and density of annual growth rings of trees vary with temperature, precipitation, and other climatic factors. With an estimate of age (provided by the number of rings), tree growth rings can serve as a climate proxy.

Glaciations

Periods during which the average temperature at the surface drops by several degrees and stays low long enough for ice sheets to grow larger (and for new ones to form) are called **glaciations** (also ice ages, glacial periods, glacial stages, or glacial epochs). Periods between glaciations, when ice sheets retreat and sea levels rise, are called **interglacials** (also interglacial stages or interglacial periods). During the past 1.6 million years, Earth has experienced more than twenty major glacial-interglacial cycles (Figure 16.3). The timing of these cycles has varied, with extreme temperature minima (low points) occurring roughly every 100,000 years over the past million years and every 20,000 to 40,000 years before that. Glaciations have been especially prevalent during the **Pleistocene Epoch** (the past 2 million years), but the rock record contains evidence of glacial ages that occurred as long ago as 2.3 billion years.

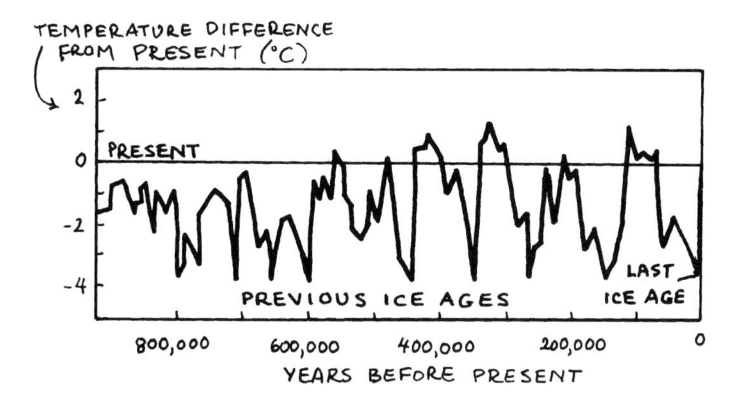

Figure 16.3.

Scientists can determine the timing and extent of glaciations from a variety of evidence. For example, rocks that were scratched and grooved by glaciers reveal the direction of ice movement. Glacial landforms, especially **moraines** (piles of poorly sorted rock and sediment deposited by the glacier along its edges and at its end, or terminus), reveal the geographic extent of land ice sheets (Figure 16.4). Tree rings

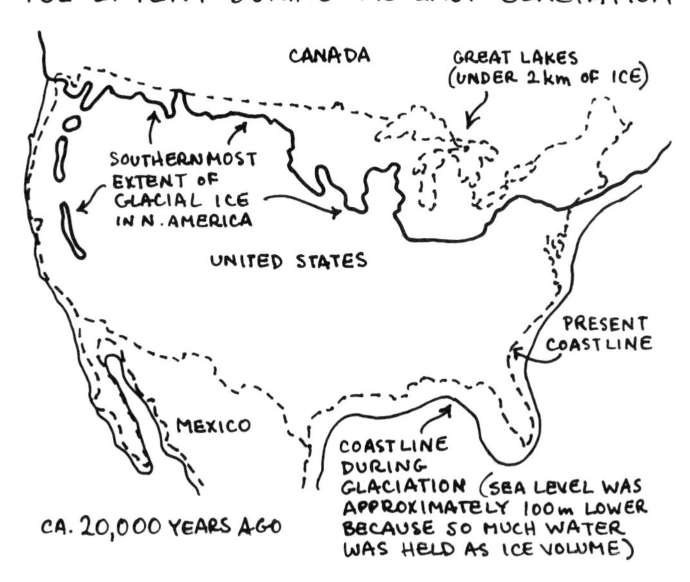

Figure 16.4.

and radiocarbon ages of trees that were felled by advancing ice tell scientists when the ice arrived in a given region. Great thicknesses of fine, wind-blown dust, called **loess** (pronounced "luhss"), deposited just south of the ice limits during glacial times, contain fossils characteristic of cold, dry weather. The loess deposits reveal that the ice age atmosphere was both colder and dustier than today's. Deposits of loess and other features and landforms that typically occur around the edges of an ice sheet are called **periglacial** features.

The most recent ice age began about 70,000 years ago. It was a time when great woolly mammoths, mastodons, long-horn bison, and saber-tooth tigers roamed North America. Early humans migrated into North America from Asia, perhaps walking across exposed continental shelf in today's Bering Strait to Alaska. This land bridge was exposed during the ice age because a large volume of water was locked up in glacial ice, causing sea levels to drop globally. Huge floating ice sheets like those in the present-day Arctic and Antarctic seas occupied large areas of the Atlantic Ocean.

The ice sheets of the most recent glaciation reached their maximum extent in North America around 24,000 years ago (see Figure 16.4). About 12,000 years ago, the climate began to warm, and by 10,000 years ago, Earth had emerged from the ice age and entered the present interglacial period. We have now passed the time of maximum warmth in the glacial-interglacial cycle. Temperatures peaked in a warm period about 6,000 to 7,000 years ago, known as the **Holocene optimum**. Since then, temperatures have been cooling gradually, with some distinctly colder fluctuations, such as the **Little Ice Age**, which lasted from about A.D. 1300 to about A.D. 1900.

When did the most recent ice age begin and when did it end? ______________

Answer: It began 70,000 years ago and ended 10,000 years ago.

The Natural and Anthropogenic Factors That Drive Climatic Change

What factors cause Earth's climate to change? The search for answers has been difficult because the climate system is very complicated, with many interacting subsystems. The factors that influence climate vary on time scales ranging from seasons, to centuries, to millions of years. Furthermore, modern human impacts on the climate system are making it increasingly difficult to separate natural from anthropogenic influences. We know that Earth's climate *will* change; the record of past climatic change is very clear. What we lack is a comprehensive understanding of *how* it will change and at what rate.

Several mechanisms cause natural climatic change. They involve the atmosphere, the lithosphere, the ocean, and the biosphere interacting in complex ways. Geographic changes resulting from tectonism—the shifting of continents, the uplift of continental crust and creation of large mountain chains, and the opening or closing of ocean basins—have a significant impact on oceanic and atmospheric circulation and therefore on global climate. Even Earth's internal processes affect climate. Large, explosive volcanic eruptions sometimes produce vast quantities of dust and tiny aerosol droplets of sulfuric acid, both of which scatter and block the Sun's incoming radiation and cause global cooling.

Earth's orbit and rotation play an important role in controlling the timing and cyclicity of variations in climate (Figure 16.5). The eccentricity (departure from circularity) of Earth's orbit; the tilt of the planet's axis of rotation; the precession (wobbling) of the axis; and the timing of perihelion (closest approach to the Sun) all have an impact on insolation, the amount of solar radiation reaching the surface at any given time. Periodic changes in climate caused by fluctuations in insolation that result from variations in Earth's orbital and rotational characteristics are called **Milankovitch cycles**, after the Yugoslavian mathematician who first recognized this phenomenon and its impact on the timing of glacial-interglacial cycles.

When processes like plate tectonics, volcanism, or orbital variations drive or profoundly influence the global climate system, it is referred to as **climate forcing**. The external forcing influences on climate are modified by the filtering action of the atmosphere—that is, the greenhouse effect (introduced in chapter 3). Acting somewhat like an actual greenhouse, radiatively active (greenhouse) gases in the atmosphere absorb outgoing infrared radiation, trapping heat and raising the surface temperature (Figure 16.6). Water vapor (H_2O) in the atmosphere is the most

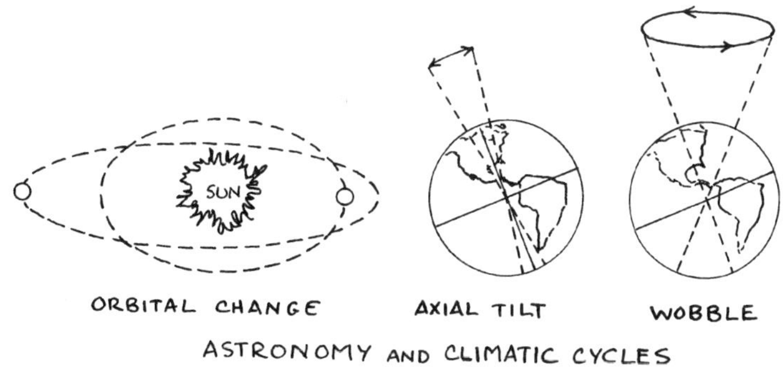

Figure 16.5.

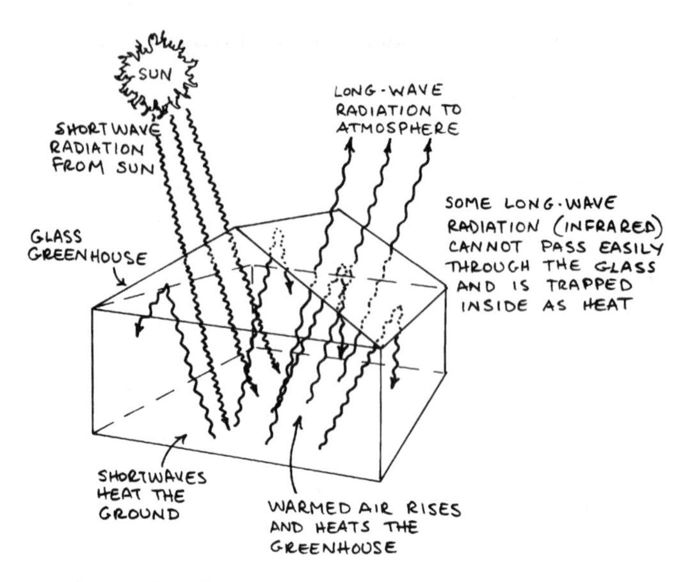

Figure 16.6.

important natural greenhouse gas. A number of naturally occurring minor gases, including carbon dioxide (CO_2) and methane (CH_4), also contribute significantly to greenhouse warming. During the past decade or so, the role of anthropogenic emissions and the possibility of accelerated global warming have been subjects of extensive scientific and political debate.

The rate of emission of anthropogenic GHGs, especially carbon dioxide, has increased dramatically since the Industrial Revolution, causing changes in the chemistry of the atmosphere. Systematic measurement and monitoring of atmospheric carbon dioxide began in 1958 with an experiment at Mauna Loa Observatory in Hawaii. This experiment was originally established to measure natural fluctuations in carbon dioxide resulting from seasonal variations in photosynthetic activity. These fluctuations account for the zigzag pattern on the graph in Figure 16.7. However, researchers soon realized that they were recording a steady increase in the level of atmospheric CO_2. It was quickly determined that emissions of other GHGs, such as methane (CH_4) and nitrous oxides (NO_x), have also increased in recent decades.

The main concern is that anthropogenic emissions of greenhouse gases will contribute to an **enhanced** or **accelerated greenhouse effect**. The gases of major concern are carbon dioxide (CO_2), which accounts for approximately 55 per-

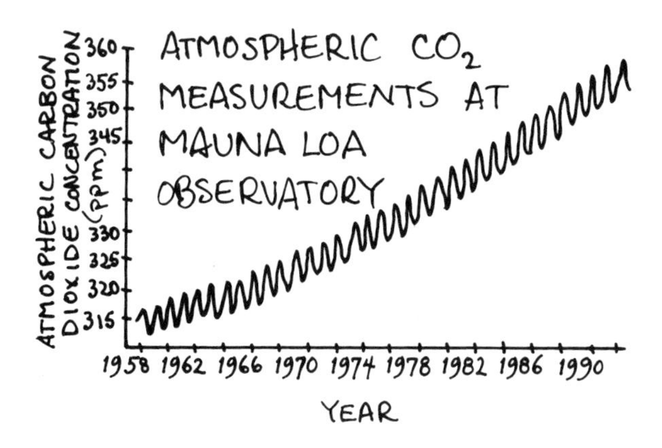

Figure 16.7.

cent of the enhanced greenhouse effect; methane (CH_4), 16 percent; and nitrous oxide (N_2O), 5 percent (Table 16.1). Sources of some confusion are **chlorofluorocarbons** (**CFCs**), a group of chemicals that are human-made and do not occur naturally. CFCs are radiatively active gases, contributing approximately 10 percent of the enhanced greenhouse effect. CFCs are also implicated in the depletion of the ozone layer, which is discussed in greater detail below. To further complicate things,

Table 16.1 Some Important Greenhouse Gases, Common Sources, and Relative Contributions to the Enhanced (Anthropogenic) Greenhouse Effect

Greenhouse Gas	Sources	Contribution to Enhanced Greenhouse Effect
Carbon dioxide (CO_2)	Fossil fuel burning, land conversion, deforestation	~55%
Methane (CH_4)	Rice culture, cattle, biomass burning, landfills, peatlands	~16%
Nitrous oxide (N_2O)	Fertilizers, land conversion, transportation	~5%
Chlorofluorocarbons (CFCs)	Refrigerants, aerosols, industrial solvents	~10%
Tropospheric ozone (O_3)	Secondary pollutant (from photochemical reactions)	~14%
Water vapor (H_2O)	Irrigation, industrial processes, enhanced evaporation	Unknown

ozone itself is a greenhouse gas that is accountable for about 14 percent of the anthropogenic greenhouse effect. Emissions of water vapor—the most important naturally occurring GHG—also result from many human activities, but the contribution (or potential contribution) of water vapor to the anthropogenic greenhouse effect is not yet known.

The preindustrial concentration of CO_2 in the atmosphere is thought to have been approximately 280 ppmv (parts per million by volume); the present level is 355 ppmv, increasing at a rate of about 0.5 percent per year. The human impact on the carbon cycle has been to accelerate the flux of carbon to the atmosphere by remobilizing carbon stored in long-term rock reservoirs in the form of fossil fuels. Past and present, the main contributors of CO_2 emissions from fossil fuel use are the economically developed countries of the world (see Table 16.1). Deforestation, the burning of wood and other biomass, and the conversion of land from forest to agricultural land or rangeland contribute significantly to atmospheric CO_2 levels. If a tree dies, the carbon stored in its biomass will be released naturally to the atmosphere through the biologic process of decay. Thus, land conversion, deforestation, and slash-and-burn agriculture are among the most important sources of GHG emissions from less economically developed countries.

Methane is of particular concern as a greenhouse gas. Although it is much less abundant than carbon dioxide in the atmosphere, it is approximately twenty-five times more effective as a radiatively active gas, and it is building up rapidly in the atmosphere. The preindustrial level of atmospheric methane is thought to have been approximately 790 ppbv (parts per billion, by volume), while the present level is approximately 1,720 ppbv, increasing at a rate of 1.1 percent per year. Methane is generated by the decay of biomass in anaerobic (oxygen-poor) environments. Swamps, wetlands, and peatlands are natural sources of methane. Anthropogenic sources include rice cultivation and the digestive processes of domestic livestock, especially cattle.

Climatologists and other scientists are working hard to understand how different parts of the Earth system work together to determine global climate and how the system can be affected by human actions. However, there are still many sources of uncertainty. Let's look at the possible impacts of global climatic change on the environment and on people.

What is a Milankovitch cycle? ______________

Answer: Periodic variations in climate caused by fluctuations in insolation, which in turn are caused by variations in Earth's orbital and rotational characteristics.

Global Warming and Its Possible Consequences

Throughout this chapter, the term *climatic change* has been used instead of the more popular *global warming.* Global warming is actually a misnomer because global climatic change will *not* result in uniform warming of the planet's surface. In this section, we examine the predicted consequences of climatic change. Many of these predictions are based on the work of the Intergovernmental Panel on Climatic Change (IPCC), a working group of scientists and policy makers that was established with a mandate to keep the world community informed about the best and most timely research on the climate system and climatic change.

The final years of the twentieth century were some of the warmest on record in North America (Figure 16.8). Then, in the summer of 2000, a disturbing discovery was made: the ice cap that formerly covered the North Pole year-round had undergone significant melting. Does this mean that the climate has started into a warming trend caused by human-generated atmospheric changes? It is impossible to be certain, but the observations do seem to lend credence to the predictions of climatologists and the IPCC.

If you were given the task of measuring the surface temperature of this planet to come up with a global average for one year, how would you go

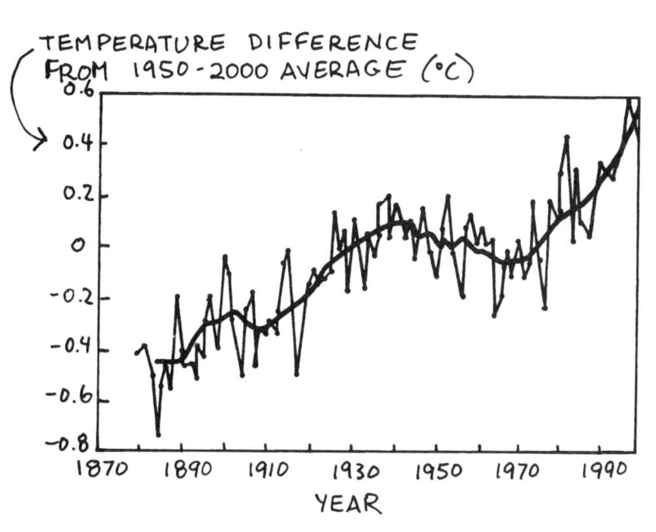

Figure 16.8.

about it? It seems simple, but how many measurements would you need to take? How closely spaced would they need to be? How many would be over land and how many over the ocean? Would you measure right at ground level or a bit higher up? How high? How often would you need to take measurements: once each day or once each hour? Measuring the surface temperature of Earth is an exceedingly difficult task; that's why it is such a contentious issue and a source of scientific uncertainty in discussions about climatic change.

The most robust analyses of surface temperatures for the past 100 to 150 years suggest that average global surface temperatures have increased by about 0.6°C over the past century (see Figure 16.8). This appears to be an actual warming trend distinct from shorter-term fluctuations in temperature that are also evident in the data set. The IPCC predicts that average surface temperature will continue to increase, reaching about 3°C to 5°C warmer than the present by about the year 2050. It may not sound like much, but this rate of change is considerably faster than anything known to have occurred in the climate system as a result of natural phenomena.

Extremes and the distribution of temperature are also expected to change. Some places will get warmer, but some may get cooler on average. Low-temperature extremes are expected to get colder, and high-temperature extremes will get hotter. These changes in surface temperature will cause changes in atmospheric and oceanic circulation. A source of considerable concern is the potential impact on the oceanic thermohaline circulation (chapter 3). If water in the North Atlantic becomes warmer (and fresher from the melting of sea ice in the Arctic), it could fail to sink and thus fail to initiate the deep oceanic circulation system that drives the global climate system. This could launch a faster and more profound shift in global climate. (This was the premise for the movie *The Day After Tomorrow.*) A temporary shutdown of the thermohaline circulation system may have started the Little Ice Age in the 1400s that devastated agriculture in much of Europe. Changes in oceanic circulation would also have a major impact on nutrient supplies for the world's fisheries.

Shifts in the major atmospheric and oceanic circulation systems are expected to lead to changes in precipitation (more in some places, less in others) and changes in the frequency and intensity of storms. Changes in the geographical patterns of temperature, precipitation, and wind systems will inevitably result in changes in the characteristics, size, and geographic distribution of biomes, ecosystems, habitats, and growing seasons. Will organisms be able to migrate quickly enough to keep up with the shifts in their habitat? Some may; others may not.

We are likely to experience more precipitation overall as a result of increased rates of evaporation due to warming of the ocean. Some locations that are currently arid or semiarid, such as the Sahel region of Africa, may receive more precipitation, which could be a good thing. Agricultural growing seasons may become longer in some areas such as Siberia; this, too, would appear to be a potential benefit of climatic change. Furthermore, it is possible that the increased carbon dioxide in the atmosphere may have a fertilizing effect on some plant types. In some areas, though, such as northern Canada, a longer growing season will be of no use, as it would take hundreds or even thousands of years for arable soil horizons to form in this previously glaciated terrain.

As warming occurs, changes in the polar ice caps are anticipated; however, this is a source of some controversy among scientists. If the temperature becomes warmer, you might expect melting of the ice caps to occur. However, some predictions call for increased precipitation in polar regions, which would still experience below-freezing temperatures for much of the year. Glaciers grow by forming new ice from snowfall that exceeds melting in a given year. This suggests that polar ice caps might grow, rather than shrink, in response to global warming.

In either case, it is anticipated that sea level will rise. Some of the increase may come from the melting of land ice in Antarctica (the melting of sea ice, which is less voluminous, is unlikely to play a major role in sea level rise). But the major contributor to sea level rise will be the warming of ocean water. Water, like most other substances, expands when it is warmed. If the ocean warms by, say, 0.5°C, every water molecule in the ocean will expand just a tiny bit. Overall, the entire volume of the ocean will increase. Sea levels worldwide are expected to rise by as much as 1 m (3.3 ft). In areas where coastlines are cliffed, a 1-m rise in sea level would not make a significant difference. In areas where the coast rises gradually, however, a 1 m rise in sea level would cause a significant incursion of seawater, leading to flooding of coastal lands and saltwater contamination of coastal water supplies. For cities located on coasts and estuaries (Venice, New York, New Orleans, Shanghai, Bangkok) and low-lying regions (Florida, Netherlands, Bangladesh), such an increase in sea level would be devastating.

A 1-m sea level rise would cause most of the state of Florida to be below sea level. Some coastal cities and island nations would be completely submerged.

Some of the anticipated impacts of global warming may have self-reinforcing effects. For example, warming may cause an increased carbon flux to the atmosphere from soil and peat gases. This would contribute to further warming in a positive feedback cycle (chapter 2). Warming of ocean water could cause the release of methane gas hydrates from where they are currently frozen in seafloor sediments. Releasing the methane from these deposits would contribute a massive volume of highly effective greenhouse gas to the atmosphere, possibly causing further warming—another positive feedback effect.

Finally, after all the predictions have been made, we must consider the potential impacts of climatic change on people and socioeconomic systems. Even without predicting dire consequences like the inundation of major cities, it is clear that the kinds of changes outlined above would seriously affect our economy and well-being. Industries like fisheries, agriculture, and tourism would be hit hard by changes in temperature, precipitation, and sea level. The risk of damage from droughts, floods, and storms would increase. The impacts of warming on already scarce freshwater resources could be devastating. Human health would be at risk because of the increase in diseases like malaria and West Nile virus carried by insects that thrive in hot climates. Interestingly, the insurance industry has taken the potential economic impacts of global change very seriously, commissioning many scientific studies on the topic.

What are some of the potential human health-related impacts of global warming? Can you think of some other health impacts in addition to those listed in the text?

____

Answer: In the text: increased incidence of diseases associated with insects that thrive in hot climates, such as malaria and West Nile virus. Others: diseases related to lack of sanitation and lack of clean drinking water as freshwater resources become more scarce, such as diarrhea and other intestinal diseases; increase in hot-weather problems, such as heat stroke and dehydration.

Modeling the Global Climate System

How do scientists and the IPCC take raw data and observations and turn them into predictions like the ones discussed above? They use a computer-based approach called **global circulation modeling** (or **GCM**) to determine the rate, magnitude, and distribution of impacts. GCMs are three-dimensional mathematical models of the climate system linking the various environmental processes and factors that interact to influence global climate (Figure 16.9). Models are important tools in environmental science because they allow us to examine complex natural systems. The original goal of GCMs was to improve weather forecasting capabilities. Now they are commonly used to predict future climatic changes by varying the inputs such as the rate of accumulation of methane in the atmosphere or the amount and

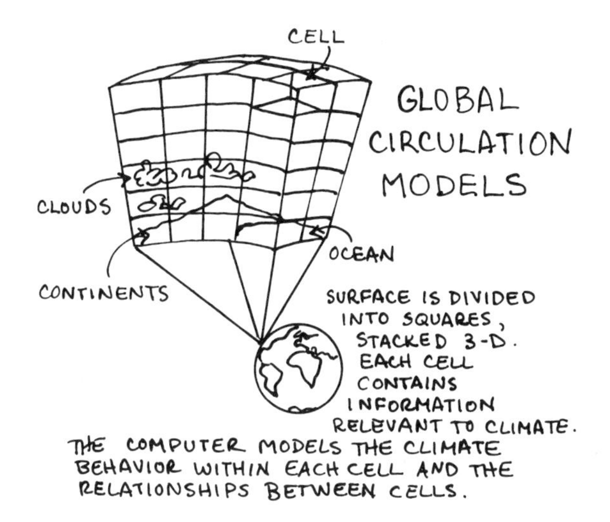

Figure 16.9.

type of cloud cover. The sophistication and accuracy of GCMs is constantly improving as computer capabilities increase.

But GCMs are only as good as our basic understanding of the climate system, and parts of the system still are not thoroughly understood. For example, clouds play a dual role in the climate system; they warm the surface by trapping infrared radiation, but they also cool the surface by reflecting incoming solar radiation. If the global temperature gets warmer, the rate of evaporation of ocean water will increase, perhaps causing the amount of cloud cover to increase. Will a change in cloud cover have an overall warming effect or a cooling effect? No one knows for certain. Other uncertainties in climate modeling include the response of the polar ice masses to global warming; the response of soils and soil gases to warming; the impacts of vegetational changes on the moisture content of the atmosphere; and the levels of future anthropogenic contributions to GHGs in the atmosphere.

Will the warming trend in surface temperatures continue? How hot will the surface temperature eventually become? How much of the change is natural, and how much is attributable to human actions? If the global climate does become warmer, how severe will the anticipated effects be on humans and ecosystems? These questions have not yet been answered conclusively. Scientists are used to dealing with uncertainty, but it causes significant challenges for policy makers who are trying to deal with the socioeconomic and risk management aspects of global change. We will end this chapter with a look at international policies and agreements designed to mitigate global change. First, let's look at the other aspect of global change that is of particular concern today: stratospheric ozone depletion.

Why are models useful in environmental science? ______________

Answer: They allow scientists to take an extremely complex system, like the climate system, impose certain restrictions on it, and study changes in the system more easily than they could study the system in nature.

Stratospheric Ozone

The **ozone layer** is a concentration of ozone (O_3) in the stratosphere (chapter 3). The main concentration of ozone is in a zone approximately 25 to 35 km (15.5 to 22 mi) above the surface. It is not a layer of "pure" ozone; the atmosphere as a whole is only about 0.000007 percent ozone—ozone is just a trace gas in the atmosphere. Approximately 90 percent of the ozone in the atmosphere resides in the stratospheric ozone layer. The remainder is in the troposphere. As you will remember from chapter 14, tropospheric (ground-level) ozone is a major component of photochemical smog, causing eye irritation and respiratory problems. Tropospheric ozone is also a greenhouse gas. If ozone at ground level causes all these problems, why are we so concerned about preserving ozone in the stratosphere?

The answer is that stratospheric ozone absorbs and blocks incoming short-wavelength, ultraviolet radiation that is hazardous to life. Without the ozone layer, life on Earth would be confined to deep water or subterranean habitats. An **atmospheric blind** blocks particular wavelengths of electromagnetic radiation from passing through the atmosphere, like a blackout curtain stops visible light from passing through a windowpane (Figure 16.10). An **atmospheric window**, however, allows radiation to pass through or be transmitted by the atmosphere. Our atmosphere provides a window for visible light; that's why our eyes evolved to see in the visible part of the electromagnetic spectrum. Greenhouse gases provide an atmospheric window for incoming short-wavelength radiation, but they act as a blind for outgoing longer-wavelength radiation—that's how the greenhouse effect works. Ozone in the stratosphere functions as an atmospheric blind in the ultraviolet part of the spectrum, blocking almost all of the incoming UV-C and most of the UV-B radiation, the wavelengths that are the most harmful to life-forms.

Ozone, which consists of three oxygen atoms bonded together, began to accumulate in the atmosphere only after free molecular oxygen (O_2) became abundant. The stratospheric ozone layer probably began to function as an ultraviolet filter when it had reached about 2 percent of its present concentration. According to models of the evolution of atmospheric chemistry, this event happened about 600 million years ago. Interestingly, this is also a time when an enormous explosion of

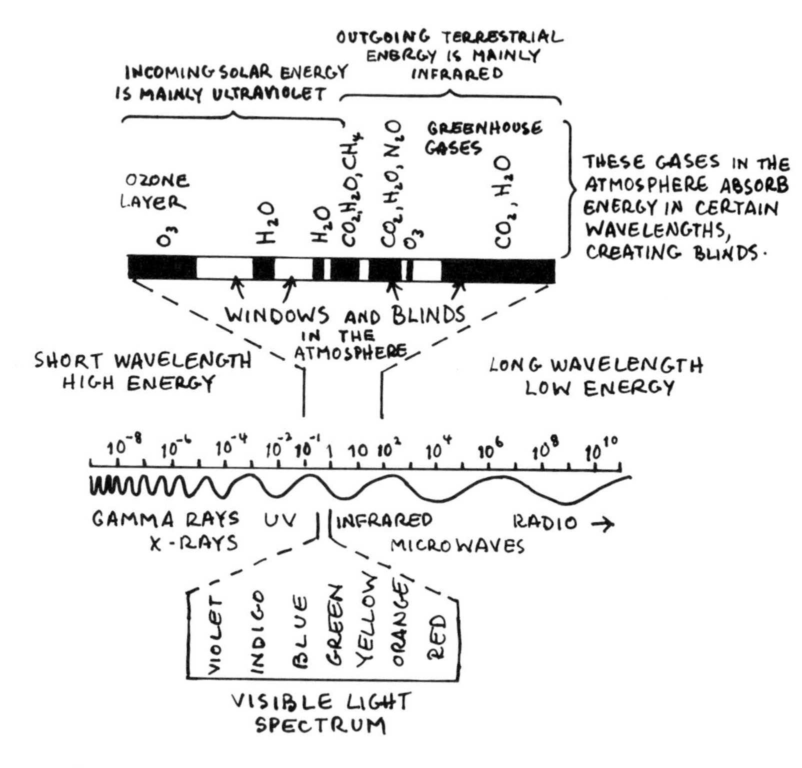

THE ELECTROMAGNETIC SPECTRUM AND ATMOSPHERIC WINDOWS AND BLINDS

Figure 16.10.

life-forms is recorded in the fossil record. The ozone layer made it possible for life on Earth to emerge from the shadows, and it continues to make life possible today.

Ozone is destroyed and re-created in the atmosphere by natural processes. It is assumed that the amount of ozone in the stratosphere was relatively constant and that a steady state existed among O, O_2, and O_3 in the atmosphere prior to human intervention. Ozone molecules are created naturally through the interaction of oxygen atoms (O) and oxygen molecules (O_2). In turn, ozone is broken down by interaction with solar radiation. The breakdown of ozone is facilitated by other natural processes, too, such as volcanic eruptions.

If all of the ozone in the stratosphere were concentrated at ground level, it would form a layer only as thick as three dimes.

The amount of ozone in the ozone layer can be measured using ground-based instrumentation that detects variations in the amount of incoming ultraviolet radiation (the more incoming UV radiation, the less ozone there is in the air overlying the instrument). Ozone can also be measured using instruments on satellites. The main instrument used for this purpose, the Total Ozone Mapping Spectrometer (TOMS), is carried on the *Nimbus* 7 spacecraft. It measures total ozone in a column of air, and scientists have been measuring stratospheric ozone using this type of instrument since 1978.

A unit of measurement that is commonly used to quantify the amount of ozone in the ozone layer is the **Dobson unit**, or DU, defined such that 100 DU is equivalent to a 1-mm-thick layer of pure ozone at surface temperature and pressure. The worldwide average in the ozone layer is approximately 300 DU. If you could collect all of the ozone that is diffused throughout the stratospheric ozone layer and bring it to Earth's surface, it would form a layer of pure ozone just 3 mm thick, which means ozone that is equivalent to the thickness of three dimes is the only thing blocking harmful ultraviolet radiation from reaching the surface.

What is ozone? _______________

Answer: O_3—that is, three oxygen atoms bonded together; it is present as a naturally occurring trace gas in the atmosphere.

What natural processes are known to cause the depletion of stratospheric ozone?

____

Answer: Volcanic eruptions and interaction with solar radiation.

The Ozone Hole

Concerns over the impact of anthropogenic emissions on the ozone layer first surfaced in the 1960s and 1970s when the possibility was raised that airplanes, rockets, and supersonic jets flying through the stratosphere might cause ozone depletion. The concern at that time stemmed primarily from nitrogen oxides in the exhaust of aircraft. Shortly thereafter, with the development of the space shuttle, an additional threat materialized. It was feared that chlorine—a trace gas in the shuttle's exhaust gases—could initiate ozone depletion. It was quickly determined that jet contrails and exhaust from the space shuttle were not a major concern, but chlorine itself became a subject of much interest and research.

Much of the anthropogenic chlorine (Cl) in the atmosphere derives from the group of chemicals known as chlorofluorocarbons, or CFCs. They are used in refrigerants, aerosol spray propellants, industrial solvents, cleaning fluids, Styrofoam, foam-blowing agents, fire retardants, and a variety of other applications. Because they are relatively stable, CFCs survive in the atmosphere long enough to migrate to the stratosphere. Once there, they are subjected to intense solar radiation, which causes the molecules of CFCs to break down, releasing an atom of chlorine. The chlorine acts as a chemical facilitator, or catalyst, in the breakdown of ozone molecules.

Here is how it works (Figure 16.11). A chlorine atom liberated from a CFC molecule "pulls" one of the oxygen atoms away from an ozone molecule, forming a molecule of chlorine monoxide (ClO) and leaving behind a molecule of oxygen, O_2. This is the basic reaction that occurs:

$$Cl + O_3 \rightarrow ClO + O_2$$

Later, when the ClO encounters a free oxygen atom (O), it will react to form another molecule of O_2, freeing the chlorine atom once again, like this:

$$ClO + O + \text{sunlight} \rightarrow Cl + O_2$$

Because the chlorine atom is a catalyst and is not itself altered in the process, it can be recycled to react again and again with ozone molecules. With an atmospheric residence time of about forty years, one atom of chlorine may destroy tens of

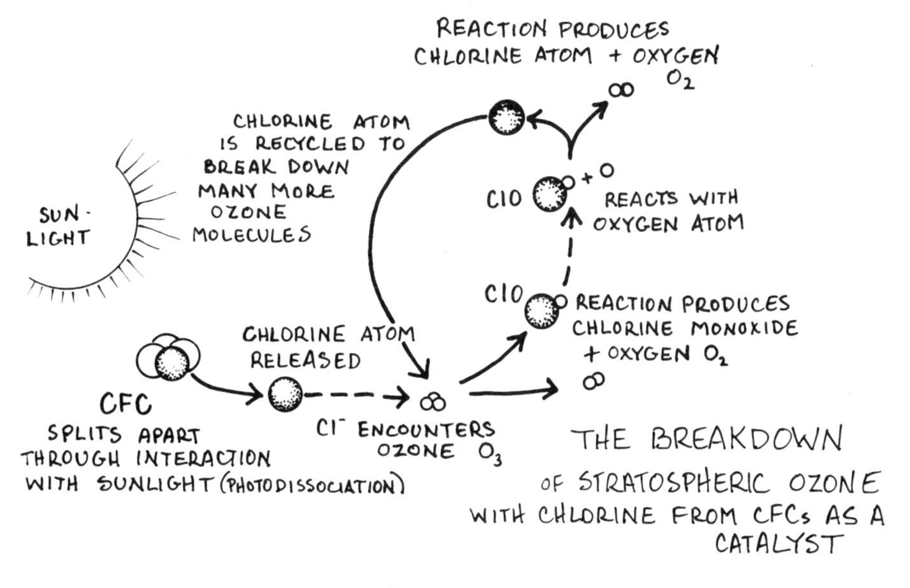

Figure 16.11.

thousands of ozone molecules. Bromine (Br), another element released by the breakdown of anthropogenic gases in the upper atmosphere, is less abundant but even more efficient than chlorine in breaking apart ozone molecules.

Do you remember the definition of residence time? Look in chapter 2 for the answer.

Although scientists were aware that elements like chlorine and bromine (from a group of elements called **halogens**) had the potential to destroy ozone molecules, the real concerns began when stratospheric ozone depletion was actually measured and documented. In 1985, it was announced that ozone levels in Antarctica had declined by as much as 50 percent since the 1970s; these measurements and the continuing depletion of ozone have since been confirmed.

Ozone depletion is both seasonally and spatially controlled. The depletion is most intense during the spring months and most pronounced over the poles, especially the South Pole. The phenomenon of ozone depletion over the South Pole has come to be known as the **ozone hole**. The depletion continues to become more severe as the years go by (Figure 16.12). At its lowest, the ozone concentration over Antarctica has dropped to less than 90 DU. The area of depletion is also increasing in extent. At its largest, the area of intense depletion covered more than 30 million km^2 (12 million mi^2), extending well over southern South America and Australia. In 1992, particularly low levels of ozone were observed, reflecting the added influence of the 1991 volcanic eruption of Mt. Pinatubo in the Philippines.

It is not absolutely clear why ozone depletion is most intense over the South Pole. It may be that tiny ice particles in extremely cold polar clouds enhance the ozone-depleting capacities of chlorine and bromine. Ice-mediated reactions in polar clouds during the cold winter months may be responsible for converting the chlorine and bromine into reactive forms. When more sunlight returns in the spring, the ozone-destroying reactions begin, with chlorine and bromine acting as catalysts. Circulating stratospheric winds, called the **polar vortex**, also play a role in ozone depletion by isolating extremely cold air masses over the poles. This would explain why ozone depletion is the worst over the South Pole: it is colder and the polar vortex is stronger than over the North Pole. Depletion of midlatitude stratospheric ozone is much less intense, although it, too, has been decreasing by about 0.5 percent per year since 1978.

That ozone depletion has occurred, especially over the South Pole, and that the depletion is increasing both in intensity and extent, is not in dispute. However, the significance of the observed depletion is less clear, especially in the context of long-

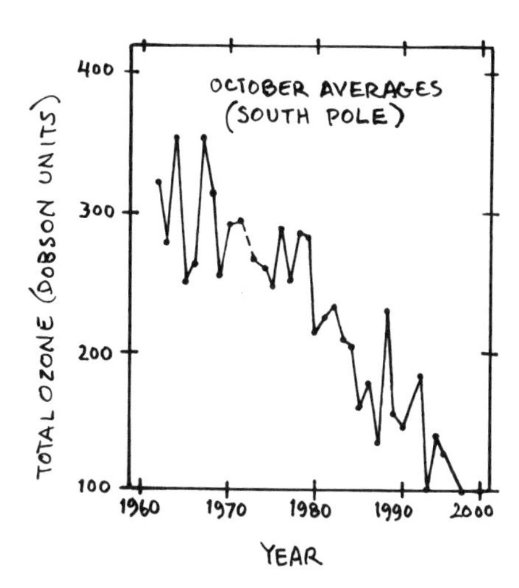

BASED ON WORLD METEOROLOGICAL ORGANIZATION (WMO) DATA

Figure 16.12.

term changes in stratospheric ozone. Unlike climate, there are as yet no proxies that would allow us to determine past concentrations of stratospheric ozone. There is also some debate about the importance of natural causes of ozone depletion, such as volcanic eruptions, in comparison to anthropogenic causes. Research continues into the various anthropogenic chemicals that cause ozone depletion; CFCs are not the only group of **ozone-depleting substances** (**ODSs**) to be implicated.

The main consequence of ozone depletion is an increase in the flux of ultraviolet radiation passing through the atmosphere. For every 1 percent decrease in stratospheric ozone, there will be a 2 percent increase in UV radiation reaching the surface. This is expected to result in increased incidence of skin cancers, particularly melanoma. Other potential impacts are more speculative. For example, an increased UV flux may cause a decrease in oceanic phytoplankton productivity, which could affect the entire oceanic food chain. Some studies have demonstrated decreasing crop yields, accelerated deterioration of some materials, and harmful effects on the human immune system from increased exposure to ultraviolet radiation. As with global warming, there are potential socioeconomic impacts as well, such as the increased cost of health care and the economic impacts on sun-based tourism.

What is the role of chlorofluorocarbons in stratospheric ozone depletion?

Answer: Chlorofluorocarbons break down in the presence of sunlight in the stratosphere, releasing chlorine; the chlorine acts as a facilitator, or catalyst, for ozone-depleting reactions. Once in the stratosphere, chlorine atoms can be recycled over and over, potentially causing thousands of ozone molecules to break down.

What is the role of the polar vortex in stratospheric ozone depletion?

Answer: The polar vortex, which is particularly strong over the South Pole, causes very cold air to circulate and become isolated over the pole; this allows the formation of tiny ice crystals, which provide sites on which ozone-depleting reactions occur.

International Policy Aspects of Global Change

Policy makers have been somewhat resistant to dealing with global change. Scientific uncertainty is a big part of the problem; policy makers want solid answers—where, when, how fast, and how much? But scientists are not yet prepared to give definitive answers for all these questions. Perhaps the most challenging roadblock to international consensus is the time lag inherent in global change processes. It will take centuries before atmospheric CO_2 levels return to normal, even if anthropogenic emissions cease. The delay for ozone-depleting substances is somewhat shorter but still daunting; the chlorine and bromine that are now present in the stratosphere will go on depleting ozone for at least forty more years even though CFC emissions have been greatly reduced (Figure 16.13). Most government officials understandably have a difficult time dealing with delayed impacts. For decision makers in less economically developed countries, it is particularly challenging to deal with issues far in the future when there are more pressing problems, such as hunger, in the present.

Another major policy obstacle is the issue of responsibility. Many developing countries are of the opinion that global change has been caused by wasteful, energy-intensive development in industrialized nations. Less economically developed countries will endure the biophysical costs and consequences of global change along with the rest of the world, but they must also forego some of their present and future economic development to join in the efforts to mitigate these consequences. A related issue concerns the development of replacement technologies that will enable industry to proceed with fewer damaging emissions. How will these technologies be developed and by whom? And under what conditions will they be made available to developing nations?

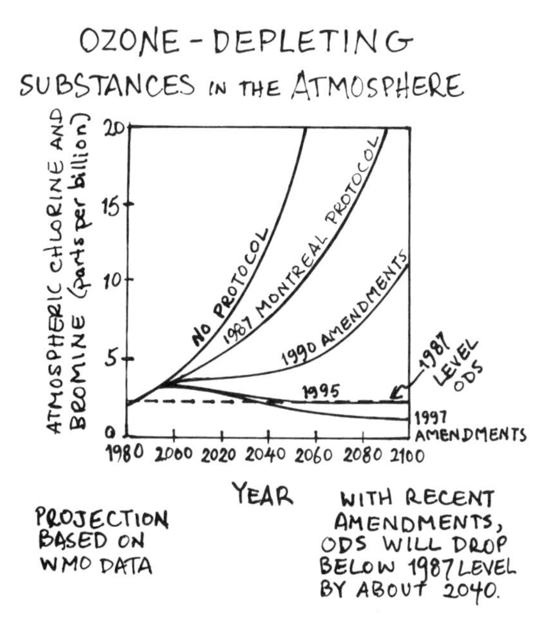

Figure 16.13.

If the predicted impacts of global warming and ozone depletion do occur, the consequences would be very severe indeed. Clearly, the world community is in a risk management situation. We must decide whether the risk is acceptable or simply too great. It is wise to adopt the **precautionary principle**, which states that in a risk management situation in which the potential consequences are unacceptably severe, measures must be taken to mitigate the risk even if the probability of occurrence of the worst-case scenario is low.

There are three basic options for dealing with global change: (1) Undertake large-scale interventions, such as orbiting solar shields, in an attempt to modify the functioning of Earth's atmosphere and climate system. This option carries a huge price tag—probably in the trillions of dollars—as well as the risk of unforeseen environmental consequences. (2) Do nothing, and be prepared to adapt to the changes. Adaptations could include building seawalls around low-lying areas like Florida. The risk inherent in this approach is that we may wait too long, and it will be too late to prevent or mitigate the worst impacts. Some small island nations are at risk of being completely flooded and would have neither the time nor the funding to attempt such adaptations. (3) The final option is to take steps to reduce anthropogenic impacts on atmospheric chemistry by cutting back on emissions of ozone-depleting substances and greenhouse gases. This is the most cautious option and probably the most achievable. It has received the most serious consideration by the international community and has been formalized through the **Montreal Protocol** on ozone-depleting substances and the **Kyoto Protocol** to limit emissions of greenhouse gases.

The Kyoto Protocol has had a controversial history so far. The original United Nations Framework Convention on Climate Change was first signed in 1992 at the Earth Summit in Rio de Janeiro. Its goal was to stabilize GHG levels in the atmosphere to prevent further human impacts on climate. Several international meetings later, the Kyoto Protocol was proposed in 1997. So far, more than a hundred countries have ratified the agreement, which calls for reductions in GHG emissions. The U.S. Senate has not ratified the agreement because of fears about economic disruption. The United States also disapproves of the principle of **common but differentiated responsibility**, which states that all nations have a responsibility to participate in efforts to mitigate the effects of global atmospheric and climatic change, but some nations have a greater capacity and should carry a greater burden of the responsibility. Early in 2004, however, the Kyoto Protocol was finally ratified by Russia. This means that a large enough percentage of the world's GHG production is represented by the ratifying nations that the agreement will now come into force as international law. However, many people fear that without the participation of the United States, the single largest contributor of GHGs to the atmosphere, the agreement will be ineffective.

The Montreal Protocol, in contrast, appears to be an environmental success story in the making. One feature that makes this agreement unique is that it has built-in provisions for regular assessments of the status of scientific understanding of ozone depletion and of the effectiveness of control measures. The agreement has already been assessed and revised several times. In accordance with the agreement, the production of CFCs and other ozone-depleting chemicals was essentially phased out by the year 2000 (see Figure 16.13). Even with reductions in CFC production and use, ozone depletion will likely continue for many decades, probably at least until the year 2030. After that, the level of ozone in the stratosphere is expected to slowly return to normal, reaching its pre-1978 level sometime after 2050. Signing the Montreal Protocol was an exceedingly difficult political decision for countries that are major producers of CFCs; in the United States alone, the production of CFCs is a multibillion-dollar industry. The truly positive outcome of this story is that scientific research and political will, with inputs from industry, led to a consensus and an international agreement to act on behalf of all the world's inhabitants.

What is the name of the international agreement to limit the production of ozone-depleting substances? What is the name of the international agreement to limit greenhouse gas emissions? ______________

Answer: Montreal Protocol; Kyoto Protocol.

Spaceship Earth

We have now come to the end of this exploration of Earth and the natural environment. Of all the themes woven throughout the fabric of this book, one in particular is fundamental: that all Earth systems are interrelated. Human systems mirror the complexity of natural Earth systems and are intertwined with them. No human action occurs in total isolation from natural systems. Humans are not separate from the environment but integral to it. In clearly thinking through the environmental consequences of our actions, we begin to recognize the necessity for appropriate management and respect for the environment both for the current and future good. We no longer have the right to "borrow" resources and carrying capacity from our children. We owe future generations the same access to a healthy environment and the same possibility of obtaining an acceptable quality of life for which we ourselves have hoped and worked.

The central challenge now facing the world community is to find pathways through which to achieve this vision of a sustainable future. Humanity does have a future on this planet; it is not too late for us to ensure that our children will inherit Earth in a reasonably healthy state. But achieving this will require creative solutions to a variety of complex problems on a wide range of scales from local, to regional, to global. This brings us full circle to the realization that the environmental problems created by us must be solved by us through our ingenuity, our ability to think

This view of the rising Earth greeted the *Apollo 8* astronauts in 1968 as they came from behind the Moon after the lunar orbit insertion burn.

creatively, and our willingness to respond to problems in a socially and environmentally responsible and even visionary manner.

I hope this book has provided you with a useful introduction to the science of the natural environment. Scientific understanding of this complex and beautiful world grows with each passing day. Through this understanding of the natural world, our impacts on it, and its influence on human lives and activities, we can begin to design policies, invent technologies, and forge agreements that will allow the world community to chart a course toward a sustainable future. Although this book has ended, I wish you many happy years of discovery and learning about the environment and Earth.

We travel together, passengers on a little spaceship . . .

—Adlai Stevenson

SELF-TEST

These questions are designed to help you assess how well you have learned the concepts presented in chapter 16. The answers are given at the end. If you get any of the questions wrong, be sure to troubleshoot by going back into that part of the chapter to find the correct answer.

1. Orbital variations of Earth, as described by Milankovitch, affect the timing of glacial-interglacial cycles because they ________________.
   a. control the amount of greenhouse gases in the atmosphere
   b. have an impact on solar insolation
   c. decrease the amount of ozone in the stratosphere
   d. All of the above.
2. In North America, the final years of the twentieth century were ________________.
   a. slightly colder than the previous few decades
   b. wetter and milder than the first part of the century
   c. some of the warmest on record
3. The IPCC predicts warming of average global surface temperatures of ________________ by the year 2050.
   a. 1°C to 2°C
   b. 3°C to 5°C
   c. 10°C
   d. 15°C

4. The ______________ states that in a risk management situation in which the potential consequences are unacceptably severe, measures must be taken to mitigate the risk, even if the probability of occurrence of the worst-case scenario is low.

5. A period between glaciations when ice sheets retreat and sea levels rise is called a(n) ______________.

6. A(n) ______________ is a poorly sorted pile of rock and sediment deposited by a glacier along its edges and terminus.

7. Direct instrumental measurements of temperature and precipitation have only been recorded on a regular basis for the past four decades. (True or False)

8. Dusty atmospheres are generally indicative of periods when the climate is particularly hot and desertlike. (True or False)

9. The cool period that lasted from about A.D. 1300 to about A.D. 1900 is called the Little Ice Age. (True or False)

10. What are the most important anthropogenic greenhouse gases? What is the most important gas in the natural greenhouse effect?

11. How do corals and lake-bottom sediments act as climate proxies?

12. What does it mean when we say that developing nations are in a position of double jeopardy with regard to global climatic change?

13. How do glacial landforms and deposits help scientists determine the timing and extent of past glaciations?

14. Why is climatic change a better term than global warming to describe the possible effects of anthropogenic greenhouse gas emissions on Earth's climate system?

ANSWERS

1. b 2. c 3. b 4. precautionary principle

5. interglacial (or interglacial stage or interglacial period)

6. moraine 7. False 8. False 9. True

10. Anthropogenic GHGs: carbon dioxide, methane, nitrous oxide, CFCs, ozone. Natural greenhouse effect: water vapor.

11. Like trees, corals add annual growth layers that reflect conditions such as the temperature and composition of seawater. Lake-bottom sediments are sometimes finely layered; the layers, called varves, are deposited seasonally. The thickness and other features of varves can indicate the length and timing of seasons.

12. Developing nations will have to endure the biophysical costs and consequences of climatic change, but they will also have to forego part of their future economic development to join in the efforts to cut back on GHG emissions.

13. Scratches and grooves in rocks reveal the direction of ice movement; moraines show the geographic extent of ice sheets; tree rings and radiocarbon ages of felled trees indicate when the ice arrived in a given region; loess deposits and other periglacial features reveal the limits of the ice and characteristics of the atmosphere and climate.

14. The changes that will occur in response to warming of Earth's climate will not be uniform. Some areas will get warmer and others will get colder, for averages as well as for extremes. There will be other changes, too, such as changes in precipitation. Thus, climatic change is a more appropriate term than global warming.

KEY WORDS

atmospheric blind
atmospheric window
chlorofluorocarbon (CFC)
climate forcing
climate proxy
common but differentiated responsibility
Dobson unit
enhanced (accelerated) greenhouse effect
glaciation
global circulation model (GCM)
global change
halogen
Holocene optimum
interglacial
Kyoto Protocol
Little Ice Age
loess
Milankovitch cycle
Montreal Protocol
moraine
ozone-depleting substance (ODS)
ozone hole
ozone layer
paleoclimatology
paleosol
paleothermometer
periglacial
Pleistocene Epoch
polar vortex
precautionary principle
varve

Appendix 1
Units and Conversions

About SI Units

Regardless of the field of specialization, all scientists use the same units and scales of measurement. They do so to avoid confusion and the possibility that mistakes can creep in when data are converted from one system of units, or one scale, to another. By international agreement the SI units are used by all, and they are the units used in this text. SI is the abbreviation for Système International d'Unités (in English, the International System of Units). Some of the SI units are likely to be familiar, some unfamiliar. The SI unit of length is the meter (m), of area the square meter (m^2), and of volume the cubic meter (m^3). The SI unit of mass is the kilogram (kg), and of time the second (s). The other SI units used in this book can be defined in terms of these basic units.

Prefixes for Very Large and Very Small Numbers

When very large or very small numbers have to be expressed, a standard set of prefixes is used in conjunction with the SI units. Some prefixes are probably already familiar; an example is the centimeter, which is one-hundredth of a meter, or 10^{-2} m. The standard prefixes are:

tera	$1,000,000,000,000 = 10^{12}$
giga	$1,000,000,000 = 10^9$
mega	$1,000,000 = 10^6$
kilo	$1,000 = 10^3$

hecto	$100 = 10^2$
deka	$10 = 10$
deci	$0.1 = 10^{-1}$
centi	$0.01 = 10^{-2}$
milli	$0.001 = 10^{-3}$
micro	$0.000001 = 10^{-6}$
nano	$0.000000001 = 10^{-9}$
pico	$0.000000000001 = 10^{-12}$

Commonly Used Units of Measure

Length

Metric Measure

1 kilometer (km) = 1,000 meters (m)

1 meter (m) = 100 centimeters (cm)

1 centimeter (cm) = 10 millimeters (mm)

1 millimeter (mm) = 1,000 micrometers (μm) (formerly called microns)

1 micrometer (μm) = 0.001 millimeter (mm)

1 angstrom (Å) = 10^{-8} centimeters (cm)

Nonmetric Measure

1 mile (mi) = 5,280 feet (ft) = 1,760 yards (yd)

1 yard (yd) = 3 feet (ft)

1 foot (ft) = 12 inches (in)

1 fathom (fath) = 6 feet (ft)

Conversions

1 kilometer (km) = 0.6214 mile (mi)

1 meter (m) = 1.094 yards (yd) = 3.281 feet (ft)

1 centimeter (cm) = 0.3937 inch (in)

1 millimeter (mm) = 0.0394 inch (in)

1 mile (mi) = 1.609 kilometers (km)

1 yard (yd) = 0.9144 meter (m)

1 foot (ft) = 0.3048 meter (m)

1 inch (in) = 2.54 centimeters (cm)

1 inch (in) = 25.4 millimeters (mm)

1 fathom (fath) = 1.8288 meters (m)

Area

Metric Measure

1 square kilometer (km^2) = 1,000,000 square meters (m^2) = 100 hectares (ha)

1 square meter (m^2) = 10,000 square centimeters (cm^2)

1 hectare (ha) = 10,000 square meters (m^2)

Nonmetric Measure

1 square mile (mi^2) = 640 acres (ac)

1 acre (ac) = 4,840 square yards (yd^2)

1 square foot (ft^2) = 144 square inches (in^2)

Conversions

1 square kilometer (km^2) = 0.386 square mile (mi^2)

1 hectare (ha) = 2.471 acres (ac)

1 square meter (m^2) = 1.196 square yards (yd^2) = 10.764 square feet (ft^2)

1 square centimeter (cm^2) = 0.155 square inch (in^2)

1 square mile (mi^2) = 2.59 square kilometers (km^2)

1 acre (ac) = 0.4047 hectare (ha)

1 square yard (yd^2) = 0.836 square meter (m^2)

1 square foot (ft^2) = 0.0929 square meter (m^2)

1 square inch (in^2) = 6.4516 square centimeters (cm^2)

Volume

Metric Measure

1 cubic meter (m^3) = 1,000,000 cubic centimeters (cm^3)

1 liter (l) = 1,000 milliliters (ml) = 0.001 cubic meter (m^3)

1 centiliter (cl) = 10 milliliters (ml)

1 milliliter (ml) = 1 cubic centimeter (cm^3)

Nonmetric Measure

1 cubic yard (yd^3) = 27 cubic feet (ft^3)

1 cubic foot (ft^3) = 1,728 cubic inches (in^3)

1 barrel (oil) (bbl) = 42 gallons (U.S.) (gal)

Conversions

1 cubic kilometer (km^3) = 0.24 cubic mile (mi^3)

1 cubic meter (m^3) = 264.2 gallons (U.S.) (gal) = 35.314 cubic feet (ft^3)

1 liter (l) = 1.057 quarts (U.S.) (qt) = 33.815 ounces (U.S. fluid) (fl. oz.)

1 cubic centimeter (cm^3) = 0.0610 cubic inch (in^3)

1 cubic mile (mi^3) = 4.168 cubic kilometers (km^3)

1 acre-foot (ac-ft) = 1,233.46 cubic meters (m^3)

1 cubic yard (yd^3) = 0.7646 cubic meter (m^3)

1 cubic foot (ft^3) = 0.0283 cubic meter (m^3)

1 cubic inch (in^3) = 16.39 cubic centimeters (cm^3)

1 gallon (gal) = 3.784 liters (l)

Mass

Metric Measure

1,000 kilograms (kg) = 1 metric ton (also called a tonne) (m.t)

1 kilogram (kg) = 1,000 grams (g)

1 gram (g) = 0.001 kilogram (kg)

Nonmetric Measure

1 short ton (sh.t) = 2,000 pounds (lb)

1 long ton (l.t) = 2,240 pounds (lb)

1 pound (avoirdupois) (lb) = 16 ounces (avoirdupois) (oz) = 7,000 grains (gr)

1 ounce (avoirdupois) (oz) = 437.5 grains (gr)

1 pound (Troy) (Tr. lb) = 12 ounces (Troy) (Tr. oz)

1 ounce (Troy) (Tr. oz) = 20 pennyweight (dwt)

Conversions

1 metric ton (m.t) = 2,205 pounds (avoirdupois) (lb)

1 kilogram (kg) = 2.205 pounds (avoirdupois) (lb)

1 gram (g) = 0.03527 ounce (avoirdupois) (oz) = 0.03215 ounce (Troy) (Tr. oz) = 15,432 grains (gr)

1 pound (lb) = 0.4536 kilogram (kg)

1 ounce (avoirdupois) (oz) = 28.35 grams (g)

1 ounce (avoirdupois) (oz) = 1.097 ounces (Troy) (Tr. oz)

Pressure

1 pascal (Pa) = 1 newton/square meter (N/m^2)

1 kilogram/square centimeter (kg/cm^2) = 0.96784 atmosphere (atm) = 14.2233 pounds/square inch (lb/in^2) = 0.98067 bar

1 bar = 0.98692 atmosphere (atm) = 105 pascals (Pa) = 1.02 kilograms/square centimeter (kg/cm^2)

Energy and Power

Energy

1 joule (J) = 1 Newton meter (N.m) = 2.390×10^{-1} calorie (cal) = 9.47×10^{-4} British thermal unit (Btu) = 2.78×10^{-7} kilowatt-hour (kWh)

1 calorie (cal) = 4.184 joule (J) = 3.968×10^{-3} British thermal unit (Btu) = 1.16×10^{-6} kilowatt-hour (kWh)

1 British thermal unit (Btu) = 1,055.87 joules (J) = 252.19 calories (cal) = 2.928×10^{-4} kilowatt-hour (kWh)

1 kilowatt hour = 3.6×10^6 joules (J) = 8.60×10^5 calories (cal) = 3.41×10^3 British thermal units (Btu)

Power (energy per unit time)

1 watt (W) = 1 joule per second (J/s) = 3.4129 British thermal units per hour (Btu/h) = 1.341×10^{-3} horsepower (hp) = 14.34 calories per minute (cal/min)

1 horsepower (hp) = 7.46×10^2 watts (W)

Temperature

To change from Fahrenheit (F) to Celsius (C): °C = (°F – 32°C) ÷ 1.8

To change from Celsius (C) to Fahrenheit (F): °F = (°C × 1.8) + 32

To change from Celsius (C) to Kelvin (K): K = °C + 273.15

Appendix 2
Some Great Environmental Science Web Sites

There are literally thousands of great environmental science web sites. Here are a few of the most comprehensive, well-organized, and reliable sites, just to get you started.

To learn more about . . .	**try visiting . . .**
Our Unique Planet	• *www.nasa.gov* NASA's main Web site gives access not only to information about space but all of NASA's Earth observing activities (try clicking on "Life on Earth" to start).
The Interactive Earth	• *www.usgs.gov* U.S. Geological Survey has information about Earth history and how the Earth works. • *www.usra.edu/esse/essonline* Earth System Science online, through the University Space Research Association.
The Hydrosphere and the Atmosphere	• *www.noaa.gov* National Oceanic and Atmospheric Administration has tons of scientific information about the ocean, atmosphere, climate, and weather.
The Biosphere: Life on Earth	• *www.ucmp.berkeley.edu/alllife/threedomains.html* The great Web site of the University of California Museum of Paleontology has information about DNA, the biosphere, and the kingdoms of life; be

sure to visit the rest of the virtual museum at *www.ucmp.berkeley.edu/index.html* for great material on the history of life.

Earth's Major Ecosystems

- *www.worldbiomes.com* Explore the world's great biomes.
  www.nationalgeographic.com/wildworld/terrestrial.html offers an interactive map, in conjunction with the World Wildlife Fund that gives detailed information about biomes and ecosystems.

Habitat and Biodiversity

- *www.worldwildlife.org* World Wildlife Fund (WWF) is perhaps the best Web site for information about endangered species and their habitats.
- *www.edf.org* is the Environmental Defense Fund's Web site.

People, Population, and Resources

- *www.un.org* The main Web site of the United Nations can give access to many U.N. organizations that deal with these issues, including *www.unesco.org, www.undp.org* (U.N. Development Program), *www.unicef.org*, and *www.unfpa.org,* is the United Nations Population Fund.
- *www.who.org* is the World Health Organization, which also has information about human health and population.

Forests, Wildlife, and Fisheries

- *www.fao.org* is the United Nations Food and Agriculture Organization, which monitors forests and fisheries around the world.
- *www.worldwildlife.org* is the World Wildlife Fund.

Soils and Agriculture

- *www.fao.org* monitors the global food supply, as well as soils and agriculture.
- *www.wri.org* is the Web site for the World Resources Institute, which keeps extensive data sets on everything from food and agriculture to population, forests, water resources, and climate change.

Mineral Resources

- *www.usgs.gov* is the Mineral Resources Program at the U.S. Geological Survey, which provides information and educational materials about minerals and mineral resources.

Energy Resources

- *www.wri.org* is the World Resources Institute. It offers a lot of data about energy use, as well as conventional energy resources and alternative energy resources.

Water Resources

- *www.nwra.org* is the National Water Resources Association.
- *www.worldwater.org* has many links to Web sites that deal with global water resources.

Water Pollution and Soil Pollution

- *www.epa.gov* is the Web site of the U.S. Environmental Protection Agency, with information about the Clean Water Act, water pollution, soil pollution, and just about everything else about the environment.

Air Pollution

- *www.epa.gov* the U.S. EPA also has information about the Clean Air Act and various types of air pollution.

Cities and Waste Management

- *www.unchs.org* is U.N. Habitat, which deals specifically with human settlements.

Global Change

- *www.epa.gov/ozone/science* offers videos of the Antarctic ozone hole.
- *www.cmdl.noaa.gov* is the main Web site for the Climate Monitoring and Diagnostic Laboratory of NOAA, which has up-to-date information about climate, ozone depletion, and air quality.

Environmental Organizations

- *www.greenpeace.org* is Greenpeace.
- *www.foe.org* is Friends of the Earth.
- *www.wilderness.org* is the Wilderness Society.
- *www.sierraclub.org* is the Sierra Club.
- These are just a few of *many* thousands of environmental organizations around the globe. You can use these links to get started.

Appendix 3
Soil Classifications

U.S. Comprehensive Soil Classification System: A Summary

Group I: Soils with well-developed horizons or fully weathered minerals

Order	*Brief Description*
Oxisol	Very old, highly weathered soils of low latitudes; subsurface accumulation of mineral oxides
Ultisol	Equatorial, tropical, and subtropical soils; subsurface horizon of clay
Vertisol	Subtropical and tropical soils with high clay content; develop deep, wide cracks when dry
Alfisol	Soils of humid and subhumid climates, subsurface horizon of clay
Spodosol	Soils of cold, moist climates; B horizon accumulation of reddish mineral matter
Mollisol	Semiarid and subhumid grassland soils with dark, humus-rich A horizon
Aridosol	Soils of dry climates; low in organic matter, with subsurface accumulations of carbonate minerals or soluble salts

Group II: Soils with a large proportion of organic matter

Order	*Brief Description*
Histosol	Thick upper layer, very rich in organic matter

Group III: Soils with poorly developed horizons or no horizons

Order	*Brief Description*
Entisol	Recently accumulated soils, lacking horizons
Inceptisol	Soils with weakly developed horizons

Andisol	Soils with weakly developed horizons and a large proportion of volcanic glassy material; volcanic soils

Canadian System of Soil Classification: A Summary

Order (and U.S. CSCS equivalent)	*Brief Description*
Brunisolic (*Inceptisol*)	Forest soils with brownish B horizons
Chernozemic (*Mollisol*)	Grassland soils with dark, organic-rich A horizons
Cryosolic (*Inceptisol with permafrost*)	Soils associated with strong frost action and underlying permafrost
Gleysolic (*wet Inceptisol*)	Soils associated with periodic or prolonged saturation with water
Luvisolic (*Alfisol*)	Forest soils with clay-rich B horizons
Organic (*Histosol*)	Soils composed largely of organic materials (peat, muck, bog soils)
Podzolic (*Spodosol*)	Forest and health soils; accumulation of amorphous humified organic material with Al and Fe
Regosolic (*Entisol*)	Soils with weakly developed horizons; recent alluvium, colluvium, or sand
Solonetzic (*Aridosol*)	Clay-rich B horizon that is very hard when dry; swells to a sticky, low-permeability mass when wet

Index

A Self-Teaching Guide

Barbara W. Murck

WILEY

John Wiley & Sons, Inc.

Published by John Wiley & Sons, Inc., Hoboken, New Jersey
Published simultaneously in Canada

Photo credits: page 2, NASA Earth Observatory; page 20, Mark Emery/U.S. Fish and Wildlife Service; page 59, Jacques Descloitres/NASA Earth Observatory; page 77, Florida Keys National Marine Sanctuary; pages 83 and 236, USGS; page 114, Gary Kramer/U.S. Fish and Wildlife Service; pages 126, 140, and 165, FAO Photo; page 191, David Nimick/USGS; page 208, OMVS, Senegal; page 228, UNEP, Peter Garside, Topham Picturepoint; page 261, *Tehran Times*; page 276, NASA Goddard Space Flight Center; page 323, NASA Manned Spacecraft Center.

Library of Congress Cataloging-in-Publication Data:
Murck, Barbara Winifred, date
Environmental science : a self-teaching guide / Barbara W. Murck.
p. cm.
Includes index.
ISBN-13 978-0-471-26988-5 (pbk.)
ISBN-10 0-471-26988-3 (pbk.)
1. Environmental sciences—Programmed instruction. I. Title.

GE105.M87 2005
333.72—dc22 2004020587

Printed in the United States of America

10 9 8 7 6 5 4 3